DEPARTMENT OF ENERGY

Energy use and energy efficiency in UK manufacturing industry up to the year 2000

to be returned on or before
  t nd below

VOLUME 2

Sector Reports containing the Detailed Analyses of the Industries, their Energy Use and Potential Energy Savings

Chief Scientist's Group
Energy Technology Support Unit
AERE, Harwell
October 1984

LONDON: HER MAJESTY'S STATIONERY OFFICE

© Crown Copyright 1984
First publlished 1984

This publication is the third in the Energy Efficiency series
published by the Energy Efficiency Office. The series is
primarily intended to create a wider public understanding and
discussion of the efficient use of energy.

The publications in the series do not necessarily represent
Government or Departmental policy.

The first two publications in the series
Energy Efficiency Demonstration Scheme: A Review (£8.25)
The Pattern of Energy Use in the UK 1980 (£9.80)

ISBN 0 11 411562 1

ENERGY USE AND ENERGY EFFICIENCY IN UK MANUFACTURING INDUSTRY

UP TO THE YEAR 2000

CONTENTS

SECTOR 1.   METAL MANUFACTURING

SECTOR 2.   CERAMIC MATERIALS

SECTOR 3.   THE CHEMICAL INDUSTRY

SECTOR 4.   PAPER, PRINTING AND PUBLISHING

SECTOR 5.   FOOD, DRINK AND TOBACCO INDUSTRIES

SECTOR 6.   ENGINEERING AND ALLIED INDUSTRIES

SECTOR 7.   THE TEXTILE, LEATHER AND CLOTHING INDUSTRIES

ENERGY USE AND ENERGY EFFICIENCY IN UK MANUFACTURING INDUSTRY

UP TO THE YEAR 2000

SECTOR 1. METAL MANUFACTURING

K F Langley

Subsector	Paragraph No
Introduction	1
1.1 Iron and Steel	12
Structure of the Industry	16
Energy Use	20
Energy Conservation Potential	68
Projection of Future Specific Energy Consumption	83
1.2 Aluminium	105
Structure of the Industry	107
Trends in Production and Consumption	110
Energy Use	114
Energy Conservation Potential	142
Projection of Future Specific Energy Consumption	158
1.3 Copper	168
Structure of the Industry	171
Trends in Production and Consumption	178
Energy Use	182
Energy Conservation Potential	206
Projection of Future Specific Energy Consumption	219
1.4 Lead and Zinc	223
Structure of the Industry	226
Trends in Production and Consumption	232
Energy Use	237
Energy Conservation Potential	249
Projection of Future Specific Energy Consumption	252
References	

TABLE 1: Comparative Summary of Statistics on Metal Manufacture

METAL	UK PRODUCTION 1980 (Kilotonnes)			PROCESS ENERGY REQUIREMENT (GJ/tonne) (heat supplied)			MID 1980 PRICE £/tonne	PRINCIPAL END USES
	Primary Smelting	Secondary Refining	Finished Products#	Primary Smelting	Secondary Refining	Finished Products		
Iron & Steel	6,689	4,443	10,389	14.2+	3.6+	9.5+		All engineering and construction.
Aluminium	374	162	504	87	15	23	750	Building, transport, packaging, electrical engineering.
Copper	0	161	530	50	25	8-25	900	Electrical engineering, building.
Zinc	87	55	245*	43	-	5-50	320	Galvanising, brass, die-casting, pigments.
Lead	30	211	305	-	2-10	2-3	370	Petrol additives, batteries, building, pigments.
Nickel	19	-	23	-	-	-	2,700	Alloys.
Tin	11	-	10	-	-	-	7,300	Tin-plate, bronze.

+ 1981 data
# includes alloys
* includes consumption of metal in pigment manufacture

INTRODUCTION

1.   This sector report contains the detailed analysis of energy use and energy conservation potential in the metal manufacturing industries.

2.   Metal manufacturing is dominated by the iron and steel industry, which alone accounted for 18% of the total energy use in manufacturing industry in 1980. Aluminium production accounted for 2.4% of energy use in 1980, but is notable for its intensive consumption of electricity in primary smelting. By comparison, copper, lead and zinc are relatively small users of energy.

3.   Table 1 shows the total amounts of each of the main metals produced in the UK in 1980 under the following headings:

- primary metal production by smelting ores;

- secondary metal production by refining scrap;

- finishing or semi-fabrication of the metal into sections, plates, wire, tube, etc.

In the case of iron and steel 'primary smelting' refers to the production of crude steel by the integrated blast furnace/basic oxygen steel furnace route, and 'secondary refining' refers to production of steel from scrap in electric arc furnaces.

4.   There is virtually no primary production of copper in the UK, but a substantial proportion of our requirements of aluminium, zinc and lead are produced by smelting imported ores. Secondary refining of non-ferrous metals from scrap is also well developed.

5.   The balance between primary smelting and secondary refining has a profound effect on the overall specific energy consumption of all the metal industries, as can be seen from the process energy requirements in Table 1. Projections of future specific energy consumption are therefore sensitive to assumption about the ratio of scrap metal which is recycled.

6.   The proportion of primary metal which is imported into the UK also has a large effect on the energy consumption in the metal manufacturing sector within the UK. Significant changes in the level of imports are possible, for example steel slabs and aluminium ingots are energy intensive products with relatively low added value which may in future be imported from energy and resource-rich developing countries, with the UK industry concentrating on high added value, less energy intensive semi-fabrication. However, since the main purpose of the present study is to identify the effects of technical changes on energy consumption, major structural changes in output have not been considered.

7.   Nevertheless, some consideration of the effects of output trends on specific energy consumption is required. This is done in the present study by adopting two scenarios for future output, designated 'high' and 'low'. The scenarios are based on the central projections of the Department of Energy's 1982 Energy Projections (ref 1). Both scenarios relate to a 1½% growth rate in GDP, but differ in the contribution which manufacturing industry makes to the overall growth of GDP.

8.   In the case of iron and steel, the 1982 Energy Projections make explicit assumptions on output, which are adopted in this study. For the non-ferrous metals, however, additional assumptions are required. Aluminium demand is expected to grow more rapidly than copper, but such that the total demand for the two metals grows at the rate assumed for 'other industry' in reference 1. Lead and zinc are assumed to grow at the rate assumed for 'building materials' in reference 1.

9.   The finished, or semi-fabricated, products of the metal manufacturing sector are generally taken to be the intermediate goods which are sold for subsequent fabrication in other industries such as engineering and construction. Substantial quantities of lead and zinc are used as oxide in pigments, but since they have passed through the metallic form, their production is included in metal manufacturing; alumina, which is refined and used directly as oxide without smelting, is not.

10.   Iron castings produced in iron foundries are treated as part of the engineering sector in this study, and are not dealt with in the metal manufacturing sector.

11.   As far as nickel and tin are concerned, the total amounts produced are relatively small, and their energy requirements are therefore of little importance in the context of the overall industrial energy requirement. Furthermore, given the high intrinsic values of these metals, the energy required for processing them is of no significance to the economics of their production and use.

TABLE 2: Output Assumptions for Metal Manufacturing Sector (kilotonnes)

Subsector	1980	1990 High	Low	2000 High	Low
Crude Steel	11,277	16,500	13,700	18,900	12,500
Aluminium	504	600	590	760	700
Copper	530	580	560	670	620
Zinc and Lead	550	640	610	800	700

ENERGY USE AND ENERGY EFFICIENCY IN UK MANUFACTURING INDUSTRY

UP TO THE YEAR 2000

1.1 The Iron and Steel Industry

FIGURE 1: A. Production of Crude Steel 1956-81 (million tonnes/year)
B. Specific Energy Consumption 1970-81 (GJ/tonne)

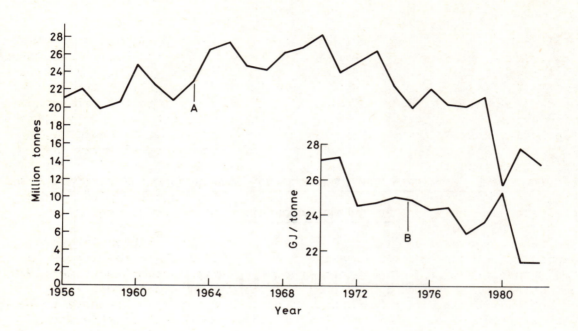

FIGURE 2: UK Iron and Steel Production 1981: Principal Processes
and Material Flows (million tonnes)

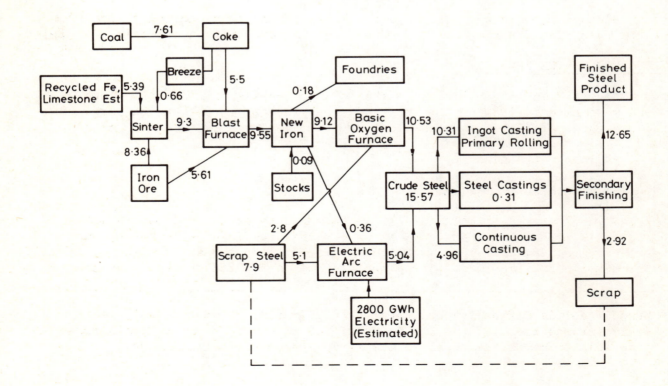

- 4 -

1.1   THE IRON AND STEEL INDUSTRY

INTRODUCTION

12.   The iron and steel industry is the third largest individual industrial user of energy, accounting for around 18% of all the energy use in industry, or 6% of the total national energy requirement.

13.   In the UK, the iron and steel industry is generally recognised to be those activities covered by SIC Groups 221 (Iron and Steel), 222 (Steel Tubes) and 223 (Drawing, Cold Rolling and Cold Forming of Steel), according to the Standard Industrial Classification (Revised 1980).   Thus it includes blast furnaces and associated material preparation plant, steel melting shops and steel foundries, casting and continuous casting plant, hot and cold rolling mills including associated coating plant, re-rolling mills, tube and pipe mills and open die forges.   Coke ovens owned by iron and steel works are included in the industry's own definition, but excluded by the SIC.   Excluded are certain products which may be made by works within the industry; such as drop forgings and stampings, cold formed sections, springs and wire.   Iron foundries are generally considered to be outside the industry (ref 2).

14.   In recent years, the industry has seen a major contraction in output accompanied by a structural rationalisation.   Figure 1 shows the trends in output of crude steel* since 1956 and specific energy consumption since 1970, when output was at its peak of 28.29 million tonnes of crude steel (2).   In 1980 output was particularly low due to a prolonged strike.   The need to keep furnaces hot during the strike, combined with the more general effects of operating plant at low capacity led to an increase in specific energy consumption, compared with the general downward trend.   Subsequent rationalisation and plant closures, accompanied by higher output from remaining plant, has resulted in considerably improved specific energy consumption in 1981.

15.   Because 1980 is such an anomalous year, it has been decided to use 1981 as the base year for the analysis of the energy conservation potential.

STRUCTURE OF THE INDUSTRY

16.   Figure 2 shows schematically the major processes and material flows which were used in 1981 in the iron and steel industry.   For the purposes of the present analysis it is possible to divide the industry into two sectors:

- integrated plants

- electric arc plants.

Integrated plants use mainly 'new' iron obtained by reducing iron ore with coke in a blast furnace; the coke is partly produced in on-site ovens, and partly by the NCB.   The iron is converted to steel in a Basic Oxygen Steel-making (BOS) furnace.   Scrap steel accounts for 25% of the input to the

---

* Crude steel is defined as the total of usable ingots, continuously cast semi-finished products and liquid steel for castings, based on the general principle of measuring steel at its first stage of solidification.

TABLE 3: Major Units of Plant in 1981

Site	Blast* Furnaces (in blast)	Basic* Oxygen Furnaces (active)	Electric Arc Furnaces	Continuous Casting Plant	Liquid Steel Capability (Mt/a)	Crude Steel Production, 10^6 tonne
Llanwern	3 (2)	3 (2)	–	0	2.8	) 3.5
Port Talbot	4 (3)	2 (1)	–	0	2.9	)
Ravenscraig	3 (3)	3 (2)	–	3	2.8	1.6
Scunthorpe	4 (4)	3 (2)	–	2	3.8	2.7
Teeside	1 (1)	3 (2)	–	3	3.5	2.7
Total Integrated Plants	15 (13)	14 (9)	–	8 machines (20 strands)	15.8	10.5
BSC Non-integrated steel making (10 sites)	–	–	24	9 machines (23 strands)	4.0	)))) 5.0
Other Non-integrated steel making (23 sites)	–	–	40	12 machines (40 strands)	4.2	)))

* Not necessarily at the same time.

FIGURE 3: Imports and Exports of Finished Steel

- 6 -

BOS furnace. By contrast, electric arc furnaces (EAF) use only scrap steel, much of it generated within the industry. Open hearth furnaces are no longer used in the UK.

17.   The numbers of different sites and major units of plant in use in the UK in 1981 are shown in Table 3. It can be seen that all of the 14 BOS furnaces in the UK were located at the five integrated steel works; they accounted for 68% of crude steel production. The remaining 32% of crude steel production was produced in 64 electric arc furnaces on 30 sites. Electric arc furnaces range widely in capacity, from under 10 tonnes to over 100 tonnes. Special quality steels are usually made in electric arc furnaces, especially the smaller ones, because orders for such qualities come in smaller quantities than for bulk steel. British Steel Corporation operate all five of the integrated plants and ten of the other sites (comprising 24 electric arc furnaces), producing 85% of the crude steel produced in the UK in 1981; the remaining 15% was produced by private companies.

18.   The domestic production in 1981 of finished steel products amounted to 12.6×10^6 tonnes, of which 9.1×10^6 tonnes was delivered for domestic consumption, the remainder being exported. Note that an average of about 80% of crude steel is converted to finished product, the actual value depending largely on quality of end product. The rest of the crude steel is recycled as scrap.

19.   In the recent past, imports and exports of finished steel products have been in approximate balance, as is shown in Figure 3. 1980 is again anomalous, because of the strike.

ENERGY USE IN THE IRON AND STEEL INDUSTRY

20.   The standard broad measure of energy consumption in the iron and steel industry is the specific energy consumption, defined as:

$$\text{SEC} = \frac{\text{total direct energy consumed (heat supplied)}}{\text{crude steel produced}}$$

The industry's own measure of energy consumption includes energy used in coke making. According to this definition, the specific energy consumption has fallen from about 36 GJ/tonne in 1956 to 21.3 GJ/tonne in 1981, a reduction of 42%. This measure does not take account of the change in the mix of fuels used. Electricity consumption increased in the 1960s with the increase in the number and size of electric arc furnaces. However, the primary energy requirement of the purchased electricity is not included in the specific energy consumption. When the primary energy requirement for electricity generation is included, the 1981 specific energy consumption becomes 27.0 GJ/tonne compared with 38.5 in 1956, a reduction of 30% which is still impressive.

21.   A detailed analysis of the energy flows in steel production was given in Energy Audit No 16 (ref 3), based on 1978 data. The structural change in the industry since 1978 has had a profound effect on the efficiency of energy use. The specific energy consumption has fallen by 10% between 1978 and 1981 (cf Figure 1). Furthermore, the industry has considerably improved its collection and analysis of data on energy consumption since 1978. Hence the energy audit is substantially outdated by events. Although a similarly detailed analysis of 1981 energy flows is beyond the scope of this report, some insight into the trend in energy consumption can be obtained by comparing 1978 and 1981 data.

TABLE 4:   Energy Consumption by Fuel (GJ/tonne crude steel)
           (Source: Iron and Steel Statistics Bureau, ref 2)

Fuel	1978	1981
Coal (other than for coking)	0.2	0.2
Coke	9.6	10.2
Coke Breeze	1.8	1.1
Fuel Oil	4.4	2.7
Gas/Diesel Oil	0.1	0.1
Electricity	2.1	2.2
Gas – Coke Oven*	1.8*	1.6*
Natural	2.0	2.5
LPG	0.1	0.1
Less sales of electricity	0.2	0.3
TOTAL	21.9	20.4

* Excludes coke oven gas used at coke ovens.

TABLE 5:   Production of Crude Steel by Different Processes (10^6 tonnes)
           (Source: Iron and Steel Statistics Bureau, ref 2)

Process	1978	1981
Basic Oxygen Furnaces	11.34	10.53
Open Hearth Furnaces	1.76	–
Electric Arc Furnaces	7.05	4.93
Induction and others	0.17	0.11
Total	20.31	15.57
Cast to Ingots	16.72	10.31
Continuously Cast	3.15	4.96
Steel for Castings	0.44	0.31
Total	20.31	15.58

22.   Tables 4 and 5 show relevant data from the annual statistics for 1978 and 1981 (ref 2). Table 4 shows a breakdown of specific energy consumption by fuel type. The following trends can be observed:

- natural gas consumption rose by 25% whereas fuel oil consumption fell by 39%, which indicates significant fuel switching as well as a net saving of 1.2 GJ/tonne for oil and gas combined;

- specific coke consumption (ie per tonne of crude steel) increased by 6%.

23.   The structural changes which can be identified with these trends are:

- closure of open hearth steel-making furnaces;

- a fall in the proportion of steel made by electric arc furnaces;

- an increase in the proportion of steel made by the basic oxygen route;

- an increase in the proportion of steel cast continuously rather than by the traditional ingot casting route.

The tonnages of crude steel produced by the different routes are shown in Table 5.

24.   In the following paragraphs a brief technical description is given of energy use in the various individual processes in the iron and steel industry. These are grouped under three main headings:

- integrated iron and steel making;

- electric arc steel making;

- finishing operations.

25.   It should be noted that the energy used in coke-making does not properly fall within the iron and steel industry according to the standard industrial classification (1980) but is part of the energy supply industry (SIC Division 1). Thus in Table 4 the coke oven gas used at the coke ovens has been removed from the original ISSB data (ref 2). However, in any analysis of integrated iron and steel making it is usual to include the energy use in coke-making. This practice is followed in this report in order to facilitate comparison with other analyses. In the subsequent calculation of a specific energy consumption, however, coke-making is omitted.

Integrated Iron and Steel Making

26.   In an integrated steel works, all of the processes required for the conversion of iron ore to 'semi-finished' steel are balanced such that the output of the various intermediate materials and processes are matched to the required output of steel. Thus, the amount of coke produced in the coke ovens is, in principle at least, matched to the requirements of the blast furnaces, while the iron produced in the blast furnaces is matched to the capacity of the steel furnaces. Table 6 shows the production of intermediate materials,

and their energy requirements, for 1978 and 1981. In practice, coke ovens, sinter plants and blast furnaces are most efficient when operated continuously at a steady rate. Thus, within limits, some stocking of intermediate products (coke, breeze, iron) is required to allow for fluctuations in steel-making.

27. For each individual process in integrated iron and steel making, the net energy consumed per unit of product can be calculated by subtracting the energy value of any surplus fuel by-products used elsewhere (eg coke oven or blast furnace gas) from the energy input to the process. Column 2 of Table 7 shows the net energy consumption estimated for each process in 1978, in terms of GJ per tonne of product (ie coke, sinter etc); the corresponding data for 1981 are shown in Table 8. Columns 3 and 5 show the amount of each intermediate product required to produce one tonne of iron and crude steel respectively, taken from reference 2. Columns 4 and 6 show the net energy consumption per tonne of iron and crude steel respectively.

28. Coke making The major energy input to an integrated steel works is in the form of coal for coke making. The average yield of coke ovens was 0.724 tonnes of coke per tonne of coal in 1981). Allowing for the lower calorific value of the coke, the yield in energy terms is about 70%. Other fuel inputs are relatively minor. About 10% of the coal energy is released as surplus coke oven gas (calorific value ca. 20 MJ/Nm^3) which may be used elsewhere in the plant. Other products, benzole and tar, account for a further 6% of the coal energy input. The net energy consumed in the coke making process represents about 14% of the energy in the coal. The largest single loss is the sensible heat in the hot coke, for which heat recovery is technically feasible. The consumption of coal per tonne of coke fell by about 3% between 1978 and 1981.

29. Sinter production About ¾ of the iron ore used in a blast furnace producing iron for steel making is charged in the form of sinter, rather than directly as ore. Sinter is an agglomerate formed from iron ore, recycled iron particles (magnetically separated from steel works and iron works slags), scale from soaking pits (cf. paragraph 48) and mill scale, totalling about 90% of the input, and 10% limestone, by roasting on a continuous grate using coke breeze as the fuel. Carbon dioxide and sulphur dioxide are driven off resulting in a high degree of beneficiation and in a material with physical characteristics which, together with the coke, supports the burden of the blast furnace. Its surface structure also improves the ability for gas/solid reactions over that of sized lump ore. The coke breeze is totally consumed in the sinter production; the amount required fell by 30% between 1978 and 1981 (ref 2).

30. Blast Furnace In the blast furnace coke combines with the oxygen in the blast to form carbon monoxide which in turn reduces the iron ore/sinter to iron. Other impurities in the ore combine with limestone to form a slag which floats on top of the molten iron and can be removed separately from the furnace. Air is introduced to the furnace at temperatures in the order of 1,000°C through tuyères; sometimes it is enriched with oxygen to reduce the volume of inert nitrogen through the furnace. Fuel oil can be injected through the tuyères to replace up to about 20% of the coke; the economics of this practice are determined by the relative costs of fuel oil and coke. The blast through the tuyères, which is driven by blowers powered by steam, is pre-heated in stoves using recycled cleaned exhaust gas from the top of the blast furnace.

TABLE 6: Energy Consumption in Iron Making 1978-1981

	1978	1981	ISSB Reference Table	Page
Coke Ovens				
Coal Consumption 10^6 tonne	10.17	7.61	8	14
Coke Production 10^6 tonne	7.27	5.59	"	"
(Coal:Coke)	1.40	1.36	"	"
Coke oven gas produced (gross	7.99	7.77	"	"
per tonne of coke (GJ) (surplus	4.78	4.32	"	"
Sinter Production				
Total Production 10^6 tonne	15.8	11.5	13	19
Coke breeze/tonne sinter (kg)	82	55	"	"
Blast Furnaces				
No. in existence < 8 m. dia.	19	2	14	20
No. in existence > 8 m. dia.	20	15	"	"
Average no. in blast	29	11	15	21
Iron Production:				
Steelmaking)	10.94	9.39	"	"
Foundries) 10^6 tonne	0.42	0.075	"	"
Ferromanganese)	0.07	0.084	"	"
	11.43	9.55	"	"
Output/Furnace annum 10^6 tonne	0.39	0.87	"	"
Coke consumed 10^6 tonne	6.71	5.51	8	14
Fuel Oil consumed 10^6 tonne	0.314	negligible	16	22
Coke rate tonnes/tonne iron	0.587	0.577	"	"
Consumption of coke + fuel oil GJ/tonne iron	18.0	16.6		
Blast furnace gas: gross production (GJ/tonne iron)	7.3	6.1		

TABLE 7: Net Energy Consumed in Integrated Iron and Steel Making
(1978 data)
(excluding finishing operations)

1 Process	2 Net Energy Consumed GJ per tonne of product	3 Requirement of Product per tonne of iron	4 Net Energy Consumed GJ/tonne iron	5 Requirement of Product per tonne crude steel	6 Net Energy Consumed GJ per tonne crude steel
Coke Oven	6.5	0.587	3.82	0.510	3.32
Sinter Plant	2.4	1.320	3.17	1.143	2.74
Blast Furnace	14.4	1.00	14.40	0.866	12.47
BOS Furnace	0.4			1.000	0.40
		Total	21.39	Total	18.93
		Excluding coke-making	17.57	Excluding coke-making	15.61

TABLE 8: Net Energy Consumed in Integrated Iron and Steel Making
(1981 data)
(excluding finishing operations)

1 Process	2 Net Energy Consumed GJ per tonne of product	3 Requirement of Product per tonne of iron	4 Net Energy Consumed GJ/tonne iron	5 Requirement of Product per tonne crude steel	6 Net Energy Consumed GJ per tonne crude steel
Coke Oven	6.0	0.577	3.46	0.494	2.96
Sinter Plant	1.9	1.044	1.98	0.908	1.73
Blast Furnace	14.0	1.00	14.00	0.866	12.12
BOS Furnace	0.4			1.000	0.40
		Total	19.44	Total	17.21
		Excluding coke-making	15.98	Excluding coke-making	14.25

31.   The main energy input to the blast furnace is in the form of coke. Approximately 63% of this energy is carried through to the BOS steel furnace by the hot metal, partly as the energy required for transformation of the ore (42%) and partly as sensible and chemical heat (21%), the latter from carbon and metalloids. A further 20% of the input energy is recovered as clean surplus gas (calorific value averaging about 3.2 MJ/Nm^3), which is available for use elsewhere in the works. The remaining 20% or so of the input energy is lost by a variety of routes.

32.   From Table 6 it can be seen that the coke and fuel oil supplied to blast furnaces in 1978 was equivalent to 18.1 GJ/tonne of iron; in 1981 no fuel oil was used, but the coke consumption was 16.6 GJ/tonne. However, part of this consumption was offset by a fall in the production of blast furnace gas. Blast furnace gas production fell from 7.3 GJ/tonne of iron in 1978 to 6.1 GJ/tonne of iron in 1981. Part of the blast furnace gas is used at the blast furnace to heat the air for the blast, but no breakdown of the proportion is given in the published statistics. According to industry sources, the net energy consumption in the blast furnaces fell from about 14.4 GJ/tonne of iron in 1978 to 14.0 GJ/tonne of iron in 1981.

33.   Oil injection was discontinued in 1980 and is unlikely to be resumed at present oil costs. Other fuels can be injected through the tuyères. Tar and natural gas have both been used, particularly in the USA and many years ago experiments with pulverised coal injection were successfully carried out in the UK. At present BSC are re-investigating the use of coal, preferably without the need for pulverising, which is a significant operational cost. The economic advantage of coal injection would stem from the difference in price between coking coal (which is relatively expensive) and ordinary grades of coal, and a reduction in the coking process. However, the economic balance is complex because fuel injection affects the overall energy balance of an integrated steel works.

34.   Basic Oxygen Steel Making The iron produced in the blast furnace is transferred in molten form to the BOS furnace, using special insulated wagons known as 'torpedoes', because of their shape. The most effective way of reducing the heat loss from the hot metal is to ensure a minimum track or turn-round time for the torpedoes, which in turn is facilitated by having the minimum number in service. However, in spite of these precautions there is still a significant loss. Total losses of temperature are in the range 120-150°C. Ways of minimising these losses are under study at BSC.

35.   Cold scrap steel is also charged to the BOS furnace of which about 60% is recycled from downstream finishing processes. The hot metal, because of the chemical heat released when the oxygen is introduced, can melt about 24-28% of cold scrap. The level of silicon in the hot metal has an important effect on determining the range.

36.   Oxygen is injected into the BOS furnace via a lance. The energy consumed in the production of this oxygen represents about 480 MJ per tonne of crude steel. No additional heat is required for the operation of the BOS process, because the combination of the oxygen with metalloids in the hot metal generates the required heat. Electricity is used only for mechanical

power.   The net energy consumption of BOS furnaces is 0.4 GJ/tonne crude steel.

37.   It is possible to increase the proportion of scrap which can be charged to the BOS furnace above the limit set by exothermicity of the process, by providing additional heat.   In principle, 100% scrap could be processed in a BOS furnace by injecting a fuel.   BSC are investigating the use of lump coal in a conventional BOS vessel.   The indications from current research are that approximately 5 tonnes of scrap can be melted for each tonne of coal injected into the furnace, which is equivalent to about 6 GJ/tonne of scrap.   This is comparable to the energy requirement for scrap meltng in reverberatory furnaces in the non-ferrous metals industries (cf. sector reports 1.2-1.4).

38.   A gas of valuable calorific value (8.0 MJ/Nm^3) is evolved from the top of the BOS furnace.   The quantity of BOS gas produced per tonne of crude steel is relatively small compared with coke oven and blast furnace gases, and is not recovered in any plants in the UK.   In Japan, however, the gas is recovered from every BOS plant but one.   At approximately 700 MJ per tonne of steel this single loss represents about 3% of the specific energy consumption as defined in paragraph 9.   BSC is currently undertaking a project on BOS gas recovery at Scunthorpe.

39.   <u>Open Hearth Furnaces</u>   Since 1979 no steel has been made in the UK using open hearth furnaces.   Open hearth furnaces used roughly equal amounts of new iron and scrap steel.   In 1978 the total crude steel output was 1.764 M.te by this process.   The energy input per tonne of crude steel was estimated in the Energy Audit (3) to be 5.6 GJ, mainly as fuel oil or gas.   Allowing for the energy required in 1978 to make the 53% new iron, the total consumption per tonne of crude steel by the open hearth furnace route can be estimated as 15.1 GJ.   Thus the higher energy consumption of the open hearth furnace compared with a BOS furnace was offset by its greater capacity for converting scrap steel.   Open hearth furnaces were displaced by the BOS convertor because the latter is a faster and more economic process.   The scope for increasing the scrap rate of the BOS process has been discussed previously (paragraph 29).

Electric Arc Furnace Steel Making

40.   Electric arc furnace steel making uses mainly cold scrap steel together with some cold scrap cast iron as its raw material.   In principle, the process can be operated with 100% scrap steel and cast iron.   In practice, however, between 5 and 10% of the feedstock metal is in the form of new iron (cf Figure 2).   The 1978 value was 5.0%, and in 1981 it was 6.4%.   Adding new iron helps to correct the balance of alloying elements present in the scrap. The carbon content of the new iron also reduces the amount of carbon which would otherwise have to be added as coke to maintain reducing conditions within the arc furnace.

41.   The main energy input to the electric arc furnace is electricity, amounting to some 2.05 GJe/tonne of crude steel.   A further 0.35 GJ/tonne is used in the form of coke giving a total energy requirement of 2.4 GJ/tonne crude steel.   When the primary energy used to generate the electricity is included, the total energy consumption in steel making is 8.5 GJ/tonne of

crude steel. Comparable data is not available for 1981, but it is assumed that there has been no change in the energy consumption of the EAF process.

42. Apart from the thermal energy lost in generating the electricity the main energy losses from the EAF process are sensible heat in the exhaust gas (0.8 GJ/tonne) and loss of iron (0.5 GJ/tonne). The scope for reducing the energy consumption by simple heat recovery is thus limited. However, new designs of electric arc furnace are being considered which enable the exhaust gases to preheat the scrap charge. Additional preheating by natural gas or LPG could be used to further reduce the electricity requirements, but is not economic at current or projected future fuel price relativities. BSC operate one EAF plant with scrap preheating at present.

Average Specific Energy Consumption in Steelmaking

43. The average specific energy consumption in steelmaking excluding coke-making can be calculated from the SECs for each steelmaking process, as shown in Table 9. The SEC for each steelmaking process is the sum of the energy used directly by the steel furnace and the energy used to make the new iron which is required by the steelmaking process. Thus the electric arc furnace uses considerably more direct energy (2.4 GJ/tonne crude steel) than the basic oxygen furnace (0.3 GJ/tonne crude steel). However, because it uses mainly scrap iron and steel, the overall energy requirement is very much lower.

44. It follows that the average SEC for all steelmaking processes depends on the proportions of the different processes in the total output. Table 9 shows that the average SEC (excluding cokemaking and finishing processes) fell from 11.3 GJ/tonne crude steel to 10.8 GJ/tonne crude steel, or 4.4%, between 1978 and 1981. Although the SEC for the BOS route fell by 9%, the contribution which the BOS route made to the overall average increased by 5%. This explains why the specific coke consumption shown in Table 4 increased during the same period (cf paragraph 22).

45. It is clearly advantageous to use the maximum amount of scrap in making steel. Methods of increasing the amount of scrap which can be used in BOS furnaces have been mentioned previously. Provided that the additional energy which must be supplied to the BOS furnace to melt the scrap is less than the primary equivalent of the electricity consumed by the EAF process, then the former is likely to be economically preferable. However, as long as there is a limit to the amount of scrap which can be used within the integrated BOS process, then the lowest overall energy consumption can be achieved by using any remaining available scrap in EAF plant. It is worth noting that since 1980, exports of scrap from the UK have been around 3 M tonnes per year, which is much higher than in earlier years.

46. There is, however, a limit to the amount of scrap steel which can be bought from outside the industry, because the level of non-ferrous metal impurities, such as copper, must be carefully controlled. These impurities are difficult to remove and will degrade the quality of the steel. Steel scrap originating from, for example, motor cars or electrical appliances may contain appreciable quantities of copper, which would build up in steel which is recycled many times. This factor alone will dictate the necessity of maintaining a large proportion of 'new' iron in steel production.

TABLE 9:   Average Specific Energy Consumption for Steelmaking
(excluding coke-making and finishing processes) in 1978 and 1981

Production Process	1978				1981			
	New Iron Requirement (r) (tonnes/ tonne crude steel)	Process SEC (GJ/tonne process product)	% of Total Production	Contribution to Average SEC (GJ/ tonne crude steel)	New Iron Requirement (r) (tonnes/ tonne crude steel)	Process SEC (GJ/tonne process product)	% of Total Production	Contribution to Average SEC (GJ/ tonne crude steel)
(Iron-making)	-	(17.6)	-	-	-	(16.0)	-	-
Basic Oxygen	0.866	15.6	56	8.7	0.866	14.25	68	9.7
Open Hearth	0.53	15.1	9	1.4	-	-	-	-
Electric Arc	0.056	3.4	35	1.2	0.072	3.6	32	1.2
Average Specific Energy Consumption (GJ/tonne crude steel)				11.3				10.8

Calculation of Process SECs:

SEC(BOS) = r x SEC(Iron) + 0.4   (para 36)
SEC(OH) = r x SEC(Iron) + 5.6   (para 39)
SEC(EAF) = r x SEC(Iron) + 2.4   (para 41)

Finishing Operations

47.   The processing of crude steel into the finished products sold by the steel industry involves two distinct stages:

- primary finishing, which is the conversion of the crude steel into slabs, blooms or billets (known as 'semi-finished' steel); it involves either ingot casting followed by primary rolling, or continuous casting directly into the above categories.

- secondary finishing, in which the products (such as bar, rod, rail wire and sheet) are made from semi-finished steel by further rolling and reheating.

Primary Finishing

48.   Ingot casting This is the traditional method for primary finishing. Liquid steel is poured into moulds, where it cools and partly solidifies. The ingots thus formed are removed from the moulds and then reheated and maintained at a high temperature (1,150-1,400°C depending on quality). This process, which lasts for 4-12 hours, is carried out in oil or gas fired pits, and is called 'soaking' (ie soaking in heat). It produces a near-uniform temperature throughout the ingot. A scale forms on the surface of the ingot which is usually thick enough to remove surface defects. The scale is broken off either as the surface of the ingot initially cools on removal from the pit and then totally in the first passes in the primary rolling mill where the ingot is transformed to the semi-finished products required.

49.   According to the Energy Audit (3), the energy input into the ingot casting process amounts to 1.53 GJ/tonne of crude steel input. The main requirements are fuel oil or gas for reheating and soaking and electricity for rolling. The lost metal, in the form of crops and oxides which are recycled, is typically about 13% of the ingot. This loss represents a significant energy penalty, but is inherent in the process. Some of the heat lost during soaking could, in principle, be recovered by installing heat recovery equipment; however it would be preferable if soaking could be avoided.

50.   Continuous casting This method of primary finishing avoids both soaking and most of the loss of metal inherent in ingot casting. Liquid steel is fed from a ladle to the casting machine from which the solidified steel can be withdrawn continuously through rollers and cooling water sprays. The cast steel is then cut into the required lengths of slab or bloom. The loss of metal (about 5%) is significantly lower than for the ingot casting route and only a very small input of fuel and electric power is required amounting to a total energy input of 0.18 GJ/tonne of crude steel (ref 3).

51.   To compare the energy losses associated with the lost metal in the two casting routes, consider a tonne of liquid steel being cast by each route. In the case of ingot casting, 13% of the cast steel has to be recycled, either to the sinter plant or to the steel furnace, and subsequently re-cast. The energy required to recycle it includes about 2 GJ/tonne for remelting it. In the BOS furnace this is provided by the hot iron from the blast furnace. A further 1 GJ/tonne is required for process energy supplied to BOS furnace and 1.5 GJ/tonne for casting. The amount of oxidised material requiring reduction in the blast furnace is small and is ignored in the analysis. Thus the total

TABLE 10: The Proportion of Crude Steel Cast by the Continuous Casting Process, and its Effect on the Average Energy Required for Primary Finishing, in 1978 and 1981

	1978	1981
Energy required for ingot casting (GJ/tonne)	1.53	1.53
Recycling lost metal (GJ/te)	0.6	0.6
Energy required for continuous casting (GJ/tonne)	0.18	0.18
Recycling lost metal	0.16	0.16
Proportion of crude steel cast by continuous process	15%	32%
Weighted average energy requirement for primary finishing (GJ/tonne)	1.33	1.10
Recycling lost metal	0.53	0.46

energy required to recycle the 'lost' metal is 0.13* (2 + 1 + 1.53) = 0.59 GJ/tonne. After one recycle the lost metal is reduced to $(0.13)^2$ = 0.017 tonnes which requires a further 0.01 GJ/tonne. The tonne of liquid steel is now semi-finished with a total energy consumption during primary finishing of 2.1 GJ/tonne. The continuous casting process can be analysed similarly, recycling 5% of the metal through the steel making process and allowing for 0.18 GJ/tonne fuel input to the casting machine. In this case the recycling requires an extra 0.16 GJ/tonne, giving a total of 0.34 GJ/tonne.

52.   When allowance is made for the recycling of lost metal the energy saving which results from continuous casting is thus about 1.8 GJ/tonne of semi-finished steel. For the EAF route, it is somewhat higher when the electrical energy required to recycle lost metal is accounted for in primary energy terms. Between 1978 and 1981, the proportion of crude steel cast continuously increased from 15% to 32%. Consequently, the average energy required for primary finishing fell by 0.23 GJ/tonne, a decrease of 17% (cf. Table 10). An additional saving of 0.07 GJ/tonne is due to the reduction in metal recycling.

Secondary Finishing

53.   A diverse range of secondary finishing operations are required to produce the variety of finished steel products. However, in all cases the process involves reheating and rolling the semi-finished slab or billet to form the final product (plate, sheet, bar etc) which is then cooled in cooling banks.

54.   An analysis of energy flows in secondary finishing operations is given in the Energy Audit (3). There is a large input of fuel for reheating which may be oil, gas or by-product gas available on site. The average yield of finished product is taken to be 86% of the semi-finished input, with the remaining 14% recycled through the steel making process. The energy requirement in secondary finishing is estimated to be 2.8 GJ/tonne of semi-finished steel (ref 3), which is equivalent to 2.5 GJ/tonne of crude steel.

55.   The Energy Audit (ref 3) estimate for energy consumption in primary and secondary finishing operations in 1978 was 3.8 GJ/tonne. This figure is an underestimate of the actual energy consumption in these operations, because the audited data was obtained for plant operating at full load. In addition, there are other energy-consuming activities which have not been included in the Energy Audit.

56.   In an earlier section (cf Table 9) it was shown that in 1981 the average specific energy consumed in the production of crude steel by all routes was 10.8 GJ/tonne (excluding finishing). The overall specific energy consumption of the industry in 1981 was 20.4 GJ/tonne (Table 4). **The difference between these two values, 9.6 GJ/tonne, represents the energy consumption in all finishing operations, together with energy 'overheads' for non-process uses.** This value includes consumption of surplus blast furnace gas, which is estimated to be equivalent to 1.5 GJ/tonne crude steel.

57.   A breakdown of the energy consumption by the British Steel Corporation, which accounts for 85% of all steel production, including 100% of integrated iron and steel production, for the month of October 1981 is shown in Table 11

- 19 -

TABLE 11:   Specific Energy Consumption by the British Steel Corp
in October 1981

Process	Specific Energy Consumption GJ/tonne crude steel
Coke making	2.66
Sinter making	1.37
Pellet making	0.18
Iron making	9.68
Steel making	0.75
Casting, rolling and finishing	4.70
Power generation	0.67
Boiler plant	0.75
Other activities	1.26
Losses	0.42
Total (excluding coke-making)	22.44 (19.78)

Note: data has been converted from liquid steel basis to
crude steel basis by assuming a ratio of liquid
steel:crude steel of 1.032.

TABLE 12:   Specific Energy Consumption by the Private Sector
Iron and Steel Industry in 1978 and 1981

Fuel	Specific Energy Consumption GJ/tonne crude steel	
	1978	1981
Coal (other than coking)	0.54	0.26
Coke and Breeze	0.89	-
Fuel oil	2.79	1.95
Gas oil	0.23	0.18
Electricity	2.94	3.55
Gas: - coke oven	6.06	0.12
- natural	2.76	6.53
- other	0.24	0.25
Total	16.44	12.83
Private sector crude steel make (10^6 tonnes)	3.631	2.330

(ref 5). This is described as a 'good month', with no interruption to production as a result of holidays. Because the data does not give any indication of the proportion of electric arc steel making, the data for iron and steel production cannot be compared directly. However, the data for finishing and other operations is directly comparable. This shows that primary and secondary finishing consumed 4.7 GJ/tonne crude steel (although a number of non-integrated finishing plants were excluded from the analysis). Power generation, boiler plant, other activities and losses accounted for a further 3.1 GJ/tonne crude steel. Thus a total value of 9.5 GJ/tonne crude steel for finishing and non-process energy overheads for the whole of 1981 is a reasonable estimate, compatible with the data in Table 11 for October 1981.

58.   Private Sector Steel-making Specific energy consumption by the private sector of the iron and steel industry was 16.44 GJ/tonne crude steel in 1978, when private sector production accounted for 17.9% of UK crude steel (ref 4). Specific energy consumption fell to 12.83 GJ/tonne crude steel in 1981, when production was 15.2% of UK crude steel. A breakdown of the SEC by fuel is shown in Table 12. The 1978 figures include production by one private sector blast furnace and two open hearth furnaces, which have since closed. Hence the 1981 SEC shows a higher electricity consumption, because of the higher weighting of EAF furnaces for steel making. The consumption of oil and gas fell from 12.08 to 9.03 GJ/tonne, which presumably is largely due to savings in finishing operations. Note, however, that there has been a significant switch from coke-oven gas to natural gas.

Hypothetical 'Best Practice' Energy Consumption in Iron and Steel Industry

59.   Variations in energy consumption from one plant to another are often dependent on specific conditions, such as differences in the product. Hence, the 'best practice' in terms of the plant with lowest energy consumption may not be readily applicable to other plants. Furthermore, there are no separate data for individual integrated plants published in the UK. Nevertheless, it is useful to consider what can be achieved with the most modern and efficient plant.

60.   The International Iron and Steel Industry (IISI) Committee on technology (5) defined a 'reference' integrated steel plant based on 8×10^6 tonnes per year of crude steel output, as shown in Figure 4. Two reference electric arc plants were also examined, one fed entirely with scrap, the other with 75% direct reduction iron and 25% scrap. The plant sizes are considerably largeh than any UK plants, so direct comparisons may be distorted by scale effects, and should be treated with caution.

61.   The integrated plant is based on oil-free operation with by-product gases, from coke ovens, blast furnaces and BOS plant, being used as process plant fuel gases. The balance is supplied to the boiler house to raise steam for power generation (70% of the plant's power requirements are internally generated). Process steam demand is satisfied primarily by the output from a back pressure turbine which constitutes part of the power generating equipment. Energy saving facilities include top pressure turbines and BOS gas recovery. The BOS furnace is fed with 75% hot metal; 70% of the liquid steel produced is continuously cast.

62.   The electric arc furnace plants are based on energy efficient, oxy-gas burner assisted furnaces. Energy inputs include electricity, natural gas

FIGURE 4: IISI 'Reference Integrated Steel Plant'
(Output in 10^3 t/yr)

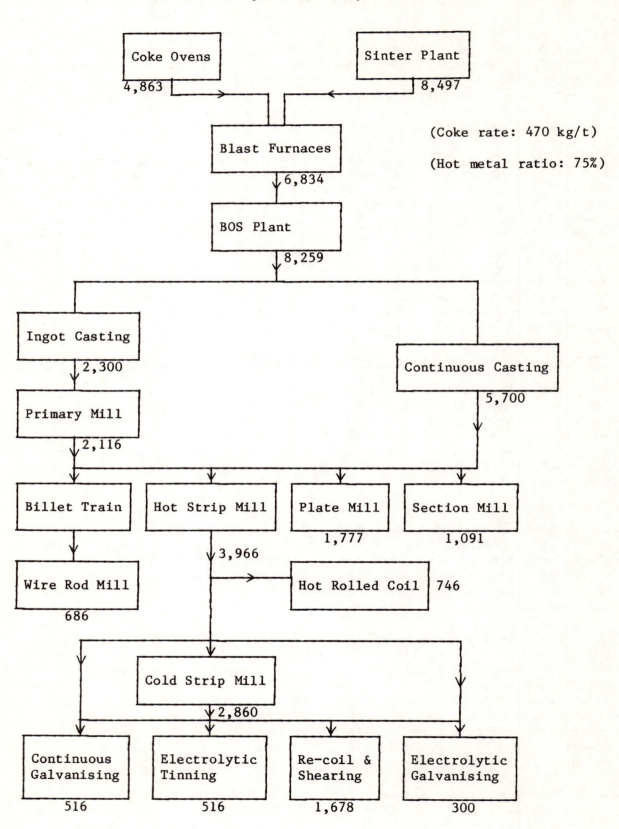

TABLE 13: Calculated Energy Consumption for IISI 'Best Practice' Reference Plants (ref 5) (including coke-making)

Reference Plant	Specific Energy Consumption, GJ/tonne crude steel		Percentage of Total Energy Inputs								
	Primary Energy	Heat Supplied	Coal	Elec.	Nat. Gas	LPG	Oxygen	Carbon	Electrode	Pellets	Unspecified
Integrated											
Iron and steel making	14.8	14.6	90.8	4.9	-	0.2	1.6	-	-	2.5	-
Finishing	4.6	4.2									
Total	19.4	18.8									
Scrap-Fed EAF											
Steel making	5.7	2.6	-	74.7	15.5	-	2.4	3.3	1.6	-	2.5
Finishing	2.8	1.8									
Total	8.5	4.4									
DRI and Scrap-Fed EAF											
Iron making	11.8		-	37.6	51.6	-	1.0	0.2	0.7	7.9	1.0
Steel making	5.9										
Finishing	2.8										
Total	20.5	15.4									

N.B. The IISI calculations are given in terms of primary energy equivalent of electricity; for the purposes of comparison in this report, the totals have been converted to heat supply terms.

- 23 -

(particularly in the case of the direct reduction route) and carbon (in the form of electrodes and as powder for slagging).

63.   For both the integrated and EAF plants, purchased electricity is accorded a primary energy equivalence based on a modern fossil fuel fired power station operating at 37.4% thermal efficiency. Self generated electricity is produced at 35.1% thermal efficiency.

64.   Table 13 shows the calculated specific energy consumption for each of the IISI reference plants in both primary energy and heat supplied terms, and the breakdown of energy inputs by fuel (and energy containing material inputs).

65.   The breakdown of the energy consumption in Table 13 indicates that the energy consumption in steel making (including coke-making) for the reference integrated plant is 14.6 GJ/tonne, which is 2.6 GJ/tonne less than the 1981 UK consumption for these processes; finishing operations and non-process energy use 4.2 GJ/tonne in the reference process, compared with 9.5 GJ/tonne estimated for the 1981 UK consumption.

66.   For the 100% scrap-fed EAF process, energy consumption in steel making is 2.6 GJ/tonne compared with the value of 2.4 GJ/tonne estimated for the UK in reference 3. The IISI reference plant presumably requires more energy supplied to it than is actually required by UK plant because the latter included 5% or more of new iron containing carbon, which has significant calorific value. The finishing operations in the reference EAF process are very simple and hence require only 1.8 GJ/tonne.

67.   Comparison of the actual 1981 UK energy consumption with the IISI reference plant indicates that there is still considerable scope for reducing the former. This can be done through a combination of better operating procedures, higher loading of plant (much of which is operating below capacity) and investment in energy saving measures. Since early 1982, BSC has been operating a monitoring and targeting programme, aimed at reducing specific energy consumption by between 7% and 12% over a three year period. By mid 1983, there were indications that the reduction in energy consumption was already within the target range. Further improvements will depend increasingly on the capital intensive measures discussed in the next section.

ENERGY CONSERVATION POTENTIAL

68.   The future trend of specific energy consumption in the iron and steel industry will depend on a complex set of factors, which include the technical scope for energy conservation and the overall requirements for new plant. In this section, the various energy conservation technologies are discussed and their cost effectiveness examined. The rate and extent of deployment of these measures will be discussed in detail in a later section.

69.   In the previous sections, it has been noted that the specific energy consumption has fallen by about 40% since 1956, and by about 7% between 1978 and 1981. Clearly much has already been done to save energy, and to some extent many of the easy options of 'good housekeeping' have already been applied. Much of the remaining potential for energy conservation lies in applying new techniques, which involve substantial investment. These are described in the following paragraphs under two headings: technically proven measures; and measures which still require research and development.

'Technically Proven' Conservation Measures

70.   Table 14 summarises those technologies which can be regarded as technically proven. This term is used to indicate that the technologies are actually in use, if not in the UK, then in other countries. However, their cost effectiveness in UK conditions must be examined in detail.

71.   Low temperature heat (30-40°C) in the cooling water from furnaces could, in principle, be recovered by using this cooling water as boiler feedwater. This would require cooling water of much higher quality than is generally used at present. In the case of blast furnaces, an alternative form of cooling (evaporative cooling) would release the waste heat at a higher temperature, allowing it to be used to generate process steam. Evaporative cooling is already used in some blast furnaces in Germany but it requires considerable structural changes to the areas of a blast furnace requiring cooling, and would not be applicable to existing furnaces.

72.   In modern, high pressure blast furnaces, the pressure of the gas at the top of the furnace is a potential source of energy. Various methods of recovering this energy have been developed abroad, notably in the USSR and France. After cleaning, the gas is expanded through a turbine. The shaft power from the turbine can either be used to drive the blowers on the blast furnace (similar to turbo-charging in car engines) or be coupled to an alternator to generate electricity. However, there are only two blast furnaces in the UK which have a sufficiently high operating pressure to use a turbine system gainfully.

73.   Heat recovery from coke ovens and sinter plants is technically proven overseas, although the cost effectiveness in the UK is marginal at present fuel prices. Partial heat recovery from sinter plants is being introduced on some plants in the UK, but full heat recovery may not be achieved until the end of the century.

74.   In basic oxygen furnaces the recovery of the calorific value of the gas produced at the top of the furnace is widely practised in Japan. In the UK, it has not been used in the past, mainly because of the lack of investment capital. However, it seems likely that BOS gas recovery will be introduced by the end of the century.

75.   The exhaust gases from an electric arc furnace can be used to preheat the scrap in another furnace prior to firing. One way of doing this would be through a tandem vessel arrangement with interchangeable tops, such that each vessel could be fired alternatively, with the other in preheat mode.

76.   In finishing operations it has been shown in earlier sections that continuous casting can save considerable amounts of energy compared with the traditional ingot casting method. Continuous casting was first introduced in the UK in the early 1960s, reaching 1% penetration of total steel production in 1965. By 1981, 32% of primary finishing was by the continuous casting process. Assuming the same rate of penetration, continuous casting should account for 50% of the total in 1990 and 70% in 2000.

77.   Where traditional ingot casting methods continue to be used, the energy losses can be reduced by improving heat recovery on soaking pits and reheating furnaces. Although recuperators have long been used, BSC are currently testing a prototype ceramic recuperator of an advanced design.

TABLE 14: 'Technically Proven' Energy Conservation Potential in the Iron and Steel Industry

Plant	Technology	Energy Saving Potential GJ/t.c.s.	Possible Use of Recovered Energy	Cost Effectiveness	Current Status	Prospects in UK
Coke Ovens	Coke dry-quenching (recovery of heat at 1000°C)	0.72	- Steam/electricity generation - Coal preheating	Capital intensive; 4-10 year payback	Full scale demonstration in Japan and USSR	Introduce ~ 2000
Sinter Plant	Partial heat recovery (> 500°C) or	0.1	Preheating combustion air	2 year payback	First installation in progress	Full development before 2000
	Full heat recovery (> 250°C)	0.3	" " "	Full recovery is marginal	Difficult to adapt to existing plant	Introduce ~ 2000
Blast Furnace	Evaporative cooling	0.38	Steam for process heat or electricity generation			
	Recovery of pressure energy at top of furnace	0.36	Shaft power for furnace blowers/ electricity generation	Capital intensive	Available but only applicable to high pressure furnaces	Two existing furnaces and any new ones
	Reduced gas losses	0.58	Replace fossil fuels on-site or sell offsite	" "		
BOS Furnace	Top gas: sensible heat recovery	0.2	Steam raising	2-4 year payback	Partial combustion system used in Japan	Full deployment by 2000
	Top gas: combustion	0.5	" "	" "		
Electric Arc Furnace	Exhaust gas recovery	0.11-0.18	Scrap preheating	2-4 year payback	First installation in operation	Full deployment by 2000
Finishing Plant	Substitution of continuous casting for ingot casting or	1.2	Saves reheating fuel		Continuous casting accounts for 32% of production	Maximum of 80% of production by 2000
	Heat recovery from soaking pit waste gases and preheat furnaces	1.0	Self use or steam raising			

TABLE 15: 'R&D' Energy Conservation Potential
in the Iron and Steel Industry

Plant	Technology	Magnitude of Loss GJ/t.c.s.	Possible Use of Recovered Energy	Current Status
Coke oven	Stack gas sensible heat recovery (~190°C)	0.26		Problems with buoyancy and corrosion
Sinter Plant	Stack gas sensible heat recovery (150°C)	~0.35		As for c.o. stack gas
Blast Furnace	Slag sensible heat recovery (1400°C)	0.32	High temperature heat	Possible fluidised bed system being developed in Sweden
BOS Furnace	Reduction of iron loss by vaporisation, etc	0.35	Improves yield of process	Detailed study of process required
	Slag sensible heat recovery	0.14	High temperature heat	As for BF
Electric Arc Furnace	Slag sensible heat and cooling water recovery	0.1	High temperature heat	As for BF

78.   If all of the measures in Table 14 were taken up, the aggregate of the
energy saving potential in integrated steel-making (avoiding double counting
of mutually exclusive measures) amounts approximately to 4.2 GJ/tonne of crude
steel and 0.2 GJ/tonne crude steel in electric arc steel-making. Assuming a
ratio of 2:1 for BOS to EAF processes, the reduction in average energy
consumption would be 2.9 GJ/tonne crude steel. This represents a 'technically
proven' conservation potential of about 15% of the 1981 specific energy
consumption; this is in addition to any further savings due to 'good
housekeeping' measures.

'R&D' Conservation Measures

79.   As well as those measures which are technically proven and commercially
available, there are additional measures which may become available, but
which still require further R&D. Table 15 summarises the different measures
which fall into this category.

80.   High temperature heat at about 1,400°C is potentially recoverable from
the slags produced in the blast furnace and steel making furnaces. Various
systems have been proposed for recovering and using this heat, including a
fluidised bed system under development in Sweden.

81.   Recovery of the sensible heat in waste gases from coke ovens and sinter
plants would provide heat at an intermediate temperature of about 150-190°C,
but there are problems in recovery of this heat due to the corrosiveness of
the gas and the loss of buoyancy which results from cooling it.

PROJECTIONS OF FUTURE SPECIFIC ENERGY CONSUMPTION

82.   In order to estimate future trends in specific energy consumption, it is
necessary to take account of:

- process changes, especially those which change the proportion of
  scrap used to make steel;

- investment in measures which explicitly save energy, including
  improved operational procedures (ie management measures).

These in turn will depend to a large extent on future levels of output.
Hence, in considering the improvements which may occur in SEC, it is necessary
to consider a range of future output, by postulating two scenarios, designated
as high and low. The high scenario assumes that output rises to around 18
million tonnes by 2000; the low scenario assumes that output will remain
around 12 million tonnes by 2000.

Effect of Output on Improvements in Energy Efficiency

83.   Future levels of output can be expected to affect energy efficiency
because they influence the rate at which existing plant is replaced. Thus, in
general it might be expected that a high output scenario would provide more
scope for installing new or replacement capacity. However, in the iron and
steel industry current levels of capacity are likely to be sufficient to meet
demand for some time, even on an optimistic view of output growth. On the
other hand, if output were to continue at the current low levels, additional
closure of capacity would be inevitable. In the recent past, capacity closure

has resulted in an improvement of energy efficiency because the more
inefficient plant has been retired. However, most of the remaining plant in
the industry is relatively modern, and the scope for improvements in energy
efficiency due to further closures is consequently less than has been observed
recently.

84.   It is assumed that, on balance, energy efficiency would improve more
rapidly in a high output scenario than in a low output scenario, because of
more rapid uptake of those measures which involve investment in additional or
replacement equipment.

85.   The availability of scrap will be affected by a number of factors, which
tend to have a counteracting effect. Thus:

  - scrap generated internally in a steel works is likely to decline as
    finishing operations become more efficient in terms of yield of
    finished metal, especially through the increasing use of continuous
    casting;

  - scrap arisings due to obsolescence of capital goods and
    consumer-durables is not related to UK output of steel, but to
    consumption of steel-containing goods including imports. Thus the
    increased imports of eg cars and white goods in recent years means
    that consumption of steel has not fallen as rapidly as production;

  - net exports of scrap have risen sharply in recent years, reflecting
    an imbalance between availability of scrap and requirement in UK
    steel production. A reversal of this trend would increase the
    availability of scrap.

86.   It is difficult to assess the overall effect which these trends would
have on the availability of scrap as a proportion of future steel production.
In the projections of future SEC it is assumed that the present proportion of
50% will remain unchanged. However, to test the sensitivity of the results to
this assumption, a sensitivity analysis has been carried out with alternative
scrap availability assumptions (cf paragraph 98).

Effect of Technical Change on Energy Efficiency

87.   The effect of various different technical changes on the overall
specific energy consumption can be analysed in terms of component SEC terms
for the individual processes. Table 16 shows these terms for integrated iron
and steelmaking, electric arc steelmaking and finishing operations, together
with the key parameters which determine the ratio of new iron to scrap, ie the
requirement of new iron for BOS steelmaking and the proportion of electric arc
steelmaking in the total output.

88.   The factors which are expected to influence these parameters in the
future are discussed in the following paragraphs.

89.   Integrated Steelmaking The specific energy consumption for steelmaking
by the integrated blast furnace/basic oxygen steelmaking furnace route depends
on:

- specific energy consumption for production of new iron in the blast furnace, and

- the ratio of new iron:scrap in the input to the BOS furnace.

The scope for reducing the former, indicated by the difference between the 1981 value and the hypothetical 'best practice' value taken from the IISI study (cf paragraph 65), is about 15%. Some of the technologies which could contribute to such a reduction have been discussed in earlier paragraphs. These include increased use of blast furnace gas. In a low output scenario, the closure of excess capacity could lead to further improvements in energy efficiency similar to those observed in the period 1978-81. In a high output scenario, improvements would result from higher utilisation of capacity.

90.   The use of pulverised coal injection into blast furnaces is assumed to reduce coke consumption by about 10%. It is difficult to quantify the effect which this would have on the overall energy balance of an integrated plant. It is likely that more energy would be required in the form of coal than is displaced as coke; there would also be an increase in production of blast furnace gas but a reduction in surplus coke oven gas. It is assumed that the energy injected as coal would be 40% more than the coke displaced, with no overall change in byproduct gas. This has the effect of transferring energy consumption from the coke-making process (which is outside the definition used for the iron and steel industry in this report) to the blast furnace.

91.   The ratio of scrap:new iron in the feedstock to the BOS furnace is assumed to increase from 24% scrap at present to about 35% in the high scenario and 30% in the low scenario by 2000. Allowing for the yield of crude steel this is equivalent to a requirement of 0.74 tonnes of iron per tonne of crude steel in the high scenario and 0.80 tonnes of iron per tonne of crude steel in the low scenario. The additional energy which would be required for processing the additional scrap is assumed to be supplied in the form of pulverised coal at the rate of 0.2 tonnes of coal per tonne of additional scrap (ie approximately 6 GJ per tonne of scrap).

92.   The component SEC for production of crude steel by the integrated BOS route can be calculated by the following expression:

$$SEC_{BOS} = r.SEC_{Iron} + (0.866 - r).6 + 0.4 - X \text{ (GJ/tonne)}$$

where SEC_{Iron} is the SEC for new iron (in GJ/tonne iron) and r is the amount of iron required to make one tonne of crude steel. The second term in the expression is the amount of additional energy which must be supplied to the BOS furnace as pulverised coal. This is assumed to be in addition to the present direct energy consumption of the BOS furnace (0.4 GJ/tonne crude steel). The final term, X, is the energy saving through BOS gas recovery - equivalent to 0.7 GJ/tonne when fully utilised.

93.   Electric Arc Steelmaking The energy consumption in electric arc steelmaking is assumed to fall from 2.4 GJ/tonne at present to 2.2 GJ/tonne by 2000, due mainly to improved heat recovery and scrap preheating.

94.   The requirement of new iron in the feedstock for EAF steelmaking is assumed to be 5% (ie 0.056 tonnes iron/tonne crude steel). The total specific

energy requirement in steelmaking by the EAF route (excluding finishing operations) in 2000 in the high scenario is therefore:

$$SEC_{EAF} = 0.056 \ SEC_{Iron} + 2.2 = 2.9 \ \text{GJ/tonne crude steel}$$

and similarly for the low scenario.

95.   The proportion of EAF steelmaking is determined by the assumptions made on the total availability of scrap and the proportion of scrap fed to the BOS furnace in integrated steelmaking. For the reasons discussed in paragraphs 85-86, it is assumed that there will be no change in the overall proportion of scrap used in steelmaking - ie 50% of crude steel production. The increased use of scrap in integrated steelmaking would therefore result in a lower proportion of EAF steelmaking. For example, in 2000 in the high scenario, EAF steelmaking would be only 17% of the total compared with 32% in 1981.

96.   Finishing and Non-Process Energy Requirement The energy use in finishing processes, estimated to be about 9.6 GJ/tonne in 1981, would appear to offer the greatest scope for future savings. Comparison with the IISI hypothetical best-practice suggests that a value of 5.5 GJ/tonne crude steel would not be unreasonable by 2000, at least in the high scenario. In the low scenario, a somewhat higher value would be more appropriate, reflecting a lower level of investment in new plant.

Projections of Specific Energy Consumption

97.   The overall specific energy consumption can be calculated from the component SECs by using the following expression:

$$SEC_{overall} = SEC_{BOS} \cdot (1 - e) + SEC_{EAF} \cdot e + S_f \ \text{(GJ/tonne crude steel)}$$

where SEC_{BOS}, SEC_{EAF} and S_f are the component SECs for integrated steelmaking, electric arc steelmaking and finishing operations respectively, and e is the proportion of steel made by the electric arc process. The values for the overall SEC which result from the assumptions described above are shown in Table 16. The overall SEC falls by 23.5% to 15.6 GJ/tonne crude steel in 2000 in the high scenario, and by 17.6% to 16.8 GJ/tonne crude steel by 2000 in the low scenario.

98.   Sensitivity Analysis It is clear that the overall SEC is dependent on the assumptions which have been made about the availability of scrap. In order to illustrate the sensitivity of the results to this assumption, alternative assumptions have been tested for the 2000 high scenario case. These are also shown in Table 16. Increasing the overall availability of scrap to 60% of crude steel production, while maintaining the 35% scrap feed rate to the BOS furnaces, would allow the EAF steelmaking proportion to rise to 37%. This would result in a reduction of the overall SEC from 15.1 GJ/tonne to 13.5 GJ/tonne, which is a reduction of about 10% compared with the base case. On the other hand, if the overall availability of scrap were reduced to 40% of crude steel production, and the BOS scrap rate unchanged, then there would be no scrap available for EAF production. This would result in an increase in SEC to 16.5 GJ/tonne, which is about 9% higher than the bas case.

TABLE 16: Components of Specific Energy Consumption Projected to 2000

Process	1981	1990		2000		Alternative Assumptions for 2000 High Scenario (changed scrap availability)	
		High	Low	High	Low	60% Scrap	40% Scrap
Integrated Steelmaking							
SEC for new iron (GJ/tonne iron)	16.0	15.5	15.7	15.0	15.3	15.0	15.0
Scrap rate (% of BOS feed)	24	30	25	35	30	35	35
Requirement of new iron in steelmaking (tonnes iron/ tonne c.s.)	0.866	0.80	0.85	0.74	0.80	0.74	0.74
BOS gas recovery (GJ/t)		0.3	0.1	0.6	0.5	0.6	0.6
SEC for steel (GJ/tonne c.s.)	14.25	12.5	13.7	11.0	12.2	11.0	11.0
Electric Arc Steelmaking							
SEC (GJ/tonne c.s.)	3.6	3.1	3.2	2.9	3.0	2.9	2.9
Proportion of EAF steel	0.32	0.24	0.31	0.17	0.24	0.37	0
Finishing, etc							
SEC (GJ/tonne c.s.)	9.6	7.0	8.0	5.5	6.5	5.5	5.5
Overall SEC	20.4	17.2	18.4	15.1	16.5	13.5	16.5

Fuel Mix

99. The fuel mix used by the iron and steel industry can be determined largely from the mix of processes assumed previously, together with some additional assumptions about interfuel substitution in finishing operations. Table 17 shows the projected specific energy consumption broken down by fuel type.

100. The previous calculations of SEC for individual processes have been based on net energy consumption, with the surplus blast furnace gas transferred to finishing processes. In order to calculate the mix of fuels supplied to the industry as a whole, it is the gross consumption of coke and breeze in iron-making which is significant; internal transfers of blast furnace gas can be ignored. Thus, in 1981, coke consumption in iron-making was 16.6 GJ/tonne of iron (table 6). The adoption of coal injection into the blast furnace could reduce the coke consumption by 10%. The amount of coal required would be 1.4 times the amount of coke displaced.

101. Coke breeze consumption in sinter plant is assumed to remain constant at about 1.5 GJ/tonne iron. Coke oven gas is assumed to be 14% of coke and breeze consumption.

102. Production of surplus blast furnace gas in 1981 has been estimated to be 2.5 GJ/tonne of new iron (ie 1.5 GJ/tonne of crude steel). This is assumed to be unchanged in the future - an approximation which ignores changes in the balance of gas production caused by coal injection into the blast furnace. Although the blast furnace gas is not accounted for as a bought-in fuel (unlike coke-oven gas, which is notionally bought-in), it does reduce the amount of commercial fuel which is required for finishing operations.

103. The consumption of electricity in 1981 was 9.3% of total energy consumption. About 1/3 of this is used by electric arc furnaces, and will therefore be reduced by a reduction in the proportion of EAF steelmaking. It is expected that as the energy requirement in finishing processes falls, the amount of private electricity generation will be increased to make use of surplus blast furnace and BOS gas. This will have the effect of further reducing net electricity purchases. The impact on the percentage share of electricity in the fuel mix will, however, be attenuated by reductions in the use of other fuels. There may be some increase in the use of electric induction heating in finishing operations, but this is unlikely to be large enough to have a noticeable effect on the overall use of electricity. It is assumed therefore that electricity would represent 9% of total energy needs by 1990 and 8% by 2000.

104. The consumption of oil and natural gas are estimated by subtracting the other fuels from the total specific energy consumption, and dividing the balance approximately equally between oil and natural gas.

TABLE 17: Projected Specific Energy Consumption by Fuel Type
(GJ/tonne crude steel)

Fuel	1980	1981	1990		2000	
			High	Low	High	Low
Coal	0.4	0.2	1.6	1.2	2.0	1.6
Coke and breeze	10.7	11.3	10.0	9.8	9.7	9.9
Oil and LPG	4.6	2.9	1.3	2.1	0.4	1.1
Natural gas	3.8	2.5	1.4	2.2	0.4	1.2
Electricity (net)	2.4	1.9	1.5	1.7	1.2	1.3
Coke-oven gas	1.7	1.6	1.4	1.4	1.4	1.4
Total	23.6	20.4	17.2	18.4	15.1	16.5

ENERGY USE AND ENERGY EFFICIENCY IN UK MANUFACTURING INDUSTRY

UP TO THE YEAR 2000

1.2 The Aluminium Industry

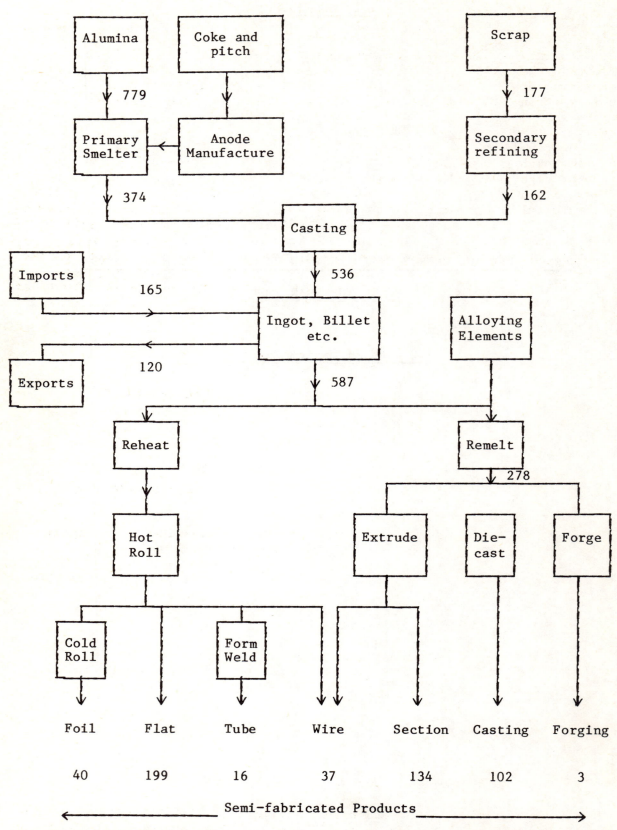

FIGURE 5: Principal Processes and Material Flows in the Aluminium Industry in 1980 (Kilotonnes)

1.2 ALUMINIUM

105. Aluminium is the most abundant metal, and the third most abundant element in the earth's crust. Yet, unlike other common metals, which have been known for thousands of years, aluminium was not discovered until 1825. The very strong affinity of the metal for oxygen makes its isolation difficult, and until the development of the Hall—Hérault reduction process in 1886, less than 50 tonnes of the metal existed in the world.

106. Although its production is energy intensive, its combination of strength, lightness, corrosion resistance and high electrical conductivity makes aluminium attractive for many applications, and it is now second only to steel in metal production. In the UK in 1980, total production of aluminium ingots was 536,000 tonnes, of which 374,000 tonnes was primary production from aluminium smelters, and the balance from recycled scrap. World—wide production of primary aluminium was 16 million tonnes in 1980. Total energy use by the UK aluminium industry in 1980 was 44×10^6 GJ (heat supplied), ie 2.4% of UK industrial energy use. In primary energy terms, the proportion is higher because of the heavy use of electricity.

STRUCTURE OF THE ALUMINIUM INDUSTRY

107. The main process steps in aluminium production are:

- extraction of alumina from ore (not carried out in the UK)

- primary smelting of alumina to aluminium

- secondary refining of aluminium scrap

- semi-fabrication of aluminium and aluminium alloy products by rolling, extrusion, etc.

Figure 5 shows these processes schematically in a simplified flow diagram, together with the quantities of materials produced in 1980.

108. Internationally, the aluminium industry is dominated by six companies, viz:

- ALCOA: The Aluminium Company of America (USA)

- ALCAN: The Aluminium Company of Canada (Can.)

- Kaiser Aluminium and Chemical Corp. (USA)

- Alusuisse (Switz.)

- Pechiney Ugine Kuhlman (Fr.)

- Reynolds Metals Co. (USA).

Two of these companies, ALCAN and Kaiser (in partnership with RTZ) are major producers of primary aluminium in the UK, while ALCOA, Alusuisse and Pechiney

TABLE 18: Plant Capacity in the UK Aluminium Industry 1980
(kilotonnes per year)

Company	Primary Production	Secondary Refining	Semi-Fabrication
ALCAN*	125	50	260
British Aluminium*	141	15	109
Anglesey Aluminium (Kaiser/RTZ)	110	–	
ALCOA	–	–	15
Pechiney	–	–	12
Alusuisse	–	–	8
Other (46 Companies)	–	c.a. 90 (20 Cos)	160 (∼ 40 Cos)

* Now merged as British Alcan Aluminium Co.

TABLE 19: International Comparisons of Aluminium Production
and Consumption

Country	1980 Production (M.tonne)		Per Capita Consumption of Aluminium (kg/yr)	Ratio of Aluminium:Steel Consumption %
	Alumina	Primary Aluminium		
USA	7.2	4.65	30	4.8
Japan	1.7	1.10	20	3.6
Canada	1.0	1.08		
W.Germany	1.3	0.73	19	4.3
France	1.3	0.43	12	3.3
UK	0	0.37	12	3.3

Ugine Kuhlman have relatively small operations in the semi-fabrication of
aluminium. Table 18 shows the plant capacities for the different sectors of
the aluminium industry in 1980, listed by company. As well as these five
major international aluminium companies, the UK aluminium industry in 1980
comprised the main UK producer, British Aluminium (a subsidiary of TI), and 46
other companies, mainly involved in semi-fabrication.

109. In 1982, British Aluminium closed its 100 kt/yr Invergordon Smelter and
the remainder of the Company was taken over by ALCAN. The Company thus formed
- British ALCAN Aluminium - now represents most of the aluminium industry in
the UK, accounting for some 60% of the remaining primary production capacity,
40% of secondary refining and 67% of semi-fabrication. The Kaiser/RTZ plant
in Anglesey is the only other primary producer, while the remainder of the
secondary refining and semi-fabrication sectors are highly disaggregated.

TRENDS IN ALUMINIUM PRODUCTION AND CONSUMPTION

110. For a major industrial nation, the UK has had a rather chequered history
in the aluminium industry. Aluminium production began in 1897, at Foyers
in the Scottish Highlands, using hydroelectricity, while bauxite was extracted
in Co. Antrim. However, although demand for aluminium grew rapidly, primary
production languished until the early 1970s, when three large smelters were
established with the assistance of Government grants. Figure 6 shows the
trends in production and consumption of aluminium in the UK since 1963.
Production of both primary and secondary aluminium remained fairly constant at
around 200 kt per year until 1970, when primary production expanded rapidly.
Demand for the metal increased at an annual rate of 4% from 1963 to 1973.
However, the effects of the general economic recession since 1973 reversed the
previous high growth rate in demand. By 1977, production and consumption were
equal, and by 1980 the industry was experiencing acute over-capacity.

111. Figure 7 shows that the increase in primary production since 1970 was
matched by a corresponding fall in imports of aluminium ingots, and a rise in
exports. By 1980 however, a high exchange rate and depressed world aluminium
prices made UK aluminium production uncompetitive internationally; it was not
possible to increase exports further in order to absorb excess production.
The aluminium industry found itself in a state of crisis, which could only be
resolved by a reduction in capacity. Since 1980, there have been widespread
doubts expressed about the future of primary aluminium production in the UK.
Because aluminium production consumes large amounts of electricity, the
industry was originally located near sources of cheap hydroelectricity,
especially in countries such as France, Norway and Canada. It is expected
that much of the future investment will be located in countries which have
both an abundance of bauxite reserves and ample hydroelectric power or
relatively low cost fossil fuels. Thus, projects planned for new smelters in
Brazil, Zaire and Australia alone will increase world capacity by 3.3 million
tonnes per year by 1990 - an increase of about 20% over 1980.

112. However, in spite of a lack of either bauxite or cheap electricity,
three of the four largest producers of primary aluminium are the USA, Japan
and W.Germany (cf. Table 19). The Japanese industry, which is heavily
dependent on electricity from oil-fired plant, has experienced a substantial
contraction since 1973, and may have to contract further; similarly the US

FIGURE 6: Aluminium Production and Consumption in the UK 1963-81

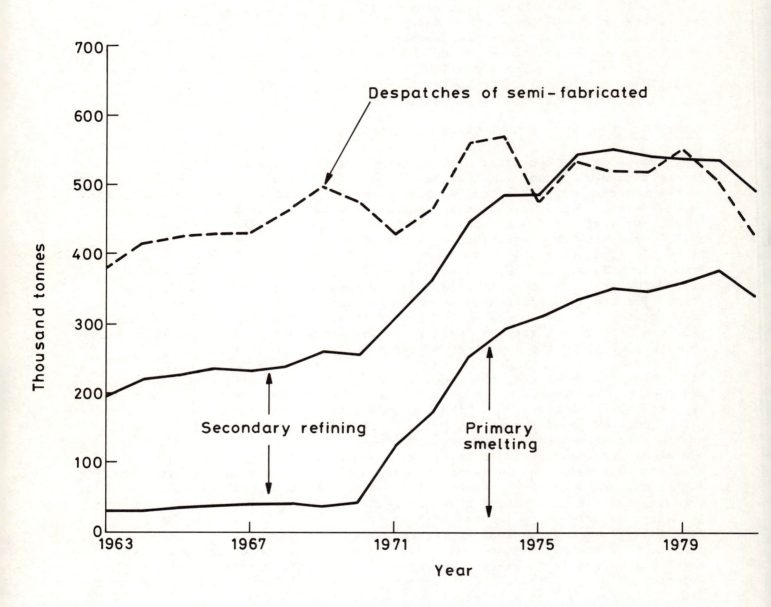

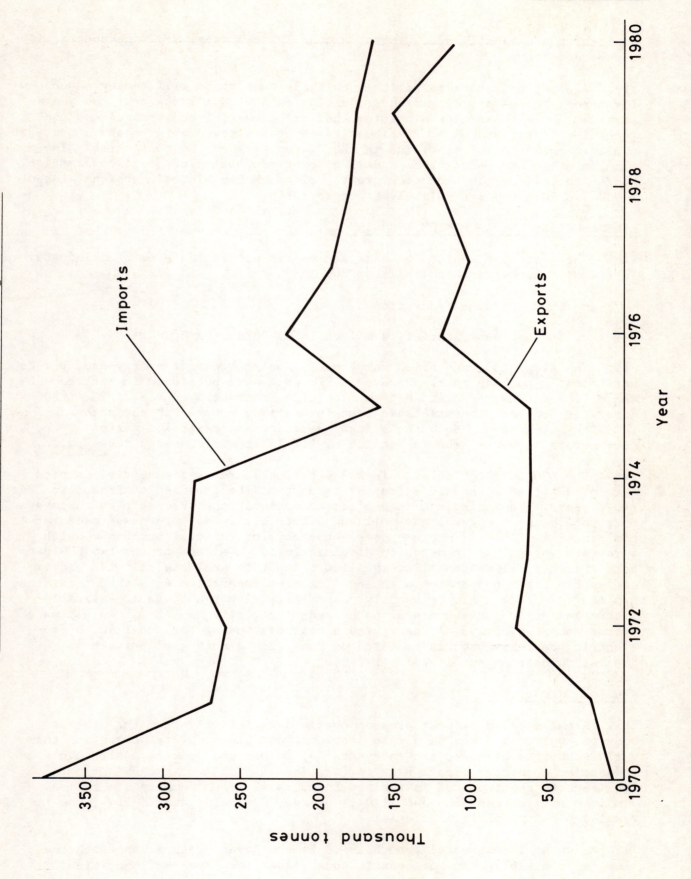

FIGURE 7: UK Import and Exports of Aluminium Ingots 1970-81

smelters are currently operating at around 70% capacity, and some contraction may be necessary.

113. Against this background it is unlikely that there will be any major new investment in primary aluminium production in the UK for the next ten years at least. The demand for aluminium should, however, increase again as the recession lifts, and allow the industry to stabilise. Table 18 shows that per capita consumption of aluminium in the UK (and France) is considerably lower than in other OECD countries, and is a lower proportion of steel consumption. Hence, there is ample scope for growth, although the historic growth rates of 7-10% (world-wide) are unlikely to be repeated.

ENERGY USE IN THE ALUMINIUM INDUSTRY

114. There are two sources of data for energy use in the aluminium industry in the UK, which are shown in Table 20:

- The Aluminium Federation

- The Business Statistics Office (BSO) Purchases Enquiries.

115. The Aluminium Federation carry out annual surveys of energy use, but do not normally publish them. However, 1974 data were published in reference 6 while 1980 data were obtained from Aluminium Federation sources. The 1980 data are, however, incomplete in certain details, notably in the secondary refining sub-sector. Both the primary smelting and semi-fabrication sub-sectors, however, show a decline of 11% in specific energy consumption.

116. The BSO data in Table 20 is based on fuel purchases enquiries carried out in 1974 (ref 7) and 1979 (ref 8) as part of the Censuses of Production for those years. They also are incomplete, especially the 1974 set, which covered only purchases of natural gas, oil and electricity. No return was made for coal in 1974. The 1979 return does, however, include coal purchases which were used for private generation of electricity. Neither of the two BSO data sets are sufficiently detailed to allow a separate analysis of total energy consumption, but comparison with the Aluminium Federation data shows that the two sources of data are reasonably compatible, although there are sizeable discrepancies in the amounts of purchased electricity reported. An estimate of the amount of privately generated electricity using both coal and hydro-electric power has been made, based on data from a survey of private generation carried out in 1977 (ref 9).

Alumina Extraction

117. Alumina (Al_2O_3) is at present obtained exclusively from the mineral bauxite, which is found mainly in tropical countries. Sixty percent of the world's bauxite reserves are in Australia, Guinea and Jamaica. Although some industrial countries import bauxite (cf. Table 18) all of the UK's alumina requirements are imported in refined form, mainly from Jamaica. A typical Jamaican bauxite contains 50% Al_2O_3, 20% FeO, 2.5% TiO_2 and 1% SiO_2, the remainder being volatile material.

118. Alumina is extracted from bauxite by the Bayer process, in which the alumina is dissolved in hot caustic soda, separated from the impurities,

recrystallised as the hydrate and finally calcined at 1,200°C. On average, four tonnes of bauxite are required to produce two tonnes of alumina, which in turn leads to one tonne of aluminium. The energy requirement for the Bayer process is estimated to be 20 GJ/tonne of alumina, or 40 GJ/tonne of aluminium. Thus UK imports of alumina have an energy content of about 15 million GJ (or 0.5 mtce).

119. World bauxite reserves are sufficient for about 300 years supply at the present rate of consumption, although if the high historic rate of growth in demand continued, this period would be reduced to about 60 years.

120. As a contingency against future constraints on bauxite supply, the major aluminium companies have examined potential alternative sources of alumina. Clays and other minerals containing 15-20% alumina are plentiful, but the extraction of the alumina would be more expensive, both in monetary and energy terms. In the UK, china clay wastes and power station fly ash, which contain about 30% alumina, have been suggested as possible alternative sources. However, it is unlikely that any alternative to bauxite will be commercially viable in the foreseeable future.

Primary Smelting

121. Primary aluminium is obtained from alumina by the Hall-Héroult electrolytic smelting process. In this process, the pure alumina is continuously dissolved in a cell containing molten cryolite ($Na_3Al F_6$), through which an electric current is passed. Molten aluminium collects above the carbon cathode, which lines the base of the cell, while oxygen is released at the carbon anode, and reacts with it to produce CO and CO_2. Pure cryolite has a melting point of 1,012°C, but various additives, such as aluminium fluoride, reduce the melting point of the electrolyte, allowing operation at 940-980°C.

122. The carbon anodes are consumed in the process at the rate of about 0.5 tonnes carbon/tonne aluminium, and must be replenished. They are usually produced from petroleum coke and pitch, baked for about one month at 1,000-1,200°C in a gas-fired muffle kiln. At one plant, however, the 'Soderberg' anode is used. This is formed continuously from a paste of petroleum coke and pitch fed directly to the electrolysis cell. Soderberg anode require less energy to make, but are less efficient in use.

123. The Hall-Héroult process is both capital and energy intensive. The largest single cost in the production of aluminium is the large quantity of electricity used to reduce the alumina, which may be as high as 50% of the value of the aluminium. Other significant energy inputs are in making the carbon anodes, and in extracting and cleaning the exhaust gases.

124. <u>Theoretical Considerations</u> The chemical reaction taking place in the cell can be represented by the following equation:

$$Al_2O_3 + 2 C \longrightarrow 2 Al + CO + CO_2$$

(25°C, solid) (25°C, solid)   (970°C, liq) (970°C, gas) (970°C, gas)

FIGURE 8:   Ellingham Plots for Reduction of Alumina

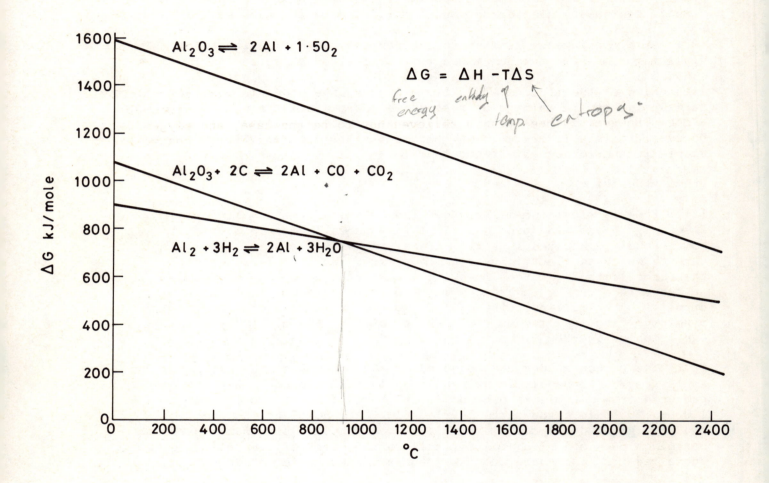

Thermodynamically, the enthalpy change for this endothermic reaction is 1,300 kJ/mole (ie 6.67 kWh/kg). This is the theoretical minimum amount of <u>enthalpy</u> which must be supplied for the reaction to proceed.

125. In general, an endothermic reaction can proceed by supplying heat provided that the <u>free energy</u> change is zero or negative. Figure 8 shows the free energy change as a function of temperature for the reduction of alumina by carbon; the reduction by hydrogen is also shown for comparison. (These plots assume that the enthalpy and entropy changes are independent of temperature, which is an approximation.) It can be seen that the free energy change does not become zero at any accessible temperature. This means that the reaction cannot proceed by the application of heat alone. The additional free energy must be supplied in the form of electricity.

126. If the carbon in the anode took part directly in the electrochemical reaction, then the electrical energy requirement would be 730 kJ/mole at 870°C. However, since this is believed not to be the case, and oxygen is liberated before it reacts with the anode, then the free energy change for the direct decomposition of alumina must be supplied. Thus, at 870°C, the minimum electrical energy requirement is 1,230 kJ/mole (ie 6.33 kWh/kg).

127. In practice, however, the electrical energy requirement is about 15 kWh/kg, owing to the resistance of the cell and other losses. The heat thus dissipated provides much more enthalpy than is required theoretically, and is sufficient to maintain the cell at the working temperature.

128. There are considerable variations in the efficiencies of smelters, as is shown in Table 21. The Lynemouth smelter, for example, at present uses 15 kWh/kg and has improved its performance by about 10% since it opened in 1974; it is expected that further improvements will be made in the near future. The other major UK smelters are not as efficient as Lynemouth, especially the now-defunct Invergordon smelter. A new small smelter at Lochaber, however, using the latest Pechiney technology, achieves 13.2 kWh/kg.

129. <u>Energy Losses</u> Under normal operating conditions, about 50% of the electrical energy to a cell is lost. This loss is due partly to electrolytic reactions other than the deposition of aluminium (for example, decomposition of the electrolyte). The amount of aluminium deposited for the passage of a given current (ie the current efficiency) is 85-90% of the theoretical maximum.

130. The major part of the losses, however, are due to resistive heating, and are functions of cell design and operation. Sources of electrical resistance include:

- the electrolyte

- the electrodes

- busbar connections.

Good cell design can minimise these losses by, for example, reducing the anode-cathode spacing or optimising the electrolyte composition.

TABLE 20:   Energy Consumption in the Aluminium Industry 1974-1980 (PJ)

FUEL	DATA SOURCE Aluminium Federation 1974 SECTOR				DATA SOURCE Aluminium Federation 1980 SECTOR				BSO 1974 Total	DATA SOURCE BSO 1979 [3] SECTOR	
	Primary Smelting	Secondary Refining	Semi-fab.	Total	Primary Smelting	Secondary Refining	Semi-fab. [1]	Total	Total	Primary Smelting	Secondary Ref. + Semi-fab.
Electricity:											
Private	7.6	0	0	7.6	9.5	n.a. [3]	0	9.5	n.a. [4]	10.6	2.0
Purchased	10.9	0.2	3.0	14.1	13.9	"	2.2	16.1	9.0	15.5	4.3
Natural Gas	0.7	1.2	4.2	6.1	1.3	:			5.0	1.0	3.7
Fuel Oil	0	0.6	2.1	2.7	0.07	:			} 5.6	0.2	0.7
Gas Oil	0.2	0.9	3.7	4.8	0.05	:	9.4	11.7	}	0.07	0.7
LPG	2.7	0	0.5	3.2	0.8	:			n.a.	1.7	0.01
Misc. Oils	0.1	0	0.1	0.2	0.1	:			0.1	0.1	0
Petroleum Coke	4.7	0	0	4.7	5.2	:	0	5.2	n.a. [4]	n.a.	0
Coal Tar Pitch	1.8	0	0	1.8	1.8	"	0	1.8	n.a.	n.a.	0
TOTAL	28.7	2.9	13.6	45.2	32.7	(2.4) [2]	11.6	46.7			11.4
Production kt	293	189	570		374	162	504			359	176/551
SEC GJ/tonne	98	15.3	23.9		87	(15)	23.0				

Notes: [1]   Aluminium Federation data for 1980 does not include returns for foil, forgings or secondary refining. 0.5 PJ has been added to allow for foil and forgings.

[2]   Secondary refining is assumed to have an SEC of 15 GJ/tonne.

[3]   Business Statistics Office data have been taken from returns of the 1979 Fuel Purchases Enquiry sub-divided by fuel size bands. Primary smelting has been separated, and the remaining total grossed up from an estimated 70% coverage, to give aggregate data for secondary refining and semi-fabrication.

[4]   n.a. = data not available.

TABLE 21: Smelter Efficiencies in the UK

Smelter	Year	Specific Electricity Consumption (kWh/kg)
Lynemouth	1974	16.5
	1980	15.5
	1983	15.0
	1985 (projected)	14.2
Invergordon	1981	16.8
Anglesey	1982	16.0
Lochaber	1982	13.2

TABLE 22: Specific Energy Consumption in Aluminium Semi-Fabrication Processes in 1980

	Electricity GJ/tonne	Fuels GJ/tonne	Total GJ/tonne	Throughput kilotonnes
Fabrication Remelt	0.49	6.22	6.71	278
Rolling	4.61	11.87	16.48	216
Extrusion	4.51	5.22	9.73	134
Casting	3.63	39.30	42.93	102

Source: Aluminium Federation

131. Heat is dissipated, partly by conduction through the cell lining and bottom, and partly through the exhaust gases. Heat loss through conduction can be reduced by increasing the insulation around the cell. This, however, makes control of the cell operation more difficult, because cell temperature can rise more rapidly when current and voltage are increased. In practice, temperature fluctuates, due to feeding of the cell with fresh alumina and 'anode effects' caused by the occasional formation of a gas film between the anode and the electrolyte, but it is essential that the temperature should not be permitted to rise above its normal level for prolonged periods because high temperatures lead to increased corrosion of the carbon sidewalls, and decreased current efficiency.

Secondary Refining

132. The process energy requirement for refining aluminium scrap is only about 15 GJ/tonne (cf. Table 20). This is about 15% of the energy required for smelting alumina - or less if allowance is made for the primary energy equivalent of the smelting electricity requirement. The proportion of scrap recycled by the industry is therefore a major determinant of the industry's overall energy requirement.

133. At present, approximately 30% of the aluminium ingot produced in the UK is from scrap. About half of the scrap supplied to secondary refiners is 'new' scrap, which arises during downstream fabrication processes. (This does not include arisings from semi-fabricators who smelt their own scrap.) The remainder is 'old' scrap recovered from products which have come to the end of their useful lives.

134. Industry sources have suggested that a further 150k tonnes per year of 'old' scrap could be recycled if the means of recovering it were available. One initiative taken by ALCOA towards this end is a campaign for collection of aluminium cans by the general public, along lines similar to municipal 'bottle banks' for the collection of glass bottles, or waste-paper collection by voluntary organisations as a means of fund-raising. Given the inherent value of aluminium scrap (about £400/tonne) one would expect such schemes to be economically attractive. A detailed assessment of the scope for aluminium recycling is being undertaken by consultants retained by ETSU in connection with another study.

135. A major constraint to aluminium recycling is the variation in alloy content in the recycled scrap. The scrap must be sorted by the refiner, and blended to give ingots of an acceptable alloy composition. It is not possible to remove alloying elements (except magnesium and zinc) from the scrap. Furthermore, scrap which is heavily contaminated with, for example, paint or lacquer, causes heavy pollution to be emitted from the melting furnaces, which are not equipped with waste gas cleaning equipment. Scrap which has a high surface/volume ratio (such as cans) tends to become oxidised during melting, again giving rise to pollution in the form of alumina in the exhaust gases, as well as a high loss of metal.

Semi-Fabrication Operations

136. Ingots of aluminium require further energy inputs in order to produce the semi-fabricated products which are the output of the aluminium industry.

These products can be classified as:

- rolled products (sheet, foil and tube)

- extrusion (wire and section)

- die-castings.

Table 22 shows the specific energy requirements of each process in 1980 together with the throughput.

137. Rolled products (sometimes referred to as wrought products) are produced by passing a suitable shaped ingot at 500°C through a succession of rolling mills to obtain the required width and thickness of sheet metal. The initial hot rolling process is generally followed by cold-rolling and annealing steps. The energy required includes thermal energy for heating the metal and mechanical energy for shaping it (ie driving the rollers). Foil is produced by continuing the cold rolling process down to a thickness of 0.005 mm. A single ingot thus yields several miles of foil.

138. Extrusions are made by forcing cylindrical billets at 450-570°C through a die of the required shape. The billets are heated either by electric induction or gas fired tunnel furnaces. A typical extrusion press uses a 2,000 tonne thrust provided by electrically driven pumps. The extruded length is stretched and cut to length, and may need to be annealed.

139. Rolling and extrusion operations are often preceded by a remelting of the original ingot/billet of metal in order either to prepare an alloy by the addition of other metals, or simply to recast it to a shape required for the process. Most secondary aluminium, and about 25% of primary aluminium ingot are remelted prior to semi-fabrication. Some secondary production is delivered to nearby foundries in liquid form, while the balance of primary aluminium is cast at the smelter into blocks suitable for rolling or extruding without remelting.

140. The theoretical amount of energy required to melt aluminium is about 1.0 GJ/tonne. In practice, however, 6.7 GJ/tonne are required to melt the metal. Much of the energy is lost because of the low thermal efficiencies of the furnaces. Furthermore, there is a considerable time during which the metal is required to be held at high temperature before subsequent processing, and the insulation of holding furnaces is not always optimum.

141. Cast products are manufactured by pouring molten metal into a mould. Sand-casting, which accounted for 13% of the tonnage of castings in 1980, uses a disposable mould made of sand, formed around a pattern. Permanent metal moulds, on the other hand, can be used repeatedly. In some cases the metal is forced into the mould under pressure, while in others the metal is poured in under gravity. Pressure die-casting accounted for 48% of the total in 1980, and gravity die-casting 39%. The main energy requirement in die-castings is for the melting of metal and holding it hot. There is a complex trade-off between energy costs and other costs, which will vary from one foundry to another. The energy efficiency of the process will therefore vary considerably. The Energy Audit (ref 6) found that pressure die-casting, starting from solid ingots, uses 28 GJ/tonne, whereas gravity die-casting uses 40 GJ/tonne. In addition there are irrecoverable metal losses equivalent to

TABLE 23: Energy Conservation Potential in the Aluminium Industry

SUB-SECTOR (energy requirement)	TECHNOLOGY	ENERGY SAVING POTENTIAL			COST-EFFECTIVENESS (payback time in years)	CURRENT STATUS	PROSPECTS IN UK
		%	GJ/tonne	Fuel Saved			
Primary Smelting (87 GJ/t)	Best practice Hall process	10	9	electricity	6	Best practice	Good
	New Alcoa process	30	27	electricity	-	Pilot plant in USA	Not before 2000
	Improved casting at smelter, or	1.5	0.8	gas		Best practice	Good/short term
	Continuous casting at smelter	4	2	gas		Prototype in Canada	Long-term
Secondary Refining (15 GJ/t)	Increased scrap utilisation (x2)	-	-	electricity		Feasibility study	Medium-long term
	Scrap preheating	25	1				
	Furnace Improvements	25	1				
Semi-Fabrication (23 GJ/t)	Furnace Improvements						
	- improved instrumentation, or	10	0.7	oil & gas	2	R&D in UK	Ten years to
	- automatic control, and	20	1.4	"	>3	R&D in UK	complete
	- scheduling, and	20	1.4	"	<1	Best practice	penetration
	- waste heat recovery	15	1.0	"	3-4	Demonstration in UK	
	- insulation	15	1.0	"	2	Demonstration in UK	
	Increased use of electric furnaces	substitutes electricity for oil and gas					
	'CONFORM' extrusion*	5	1		<2	Demonstration in UK	Rapid deployment after 1985
	Product yield improvement	5	1		<1	Best practice overseas	Good/medium term

* Applicable to only about 25% of total semi-fabrication.

about 3 GJ/tonne on a heat supplied basis (ie about 4% of the metal), and overheads for space heating, etc.

ENERGY CONSERVATION POTENTIAL IN THE ALUMINIUM INDUSTRY

142. Table 23 shows the major energy conservation technologies which have been identified by ETSU. The technologies are grouped according to the sub-sector to which they apply, with the potential energy savings expressed both as a percentage of the specific energy consumption for that sub-sector, and as an absolute reduction in the SEC.

143. The current status of each technology is classified in Table 23 according to whether it is best practice, either in the UK or elsewhere, or at the stage of demonstration or R&D. The future prospects for deployment of the technology is also indicated.

Primary Smelting

144. Since the major energy consumption in aluminium production is the electricity used in smelting, any measures which can reduce this consumption will have a major impact. There is still considerable scope for improving the energy efficiency of UK aluminium smelters. Indeed, the efficiency improved by about 10% between 1974 and 1980, and further improvements have been recorded since (cf. Table 21).

145. The average electricity consumption in smelting in 1980 was about 16 kWh/kg (ancillary processes at the smelter use a further 1.3 kWh/kg), whereas the 'best practice' was 13.5 kWh/kg for the small smelter at Lochaber. It is unlikely that this level could be achieved on the remaining two large smelters, but the closure of the relatively inefficient Invergordon smelter, together with proposed improvements and Lynemouth and possibly Anglesey should result in an average electricity consumption of 15 kWh/kg being achieved by 1990; further improvements may be possible, allowing a figure of 14 kWh/kg to be achieved by existing smelters by 2000.

146. A new process, to replace the Hall-Hérault process, is under development in the USA by ALCOA. This process reacts the alumina with chlorine to produce $AlCl_3$, which is electrolysed in a molten sodium chloride and lithium chloride electrolyte at 700°C. The chlorine is liberated at the anode and is recycled, while liquid aluminium is obtained at the cathode. The process is claimed to use 30% less energy than the Hall-Hérault process. However, the existing smelter capacity in the UK is comparatively new and it is unlikely to need replacement before the end of the century; furthermore, additional capacity is not likely to be required before 2000.

147. Savings could also be achieved in the casting operation at primary smelters, either by improving the efficiency of the existing holding furnaces, in which metal is heated prior to casting, or by the introduction of continuous casting techniques.

Secondary Refining

148. The efficiency with which scrap is melted in the secondary refining industry can be significantly improved. For example, the actual energy

required for melting scrap is about 15 times the theoretical requirement.
Secondary refiners in Europe have less than half of the energy requirement of
UK refiners, which is believed to be a major factor in the present market
conditions, where scrap prices are too high for UK refiners, many of whom have
closed since 1980. Better furnace insulation and control, and waste heat
recovery for scrap preheating could save a significant proportion of the
energy losses.

149.   Even at its present level of efficiency, however, the secondary refining
of scrap uses much less energy than primary smelting; increasing the rate of
scrap recovery is the largest single opportunity for saving energy in the
aluminium industry as a whole. There are, however, problems associated with
the collection and processing of low quality scrap. Methods for extracting
aluminium from municipal refuse (such as induction magnets) and special
furnaces for melting without high metal losses would need to be developed.
However, it should be noted that an increased scrap utilisation would tend to
displace imported primary aluminium rather than domestic production. Its
impact on UK energy consumption could therefore be small.

Semi-Fabrication

150.   The main use of energy in semi-fabrication is reheating and remelting of
the ingots. As with the secondary refining sector, the efficiency of
furnaces is poor, and offers considerable scope for improvement. Specific
measures include:

- improved instrumentation

- automatic control

- waste heat recovery

- improved insulation

- scheduling of hot metal requirements.

151.   The regulation of furnace behaviour can be improved if appropriate
instruments are used to measure fuel and air flows, metal and gas temperatures
and the chemical composition of stack gases. The waste heat contained in
furnace combustion gases can be recovered and used to preheat incoming metal
in a counter-flow arrangement. In large furnaces, there is scope for the use
of recuperators, heat wheels, etc, either to preheat combustion air or for
steam generation and space heating. The heat losses from furnace shells need
to be reduced, especially when metal is held molten after melting. The use of
new insulating materials is currently being demonstrated in the UK.

152.   A modern gas-fired reverberatory furnace with programmable controls for
firing rate and combustion air:fuel ratios, and with recuperators in the
furnace exhaust gas system to preheat the combustion air, will melt a 10 tonne
charge of aluminium with a fuel consumption of 2.5 GJ/tonne (ref 12). This
can be compared with average value of 6.7 GJ/tonne for the industry in 1980
(cf. paragraph 41).

153.   Electric induction furnaces give a melting efficiency of around
2.2 GJ/tonne. In terms of the primary energy equivalent of electricity (which

is reflected in the price paid by the consumer), this is equivalent to about
7.5 GJ/tonne. In primary energy terms, therefore, electric induction furnaces
are not as efficient as fuel-fired furnaces, especially ones with heat
recovery systems. However, electric induction furnaces do have additional
advantages, such as reduced metal loss, better metallurgical control and more
flexibility in operation, which offset the inherent disadvantage of the
primary energy efficiency of electricity generation. Furthermore, more
efficient induction furnaces should be available in the near future.

154. For small melting/holding furnaces of 1-2 tonnes capacity, electric
resistance heated crucible furnaces are more efficient than gas-fired crucible
furnaces (1.9 GJe/tonne, ie 6.3 GJ/tonne for electric resistance compared with
8.5 GJ/tonne for gas-fired). An analysis of the systems cost for melting
aluminium by the two methods found that the electric resistance furnace is 12%
cheaper than the gas-fired furnace (ref 10). Between 1975 and 1980 there were
300 resistance-heated crucible furnaces installed in UK aluminium foundries.

155. There is often a considerable mismatch in time in the supply and demand
of hot metal, especially in foundry casting. This results in considerable
energy losses entailed in holding the hot metal. Improvements in planning and
scheduling can bring about a considerable reduction in the energy required.

156. A new process for the extrusion of metals at temperatures below their
melting point, known as 'CONFORM', is currently the subject of a demonstration
project. This process was originally invented by the UKAEA Springfield
Laboratory, and has been developed by a private company for commercial applications. The first design of a commercial machine had limited application,
but an improved design should be more widely applicable. Approximately 20% of
the energy required for extrusion is saved. The process has other advantages,
including lower capital costs and the ability to use scrap directly, which
should ensure rapid deployment once it has been successfully demonstrated.

157. About 40% of the aluminium ingot used in semi-fabrication is wasted as
off-cuts and returned for remelting. An improvement in the yield of 3% would
reduce the energy used in melting by 5%. Savings of this order should be
possible by taking measures to improve the yield in processing.

PROJECTIONS OF SPECIFIC ENERGY CONSUMPTION FOR THE ALUMINIUM INDUSTRY

158. Projections of future specific energy consumption require assumptions to
be made about the following parameters:

- the specific energy consumption (SEC) for each of the three main
  processes (primary smelting, secondary refining and
  semi-fabrication);

- the proportion of total output which is met by primary smelting and
  secondary refining in the UK;

- the mix of fuels for each of the processes.

These parameters can be expected to evolve differently in different scenarios
with different total outputs. Two scenarios of output are considered,
designated 'high' and 'low'. The detailed background of these scenarios is
discussed elsewhere in this report (cf. Volume I). In the high scenario

output of semi-fabricated aluminium rises to 760 kilotonnes/year by 2000; in the low scenario, output rises to 700 kilotonnes/year by 2000.

159. The assumptions which have been made about the parameters which determine specific energy consumption are shown in Table 24. The arguments supporting these assumptions are discussed below for each scenario.

High Scenario Assumptions

160. In the high output scenario, demand for aluminium increases by 50% by 2000. Production of primary aluminium has already fallen by 100 kilotonnes per year due to the closure of the Invergordon smelter in 1982. Although it is possible that a rapid recovery in aluminium demand could lead to the recommissioning of the Invergordon smelter in the short-term, it is assumed that no new smelter capacity will be installed in the UK before 2000; by then, smelter capacity will have fallen from 74% of aluminium output to about 36%. This has a marked effect on energy consumption for the industry as a whole.

161. The specific energy requirement of primary smelters is assumed to fall by 15% by 2000. This implies that the present best practice will become the average by 2000; new technology, such as the ALCOA process, is not expected to be adopted in the UK before 2000.

162. Output of secondary refining from recycled scrap was 32% of the requirement for aluminium in 1980. It is assumed that this proportion will rise to 50% by 2000. The specific energy consumption is assumed to fall from 15 GJ/tonne in 1980 to 7 GJ/tonne in 2000. It is assumed that the specific energy consumption in semi-fabrication falls by 10% by 1990 and 20% by 2000, mainly due to more efficient furnaces.

163. Electricity will increasingly be used for mechanical power for continuous processing machinery and for induction heating and melting. As the price of oil and gas rises, it is likely that coal-firing of reverberatory furnaces will increase.

Low Scenario Assumptions

164. In a low scenario it is less likely that there would be investment in new plant and furnaces, although pressure to reduce costs would remain strong, tending to ensure the most efficient use of existing plant. Further closures of smelting capacity are possible, but are not included in the assumptions. It is assumed that smelter efficiency would fall by only 7% by 2000.

165. Secondary refining is assumed to increase to 40% of aluminium output by 2000, with the specific energy consumption falling to 10 GJ/tonne. The specific energy consumption for semi-fabrication falls by 15% by 2000.

Overall Specific Energy Consumption

166. The overall specific energy consumption for the aluminium industry is given by the expression:

$$SEC_{overall} = p \times S_P + r \times S_R + S_F$$

where S_P, S_R and S_F are the SECs for primary smelting, secondary

TABLE 24: Assumptions on Parameters Determining Future Specific Energy Consumption in the Aluminium Industry

	1980	1990 High	1990 Low	2000 High	2000 Low
Primary Smelting					
Proportion of output	74	45	45	36	38
SEC (GJ/tonne)	87	80	84	74	81
Electricity %	72	72	72	72	72
Oil %	16	16	16	16	16
Gas %	6	6	6	6	6
Coal %	6	6	6	6	6
Secondary Refining					
Proportion of output	32	40	35	50	40
SEC (GJ/tonne)	15	11	13	7	10
Electricity %	18	25	22	30	27
Oil %	42	32	36	24	31
Gas %	40	33	37	25	32
Coal %	0	10	5	21	10
Semi-fabrication					
Proportion of output	100	100	100	100	100
SEC (GJ/tonne)	23	21	22	18	20
Electricity %	18	25	22	30	27
Oil %	42	32	36	24	31
Gas %	40	33	37	25	32
Coal %	0	10	5	21	10
Overall SEC (GJ/tonne)	93	61	64	48	55
Electricity %	55	53	52	53	52
Oil %	24	23	24	20	22
Gas %	17	16	20	15	18
Coal %	4	8	4	12	8
Sensitivity to Output Assumptions					
Overall SEC with 1980 output proportions (GJ/tonne)	93	82	87	75	83

refining and semi-fabrication; p and r are the outputs of primary smelting and secondary refining as a proportion of total output.

167.   It can be seen from Table 24 that the overall SEC for the aluminium industry falls very markedly, from 93 GJ/tonne in 1980 to 48 GJ/tonne in 2000 for the high output scenario and 55 GJ/tonne in 2000 for the low output scenario. Much of this fall is due to the closure of the Invergordon Smelter. The sensitivity of the overall SEC to the output assumption can be judged by calculating what the SEC would be if there were the proportions of primary smelting and secondary refining were the same as for 1980 - ie 74% and 32% of total semi-fabrication output respectively. Under these circumstances, the SEC in 2000 would be 75 GJ/tonne for the high scenario and 83 GJ/tonne for the low scenario. Thus more than half of the projected improvement in SEC in each scenario is due to structural change rather than technological improvement in energy efficiency.

ENERGY USE AND ENERGY EFFICIENCY IN UK MANUFACTURING INDUSTRY

UP TO THE YEAR 2000

1.3 The Copper Industry

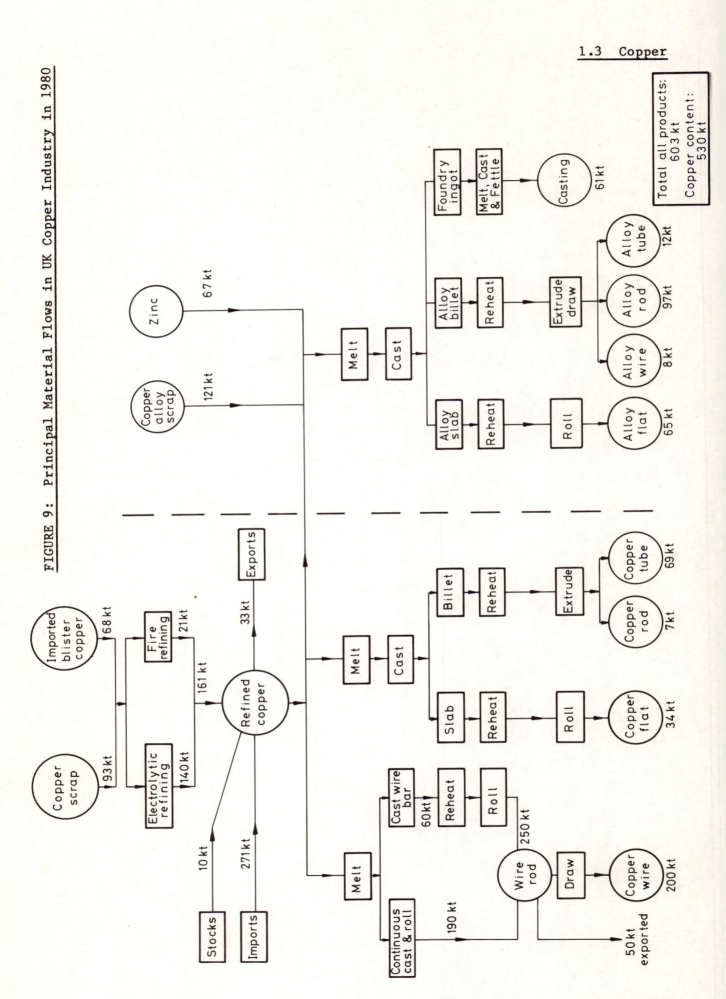

FIGURE 9: Principal Material Flows in UK Copper Industry in 1980

Total all products: 603 kt
Copper content: 530 kt

1.3 COPPER

INTRODUCTION

168. In some respects, the copper industry is similar to the aluminium industry. Thus, a comparison of the data in Table 1 shows that very similar amounts of copper and aluminium semi-fabricated products are produced in the UK, using similar techniques and for similar applications.

169. However, energy consumption is much lower in the copper industry, because there is virtually no primary smelting of copper ore carried out in the UK. (A small amount of primary copper is obtained in the UK from lead/zinc ore concentrates.) Almost all of our primary copper requirements are imported, amounting to some 271 kilotonnes of refined copper and 68 kilotonnes of unrefined copper in 1980. Furthermore, the intrinsic primary energy content of copper (50 GJ/tonne) is much lower than that of aluminium (250 GJ/tonne).

170. The UK copper industry is therefore concerned only with the semi-fabrication of copper, together with the recovery of scrap copper, which supplied some 40% of the copper used in 1980. Energy use in 1980 is estimated to be 9.10^6 GJ (heat supplied), or 0.5% of the total UK industrial energy use.

STRUCTURE OF THE COPPER INDUSTRY

171. The UK copper industry can be divided into three distinct sub-sectors:

- refining of copper scrap and imported unrefined primary metal (known as 'blister')

- semi-fabrication of pure copper

- semi-fabrication of copper alloy (brasses and bronzes).

The relationship between these sub-sectors can be seen in Figure 9, which shows the principal material flows in the copper industry and the quantities of products dispatched in 1980, taken from reference 11.

172. There are 217 separate establishments listed under SIC 2246. The largest of these, with production capacities in excess of 20,000 tonnes/year, are listed in Table 25. This shows that the secondary refining and pure copper semi-fabrication sub-sectors are relatively concentrated, with BICC and IMI accounting for ca. 85% of secondary refining, and BICC and Delta Metal Co about 66% of pure copper semi-fabrication. Copper alloy semi-fabrication is more disaggregated, especially the manufacture of alloy ingots and foundry castings, which are highly specialised.

173. Secondary refining from scrap accounts for about 20% of the refined copper used in the UK (cf. Figure 9); a further 12% or so of refined copper is obtained by refining imported 'blister' copper (impure copper obtained from roasting sulphide ores). BICC refine only high grade scrap and blister, but IMI and Elkington also refine substantial quantities of low-grade scrap. Mos of the refining done in the UK includes a final electrolytic process which produces the very high purity copper required for electrical conductors.

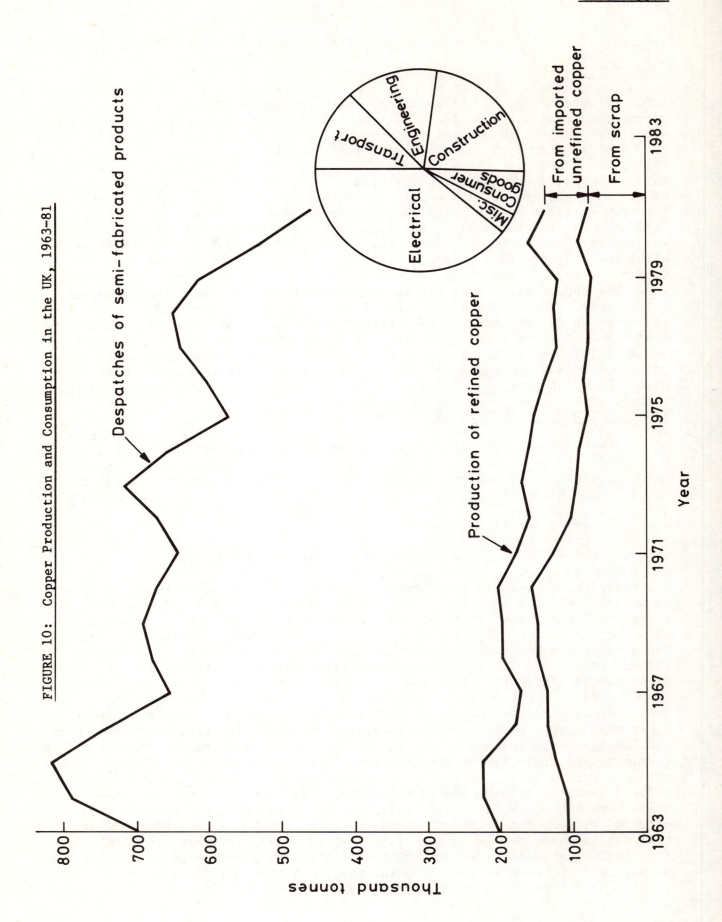

FIGURE 10: Copper Production and Consumption in the UK, 1963-81

The proportion of copper which is only fire-refined fell from 60% in 1969 to 13% in 1980.

174. Semi-fabrication of pure copper accounted for approximately 88% of the refined copper consumed in the UK in 1980, most of which was used for production of copper wire (55%) and tube (20%). Note that there is also a significant export trade in wire rod.

175. The continuous casting and rolling (CCR) of wire rod has in recent years largely replaced the traditional wire bar casting and rolling technique, with significant energy savings. From the data in Table 25 it can be seen that, in 1980, two-thirds of the nominal wire-making capacity was through continuous casting; actual production of wire rod by CCR in 1979 was estimated in reference 13 to be 75% of the total and this proportion is now believed to be higher.

176. The copper alloy industry is largely based on the recycling of copper alloy scrap. This activity is quite distinct from secondary refining of copper scrap to make pure copper; it is more akin to the secondary refining of aluminium. To make alloys of a required composition, high grade alloy scrap is carefully sorted and analysed, and blended with the necessary amounts of pure copper or zinc and other alloying elements. Semi-fabricated products are then obtained by casting slabs or billets which can be further processed by rolling, extruding or drawing as required.

177. There are a large number of foundries which cast copper alloys directly to their final shape, using sand-moulds or shell-moulds. The foundries do not prepare their own alloys, however, but buy them from alloy-ingot makers who specialise in segregating, grading and blending alloy scrap. There are estimated to be 21 alloy ingot makers, and 429 foundries, but together they account for only 10% of the total output of the copper industry.

TRENDS IN COPPER PRODUCTION AND CONSUMPTION

178. Figure 10 shows the annual consumption of copper in making semi-fabricated copper and copper alloy products from 1963 to 1981. From a peak of 815,000 tonnes in 1965, consumption has fallen by 44% to 460,000 tonnes in 1981. Part of this decline, particularly the sharp slump since 1979, is due to the general effects of recession. However, the long term trend is also downward, reflecting an increasing substitution of other materials, for example aluminium, stainless steel and plastics, in the traditional applications for copper.

179. The main end-uses of copper are shown in Figure 10 (pie-chart insert). Electrical engineering is the largest use, accounting for almost 50% of the total consumption of copper. In the 1960s, the substitution of aluminium for copper in overhead power lines contributed markedly to the downward trend, but the penetration of this market by aluminium has stabilised since the early 1970s. Technical difficulties in making end-connections with aluminium conductors has prevented the displacement of copper from the house-wiring market. In telecommunications, however, British Telecom have been moving towards aluminium cables, with 90% of new cables laid in the UK by 1985 expected to be aluminium. Optical fibres are also expected to make an impact on the use of copper in telecommunications in future years.

TABLE 25: Major Production Plants in UK Copper Industry in 1980

Product	Company	Location	1980 Capacity (kt/yr)
Secondary Refining			
Electrolytic Copper	IMI Ltd	Walsall, W Midlands	75
" " " "	BICC Metals	Prescot, Merseyside	55
" " " "	Elkington (PUK)	Walsall, W Midlands	25
Fire-refined Copper	T Bolton (BICC)	Froghall, Staffs	20
Semi-fabrications			
Wire rod (CCR)	BICC Metals	Prescot, Merseyside	130
" " "	Delta Metal Co	Enfield, Middlesex	100
Wire rod (Rolled)	F Smith (GEC)	Trafford Park	50
" " " "	Pirelli	Eastleigh, Hants	40
" " " "	E&E Kaye (PUK)	Enfield, Middlesex	20
Copper tube	Delta Metal Co	Birmingham	50
Copper and alloy tube	IMI	Kirkby, Barrshead, Leeds & Smethwick	60
Alloy rod, etc.	McKechnie Metals	Walsall, W Midlands	50

Source: World Copper Survey 1980 (ref 12)

180.   The construction industry uses large amounts of copper tube for domestic hot and cold water supplies, central heating systems, gas piping, etc, where it has displaced older materials such as lead and galvanised steel. It is now, however, facing growing competition from plastics and, in the USA at least, stainless steel. The use of thinner-walled copper tubing in recent years has reduced the overall consumption of copper, but this trend has largely run its course.

181.   In engineering, copper and its alloys, eg brass, are used for many purposes, such as valves, condensers and heat exchangers. There is a limited threat of substitution in this application by other materials, and material-saving design improvements. For example, car radiators made of aluminium and plastic have a growing market in Europe.

ENERGY USE IN THE COPPER INDUSTRY

182.   Reports on energy use in the copper industry in the Energy Audit (ref 13) and Thrift (ref 14) series have been prepared by British Non-Ferrous Metals Technology Centre, based on two sources of data:

- visits to individual company sites carried out between 1975 and 1976

- the 1974 purchases enquiry, which was carried out by the Business Statistics Office (BSO) as part of its Census of Production for that year (ref 15).

The Audit also notes certain technological changes affecting energy use in the industry, but does not attempt to quantify their effect on overall energy consumption.

183.   The only source of more recent data is the 1979 BSO purchases enquiry (ref 8). The data from both the 1974 and 1979 Censuses are shown in Table 26. These data indicate that energy consumption per tonne of copper produced decreased by 16% in the period 1974-79.

184.   The Department of Energy's Digest of UK Energy Statistics (ref 16) includes the copper and other non-ferrous metal industries with the engineering sector. Between 1974 and 1979 energy consumption by this group of industries increased by 1.8%; for all industry, excluding iron and steel, it fell by 2.6%. In the same period, the index of manufacturing output fell by 2.8%. Hence the dramatic improvement in energy efficiency in the copper industry is concealed in the aggregated data.

185.   The trend in consumption of individual fuels shown in Table 26 indicates three identifiable factors which may explain the overall decrease in energy use:

- consumption of coal decreased by 50%. This may reflect a shift from coal-firing to gas-firing in the mid 1970s.

- a reduction in the amount of low grade scrap being smelted in blast furnaces may account for the decrease in coke consumption

TABLE 26:   Energy Use in the Copper Industry (10^6 GJ)

	1974	1979
Coal	1.04	0.50
Coke	0.52	0.12
Fuel Oil	3.62	1.62
Gas/Diesel oil	–	0.25
Gas	5.16	4.77
LPG	0.25	0.18
Derv	0.20	0.12
Electricity	2.18	2.54
TOTAL	12.97	10.10
Production of Copper Semis (kilotonnes)	661	616
Specific Energy Consumption (GJ/te)	19.6	16.4

Source: Business Statistics Office Fuel Purchases
          Enquiries (ref 8, 15).

- consumption of fuel oil decreased by 55%, while electricity consumption increased by 17%. This reflects a switch towards electric induction heating.

There is at present little quantitative information available to confirm these trends, although qualitative reports from industry sources indicate that they have occurred. For example, according to reference 12, 80 coreless induction melting furnaces were installed in UK copper alloy foundries in the period 1975-1980.

186. In the following paragraphs, a detailed description of energy use in the different sub-sectors of the copper industry is given. This is based on the process analyses given in the Appendices of reference 15. Table 27 summarises these analyses of energy use. Note that the electricity data from reference 15 has been converted from primary energy to heat supplied terms. The data in reference 15 are based on measurements which were carried out on actual plants. Hence, although the plants audited were chosen to be representative of the most typical processes employed, the data are not an average of all plants; variations from one plant to another are to be expected.

187. Where appropriate, the effect of known trends since 1974/76 is commented on and estimates made of the specific energy consumption of each process in 1980, the base year for the present study.

Secondary Refining

188. Secondary refining of copper scrap involves a number of stages, depending on the quality of the scrap, as shown in Figure 12. Low grade scrap, which contains a substantial proportion of oxidised copper, is first smelted in a blast furnace with coke and oxygen-enriched air. The reactions in the furnace reduce copper and other non-ferrous oxides, while iron is oxidised, and separated as a slag. The coke rate of the blast furnace is adjusted according to the composition of the feedstock scrap. The copper obtained in this way is about 70% pure, and contains about 5% each of zinc, tin, lead, nickel and iron.

189. The next stage in the refining process is the transfer of molten copper to a 'converter' furnace, to which is added tin-rich copper scrap. The tin, together with the other metallic impurities, are oxidised and separated as slags and fumes. The oxidation of the tin supplies most of the heat required in the converter, and only a small addition of external heat is required. Tin-rich slags are recovered from the converter, from which valuable tin-lead alloy may be recovered, while copper-rich slags are either returned to the blast furnace or transferred to the reverberatory furnace used for processing medium-grade scrap. In the reverberatory furnace, medium-grade scrap is upgraded to better than 95% copper. Here, too, impurities such as lead, tin and zinc are slowly oxidised and removed as slags.

190. The copper from the reverberatory and converter furnaces is further purified in a final furnace, in which oxygen is removed by the addition of timber logs in a traditional process known as 'poling'. Slags from the melt are skimmed off and returned to the blast furnace. The molten copper is then cast into slabs which form the anodes for electrolytic refining - hence the name 'anode furnace' for this furnace. However, if electrolytic refining is not carried out (which was the case for only one plant in the UK in 1980; cf.

FIGURE 11:   Flow Diagram for Secondary Copper Refining

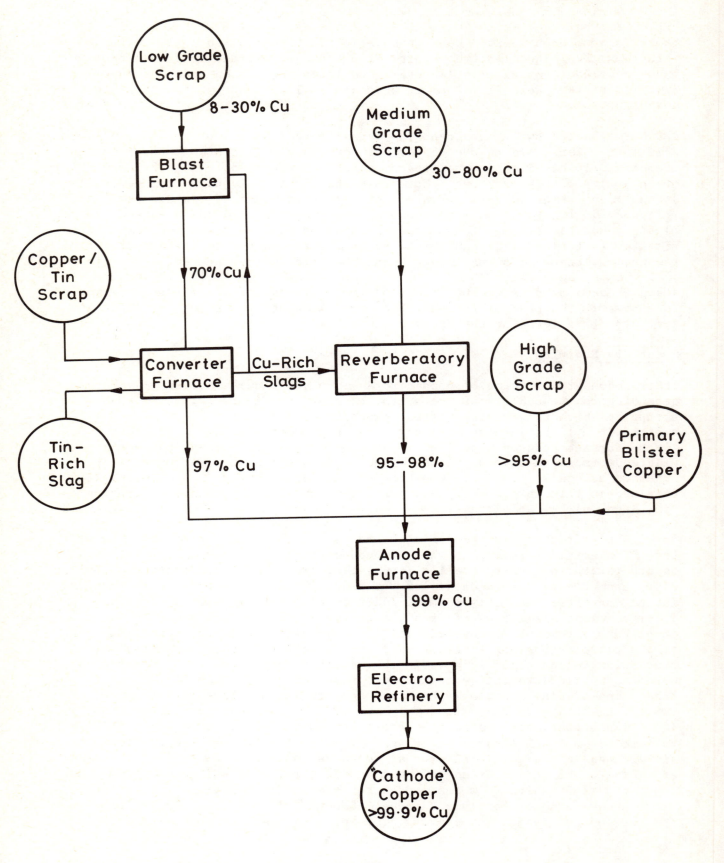

Table 25), the copper at this stage is the final 'fire-refined' product. The complete cycle of fire-refining takes about 24 hours.

191. Electrolytic refining is carried out in cells comprising alternate anode blocks and cathode sheets immersed in tanks of acidic copper sulphate solution. Over a period of several days, copper is slowly dissolved from the anode and deposited on the cathode by the passage of an electric current. Some impurities, like nickel, accumulate in the solution, while others, notably precious metals, separate as slimes, which are a valuable byproduct.

192. The energy use in secondary refining will depend on the balance of scrap available. Reference 13 estimates that the electrolysis step requires about 4 GJ/tonne of copper (heat supplied basis), of which 1 GJ/tonne is electricity. The total energy used ranges from 10 GJ/tonne for high grade scrap and blister copper which goes directly to the anode furnace, to 28 GJ/tonne for low grade scrap; the average was estimated by reference 13 to be 25 GJ/tonne. Although it is not clear to which year this refers, this figure seems excessive ($\sim$ 67%). There are no published figures on the proportion of low-grade scrap which is smelted, but it is believed to have fallen to about 60%. Hence, an average consumption of 20 GJ/tonne for secondary refining seems to be a more reasonable figure to apply to estimating energy consumption for 1980.

Semi-Fabrication of Copper and Copper Alloy

193. Up to 1975, all production of copper wire was by the traditional wire-bar casting and rolling method. By 1980, approximately three-quarters of production was by the continuous casting and rolling (CCR) technique, and this proportion has since increased further.

194. In the traditional method, special wire bars, with square sections and tapered ends, were cast by remelting refined copper in a reverberatory furnace, and casting. Wire bars are a traded commodity in their own right. They are subsequently reheated to 800°C and rolled to form wire rod, which is then drawn and annealed to produce the final wire product. Energy consumption is 8.7 GJ/tonne.

195. The continuous casting process requires only 5.0 GJ/tonne. Refined copper is melted in a more efficient shaft furnace, and the liquid copper cast into a profile section which forms the rim of a casting wheel. The bar thus formed solidifies during one revolution of the casting wheel, is cooled to the temperature at which it can be rolled, and is then fed directly through two trains of rollers to emerge as wire-rod. The wire-rod is drawn and annealed as before. The CCR process thus eliminates a reheat step, as well as using a more efficient type of furnace.

196. Copper strip and sheet are produced from slabs of copper some 4"-5" thick. As with wire bar, they are reheated and subjected to a sequence of hot and cold rolling operations, followed by annealing. The whole process, including the melting and casting of refined copper (and internally generated scrap), requires 18.5 GJ/tonne. It is now possible to cast copper strip 20 mm thick and up to 650 mm wide which can be cold rolled without reheating. The

full energy saving for this process change has yet to be evaluated, but it is expected to be at least 20%.

197. Narrow strip casting for copper and copper alloy has long been established for a range of uses which includes coinage, electrical parts and radiators. These have mainly utilised small melting furnaces using oil, gas or electricity. Oil-fired reverberatory furnaces are used for bigger, batch melting operations. For smaller or more continuous operations, electric induction furnaces are used. These generally have lower losses than oil furnaces in heat supplied terms (though not always in primary energy terms) but there is scope for improving the efficiency of reverberatory furnaces, or, where a regular shape of feed is used, replacing them with shaft furnaces.

198. In the reheat sections of rolling and extrusion mills, gas or oil firing is generally used, but for annealing and heat treatment, electric induction furnaces are most common.

199. For certain alloys which exhibit work-hardening properties, reductions by cold working have to be gradual, with repeated annealing operations between each reduction. High reduction mills which achieve the required dimensions in a single pass are under development; if successful, they would eliminate intermediate annealing.

200. Copper and copper alloy tube are both produced either by extrusion of billets into a thick shell which is then drawn to size, or by hot piercing of the billet, followed by cold drawing. In both cases, the billet is cast from the melt, and subsequently reheated. Reference 13 found that the extrusion process for pure copper has an energy requirement of 13.6 GJ/tonne. Copper alloy required rather less energy (11.6 GJ/tonne). A separate audit of the piercing process was not done.

201. A shaft furnace and continuous casting plant to produce cylindrical billets has been installed at one copper refinery (IMI, Walsall) and has eliminated the batch-wise melting and casting, with significant (ca. 30%) savings in the energy required for melting and casting. A novel process for casting hollow billets which can be directly cold worked is under development, but so far it has proved technically difficult and costly.

202. In the foundry industry, energy consumption can vary widely from one foundry to another, from 7 to 70 GJ/tonne of casting produced. This variation depends on the size of the castings, differences in product mix within a foundry, furnace configuration and degree of utilisation of furnaces. The bulk of the energy is used in melting and holding the alloy. The figure of 22.6 GJ/tonne given in reference 13 is taken as an average value.

Energy Consumption in 1980

203. For the purposes of the present analysis of energy conservation potential, an estimate of energy use in 1980 has been arrived at by modifying the energy requirement for individual processes obtained from reference 13 (Table 27) to take account of the changes observed in the industry, as discussed previously. The results of these estimates are shown in Table 28. The overall specific energy requirement obtained from the estimates is 16.8 GJ/tonne, which is in reasonably close agreement with the value of 16.4 GJ/tonne derived from the 1979 fuel purchases enquiry (Table 26).

TABLE 27: Analysis of Energy Use by Process in the Copper Industry
(GJ/tonne) in 1975/76

Process	Process Energy		Non-Process Energy	Metal Loss Energy	Total Specific Energy Requirement
	Fuel	Electricity			
Secondary Refining	9-27	1.3	–	–	10-28
Alloy Ingot Casting	–	–	–	–	5-10
Semi-fabrication of:					
– wire by wire bar casting	5.3	1.3	2.0	0.1	8.7
– wire by continuous casting	2.4	1.4	1.2	0	5.0
– copper strip	12.0	2.9	3.0	0.6	18.5
– copper tube	5.0	6.4	1.8	0.4	13.6
– alloy rod	0.9	3.9	1.4	1.0	7.2
– alloy strip	3.1	4.6	1.5	1.1	10.3
– alloy tube	3.8	4.6	1.7	1.5	11.6
– alloy castings	20.0	0.1	2.0	0.5	22.6

N.B. Data has been taken from ref 13, by converting the electricity component
      to heat supplied.

TABLE 28:   Estimate of Specific Energy Consumption in Copper Industry in 1980

Product	Specific Energy Requirement GJ/tonne	1980 Production (thousand tonnes)	Energy Use (10^6 GJ)
Scrap refining	20	93	1.86
Blister refining	10	68	0.68
Alloy Ingots	8	60	0.48
Total refining and ingot making	13.6	221	3.02
Copper Wire	5.9	200	1.18
Wire Rod	4.9	50	0.25
Copper Strip	18	34	0.61
Copper Rod	8	7	0.06
Copper Tube	12	69	0.83
Alloy Strip	10.3	65	0.67
Alloy Wire	8	8	0.06
Alloy Rod	7.2	97	0.70
Alloy Tube	11.6	12	0.14
Alloy Castings	22.6	61	1.38
Total semi-fabrication	11.1	530*	5.88
		Total Energy Use	8.90
		Overall SEC	16.8

* Copper content only

204. The types of furnaces used in the copper industry are shown in Table 29, together with an estimate of the total energy supplied to them in 1980. In this table, it is assumed that the distribution of furnace types in 1980 was the same as in reference 13. Energy consumption by type is calculated from the material throughputs required to produce the 1980 output of semi-fabricated products (cf. Figure 9).

205. It can be seen from Table 29 that approximately 90% of the energy consumption is used for heating in furnaces of various types for melting metal, reheating it for hot working and annealing it following cold working. Fossil fuel furnaces alone account for 78% of the total energy use. The remaining 10% of energy consumption is largely electricity for mechanical power (rolling, extruding, etc) and electrolytic refining.

ENERGY CONSERVATION POTENTIAL IN THE COPPER INDUSTRY

206. The following analysis of the scope for improving energy efficiency in the copper industry draws on reference 13, together with the experience gained from demonstration projects funded by the ECDPS scheme. Reference 13 concluded that 30% of the energy consumption in 1974 could be saved; other studies within ETSU suggested that the longer term potential could be as high as 46% of the 1974 figure.

207. The recently available results of the 1979 BSO purchases enquiry (ref 8) show an energy saving of 16% compared with the 1974 BSO purchases enquiry. While some of this saving may be due to a shift away from low-grade scrap refining, and fuel substitution (cf. paragraph 185), it is clear from the analysis in the previous section that the introduction of new methods, such as continuous casting techniques have led to real improvements in the efficiency of energy use.

208. Nevertheless, much of the energy saving potential identified in the Energy Audit (ref 13) remains to be fulfilled. A revised analysis is made of the energy saving potential in 1980, based on a consideration of the various changes known to have occurred in the industry since the Energy Audit was prepared.

209. The technically proven energy saving measures are listed in Table 30. The energy saving potential is expressed both as a percentage of the energy used by the particular process to which the energy saving measure is applied, and as a reduction of the overall specific energy consumption which would result from a complete adoption of the measure throughout the industry in those plants to which the measure is applicable. The applicability of the measure is shown as a percentage of total copper production.

210. Since most of the energy used in the copper industry is used in furnaces for melting metal, reheating it for hot working and annealing it, it is to be expected that improvements in furnace design and operation represent the greatest scope for energy saving. There are various measures which can be taken, analogous to those previously discussed in the context of the aluminium industry, ranging from fitting existing furnaces with improved burners and more extensive instrumentation, to installing new furnaces of better design

211. Waste heat recovery from an anode furnace has already been installed in one copper refinery, with the recovered heat used to heat the electrolysis

TABLE 29: Energy Supplied to Furnaces Classified by Type

Furnace Type	Energy Consumption in 1980 (10^6 GJ)
Blast furnaces and Converters, for smelting	1.7
Reverberatory, rotary or crucible furnaces, for melting	3.7
Induction furnaces, for melting	0.8
Reverberatory furnaces, for reheating	0.5
Fossil furnaces used for annealing	1.0
Electric heating used for annealing	0.3
TOTAL	8.0

TABLE 30:   Technically Proven Energy Saving Potential in the Copper Industry

Process	Process Energy Requirement in 1980 (GJ/te)	Energy Saving Technology	Applicability (%)	Potential Energy Savings (%)	Reduction in Average SEC (GJ/tonne)	Cost-effectiveness (payback years)	UK Prospects
Melting, reheating, annealing	13	Improved operation and control of fossil-fuel furnaces	100	10-20	1.3-2.6	1-3	Good in short-medium term Now
Melting, reheating, annealing	13	Improved design of fossil-fuel furnaces	100	10	1.3	3-5	Good Now
Melting, reheating, annealing	13	Waste heat recovery from fossil-fuel furnaces	100	10-15	1.3-2.0	Uncertain	Technical difficulties, longer term only
Foundry casting	20	Scheduling of furnaces to reduce holding of hot metal	10	20	0.4	1-2	Now
Semi-fab.	11	Continuous casting processes	50	30	1.4	< 3	Now
General	17	Improved maintenance of process machinery	100	3	0.5	<2	Now
				Total	6.2-8.2		

Note: Energy saving potential is expressed both as a percentage of the energy used by the particular process to which the measure is applied and as a reduction in the overall specific energy consumption if the measures is applied to its full potential. 'Applicability' is the proportion of total copper production which undergoes the process concerned.

tanks. In other circumstances, waste heat could be used to preheat combustion air. However, waste heat recovery still presents technical difficulties, and the economics are somewhat marginal at present fuel prices.

212. Continuous casting has already penetrated most of the coper wire sub-sector; the extension of the technique to the casting of wide-strip and other shapes suitable for further working to make final products is beginning to occur. Up to 50% of total copper and copper alloy production might be suitable for continuous casting.

213. These techniques contribute to energy saving for two reasons:

- reheating of castings for hot working is avoided

- metal losses in trimmings, etc which have to be recycled as clean scrap to the melting furnace are reduced.

The total energy savings potential of these measures amounts to some 37-49% of the 1980 energy requirement.

214. In the longer term, additional measures could further reduce the energy requirement. These measures involve the application of new technologies which are still in the development stage. They are listed in Table 31, as follows:

- high reduction milling

- conform extrusion

- seam-welded tube

- powder metallurgy.

215. __High reduction mills__ produce a much greater reduction in thickness of metal in a single pass than is possible in a conventional mill. This is achieved by an oscillating mechanism so that the incoming strip is rolled repeatedly as it passes through the machine. An energy saving of 4.7 GJ/tonne would result from the elimination of intermediate annealing between rolling operations. Further economic advantages result from the compactness of an installation.

216. __The CONFORM extrusion process__ has previously been described in the context of the aluminium industry (paragraph 156). The same principle of continuous extrusion can be applied to copper rods and sections; in some respects it is similar to continuous casting, but is applicable to smaller scale operations and more complex cross-sections of product. However, the technique is not easily adaptable for the extrusion of hollow sections for tube-making. The energy saving potential is about 30% of the energy used in semi-fabrication of those products to which it is applicable.

217. __Seam-welded tube__ Tube made from continuously cast strip which is folded into a circular cross-section and welded along the seam would reduce the amount of scrap metal arising from conventional tube-making processes, in which the central core of a billet is removed and returned to the melting stage. The energy saving from this process might be about 10% of the present energy requirement for tube-making, although this has not yet been

demonstrated. However, since the process would involve substantial investment in new equipment, and the product would need to gain acceptance over seamless tube, it is not likely to make a significant impact in the short term.

218. Powder Metallurgy Many metal products can be produced directly by compacting powder into the final shape and fusing it with a heat treatment. An energy saving would result because of the elimination of the various intermediate process stages of melting, hot working and annealing. In principle this could amount to 66% or more of the energy use in semi-fabrication, but the technique is still at an experimental stage, and it is too early to quantify the savings in detail.

PROJECTIONS OF FUTURE SPECIFIC ENERGY CONSUMPTION

219. From a consideration of the technical potential for improving energy efficiency described above, judgements about the rate and extent of uptake of the technical potential have been made, in the context of two scenarios of output, designated 'high' and 'low'. These judgements are set out in Table 32. They have been arrived at by taking account of the general factors affecting uptake of energy conservation measures identified in Volume I of this report.

220. In a high output scenario, it is assumed that SEC will fall by 15% by 1990 and 30% by 2000; this represents a take-up of most of the 'technically proved' energy saving potential in Table 30, but little (other than CONFORM) of the 'R,D&D' potential in Table 31.

221. In a low output scenario, uptake is assumed to be somewhat slower, with SEC falling 10% by 1990 and 20% by 2000.

222. In both scenarios it is assumed that electricity and solid fuel will increase their market share at the expense of oil and gas. Thus electricity will increasingly be required for mechanical power in continuous processes, and for induction heating. Coke will continue to be required for smelting, while coal will increasingly be used to heat reverberatory furnaces, in place of oil or gas.

TABLE 31: R,D&D Energy Saving Potential in the Copper Industry

Process	Process Energy Requirement (GJ/tonne)	Technology	Applicability (%)	Energy Saving Potential		Cost-effectiveness	UK Prospects
				%	GJ		
Strip rolling	25	High reduction mills	15	15	0.6	Uncertain	Long term only
Extrusion	10	CONFORM process	15	30	0.5	3	Good in medium term
Tube making	25	Seam-welded tube	15	10	0.4	Uncertain	Long term only
Semi-fab.	11	Powder metallurgy	100	ca 66	7.2	Uncertain	Fair in medium term

TABLE 32: Projections of Future Specific Energy Consumption

	1980	1990		2000	
		High	Low	High	Low
SEC (GJ/tonne)	16.8	14.3	15.1	11.8	13.4
Electricity %	25	30	28	35	32
Oil %	21	16	19	13	17
Gas %	48	39	43	30	36
Coal %	6	15	10	22	15

ENERGY USE AND ENERGY EFFICIENCY IN UK MANUFACTURING INDUSTRY

UP TO THE YEAR 2000

1.4 The Lead and Zinc Industries

FIGURE 12: Structure and Principal Material Flows in the Lead and Zinc Industry in 1980
(Units in kilotonnes)

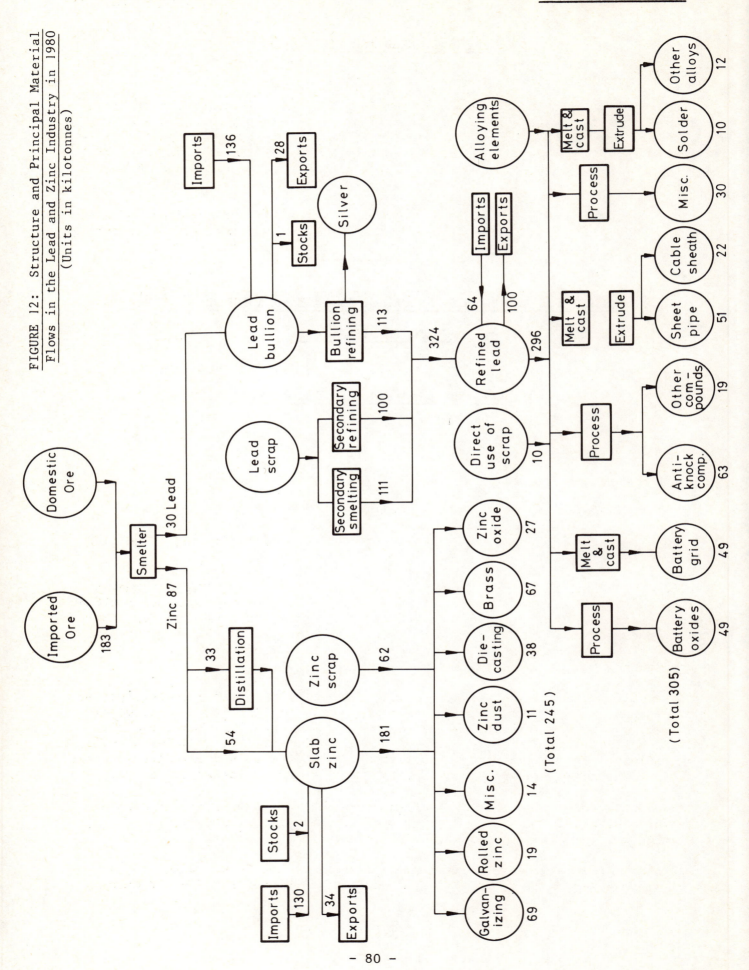

1.4   LEAD AND ZINC

INTRODUCTION

223.   Although the lead and zinc processing industries are largely separate, they are treated together in this report, mainly because the metals are obtained from the same ores in a single process.   There are some similarities in the properties of the two metals, such as corrosion resistance.   They are, however, mostly used for different purposes and are not substitutes for each other.

224.   An energy audit of the lead and zinc industries was carried out by BNF Metal Technology Centre (ref 17).   The audit adopted a convention of assigning all of the energy used in smelting zinc/lead ores to the zinc.   This convention is maintained in the present report.

225.   The total energy used by the lead and zinc industries in 1980 is estimated to have been about 13×10^6 GJ.

STRUCTURE OF THE UK LEAD AND ZINC INDUSTRIES

226.   The lead and zinc industries can be divided into the following sub-sectors:

- primary smelting of lead/zinc ores

- processing of zinc

- secondary smelting of lead

- refining and processing of lead.

Figure 12 shows the material flows in the two industries; the major smelting and refining plants in the UK are listed in Table 33.

227.   The single UK smelter at Avonmouth has a capacity of 100,000 tonnes of zinc and 40,000 tonnes of lead per year.   It operates by the Imperial Smelting Furnace process (ISF), which was originally developed in this country.   There are about ten other ISF plants worldwide, accounting for ~ 10% of world zinc production.   An electrolytic process accounts for 80% of world zinc production, but is not used in the UK.

228.   Refining processes are carried out on about 100 kt per year of imported crude lead, as well as on the UK smelter production.   The crude lead contains silver, which has a value equal to that of the lead itself - hence the term 'lead bullion' for the crude lead.   About 200 kt of lead is produced annually by refining scrap lead, mainly from batteries and demolished buildings.   About half of the scrap requires a smelting process, which is energy intensive.

229.   There are few sources of zinc scrap, other than the zinc-copper alloys recycled within the copper industry, and, consequently, there is no significant industry engaged in the secondary refining of zinc.

230.   Zinc from overseas smelters is imported into the UK as 'slab zinc'. This term covers a range of grades of purity for the metal.   Zinc produced

TABLE 33: Major Lead Refining Plants in the UK

Process	Company	Location	Maximum Capacity (tonnes/yr)
Primary Smelting	Commonwealth Smelting (RTZ)	Avonmouth	40,000 Pb (+ 100,000 Zu)
Bullion Refining	Britannia Refined Metals Ltd (MIM Ltd)	Northfleet	150,000
Secondary Refining	Britannia Refined Metals Ltd	Northfleet	30,000
" "	Associated Lead Manufacturers Ltd	Newcastle, London, Chester	75,000
" "	Chloride Metals Ltd	London, Wakefield, Manchester	70,000
" "	Capper Pass & Sons Ltd (RTZ)	North Ferriby	10,000
" "	H J Enthoven & Sons Ltd	Darley Dale	45,000

Source: Metalli Nonferrosi 1980.

by the ISF process is 98.8% pure, whereas electrolytic zinc is ordinarily 99.95%, and can be better. However, ISF zinc can be purified to 99.99% by fractional distillation.

231. The processing of lead and zinc is largely carried out by the firms which use the metals to make a final product. Thus the copper and steel industries consume large quantities of primary zinc to make brasses and galvanised steel, while battery manufacturers consume large amounts of refined lead. Thus the processing sub-sector is highly fragmented.

TRENDS IN LEAD AND ZINC PRODUCTION AND CONSUMPTION IN THE UK

232. The trends in UK production and consumption of lead are shown in Figure 13. It can be seen that there has been a downward trend in consumption since 1965. Secondary production has, however, risen by 250% over the same period. Total production of refined lead, including that refined from imported bullion, has been about 10% higher than domestic consumption since 1975 (statistics on total production prior to 1975 are not available). The UK is thus a net exporter of refined lead.

233. The decline in the consumption of lead has been due to its displacement by alternative materials, especially in plumbing and pigments, where concern over the toxicity of lead has been a marked influence. The use of lead as a petrol additive will decline, if not actually cease, for the same reason, while its use in the printing industry for type metal will be displaced by new printing technologies. However, lead will continue to be used for secondary batteries, in buildings (for flashings) and in chemical plant, where its corrosion resistance in aggressive environments is valuable.

234. The long term trend in zinc consumption (Figure 14) shows a gradual increase, fluctuating with the business cycle and interrupted by sharp slumps in 1973 and 1979. Approximately one third of domestic consumption is met by primary smelting in the UK, from imported ores. A small amount of domestic ore is also smelted (zinc content, 4 kt).

235. The most important uses of zinc are for galvanising steel, for alloying with copper to make brass and for die-casting. Die-castings are used extensively in the building and engineering industries. Zinc strip is used for dry-cell batteries, and the oxide is used as a pigment. Although there are substitutes for zinc in many of its application (eg aluminium for galvanised steel, stainless steel for brass, plastics for die-castings) there is no acute threat to the zinc industry.

236. The future trend in zinc consumption is likely to continue to reflect activity in building and manufacturing. Primary zinc production, however, depends on discrete units of plant; any future increase would require investment in additional smelter capacity. Whether or not this would be installed in the UK rather than overseas, in the countries producing the ore, depends on many factors; present conditions suggest that future smelting capacity would be sited overseas, but this could change towards the end of the century.

FIGURE 13:   Lead Production and Consumption in the UK, 1963–81

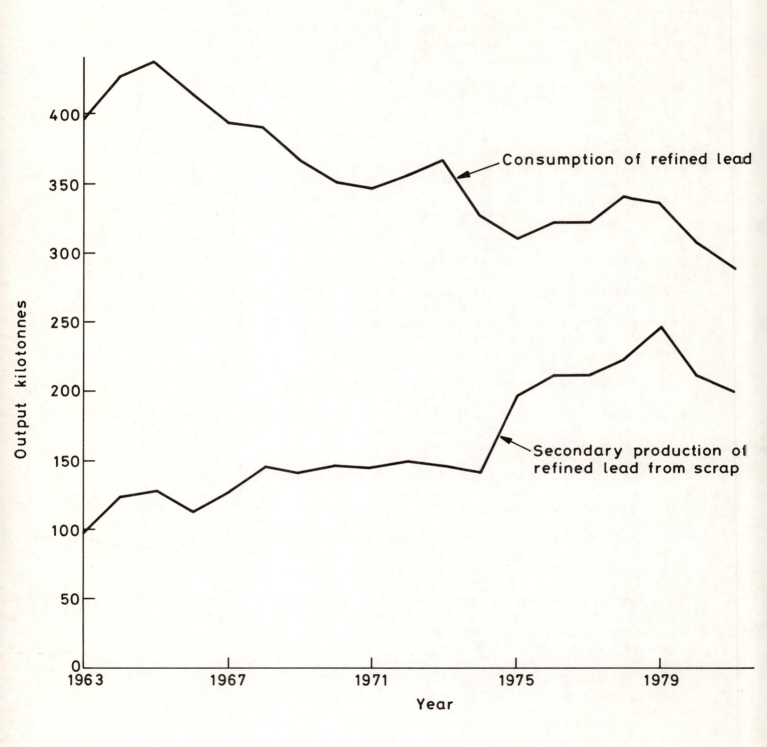

FIGURE 14: Zinc Production and Consumption in the UK, 1963-81

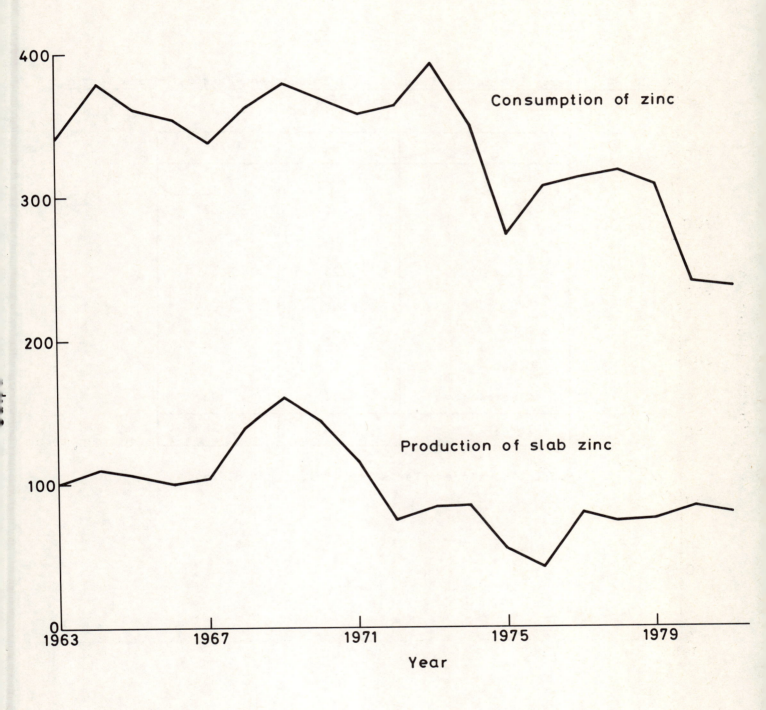

TABLE 34:   Energy Use in the Lead, Zinc and Other Non-Ferrous Metals Sector
(Units: 10^6 GJ)

	1974	1979
Coal	11.31	4.92
Fuel Oil	2.42	2.62
Gas Oil	-	1.12
Natural Gas	2.25	3.44
LPG	-	1.34
DERV	0.05	0.06
Other	0.7	-
Electricity	1.52	2.36
Total	18.25	15.86

Source: Business Statistics Office (ref 8, 18).

TABLE 35: Analysis of Energy Use by Process in the Lead and Zinc Industries

Process	Process Energy Requirement (GJ/tonne)				Production (kt)			Estimated Total Energy Consumption (PJ)		
	Coal	Elec.	Oil & Gas	Total	1974	1979	1980	1974	1979	1980
Zinc Smelting	40.3	3	9.2	52.5	84	77	87	4.4	4.0	4.6
Zinc Processing										
Galvanising	–	2.6	40.0	42.6	92	89	69	3.9	3.8	2.9
Rolled zinc	–	0.5	2.5	3.0	22	21	19	0.1	0.1	0.1
Die-casting	–	2.2	20.0	22.2	70	62	38	1.6	1.4	0.8
Miscellaneous	–	n.a.	n.a.	3.0	169	140	119	0.5	0.4	0.4
Lead Refining										
Bullion	–	0.1	2.0	2.1	137	124	113	0.3	0.3	0.2
Secondary smelting	8.3	1.0	7.0	16.3	66	130	111	1.1	2.1	1.8
Secondary refining	–	0.1	2.0	2.1	74	114	100	0.1	0.2	0.2
Lead Processing	–	n.a.	n.a.	3.0	325	336	305	1.0	1.0	1.0
TOTAL								12.7	13.1	11.8

Note: Process energy requirements are modified from reference 19. Coke consumption has been converted to the equivalent coal requirement, assuming 72% efficiency for coke-making. Electricity consumption is given in heat supplied terms (ie GJe). The calorific value of the sulphide ore is not included in the zinc smelting PER.

ENERGY USE IN THE LEAD AND ZINC INDUSTRIES

237.   There are no published data for energy use in the lead and zinc industries as such.   The Business Statistics Office data (ref 9) includes lead and zinc with other non-ferrous metals, such as tin, nickel, titanium, chromium, magnesium, etc.   (Activity 2247 of the 1981 standard industrial classification.)   Table 34 shows the energy use in SIC 2247 reported in the 1974 (ref 13) and 1979 (ref 18, 8) fuel purchases enquiries.   From the tonnages of non-ferrous metal production reported in the Annual Abstract of Statistics (ref 21), it is estimated that lead and zinc account for about 80% of the total production in SIC 2247.

238.   The data in Table 34 indicate a substantial shift from coal to oil and gas between 1974 and 1979.   Although total energy consumption fell by 13% in the period, a comparison of specific energy consumption is difficult, because of relative changes in the output of different processes within the industries.   Thus, for example, smelter production of zinc fell by 10%, whereas secondary refining of lead increased by 70% between 1974 and 1979.

239.   The energy audit of the lead and zinc industries (ref 17) is based on individual plant audits, from which specific energy requirements of individual processes were derived.   These are shown in a modified form in Table 35.   By compounding these process energy requirements with the output of each process in a given year (from ref 19) an estimate can be made of the total energy consumption by the lead and zinc industries for that year.   Table 35 shows this for 1974, 1979 and 1980.   The calculated value for 1979 can not be directly compared with the BSO 1979 data in Table 34.   However, if it is assumed that the BSO total of 15.9 PJ includes 20% for other non-ferrous metals, then the lead and zinc components can be taken to be 12.8 PJ compared with 13.1 PJ for the calculated value.

240.   The calculated value in Table 35 for 1974 is almost the same as the 1979 value, which suggests that the structural changes in output had no net effect on energy consumption.   It can therefore be concluded that the actual fall of 13% in energy consumption between 1974 and 1979 represents a decrease in specific energy consumption.   Presumably this is mainly a consequence of the change from coal to more efficient fuels.

241.   From the process energy requirements in Table 35, it can be seen that the most energy intensive processes are primary smelting, galvanising and die-casting of zinc, and secondary smelting of lead.

242.   Primary Smelting   In the ISF process for smelting zinc/lead ores a mixture of the ore concentrate and coke is fed into the blast furnace.   Zinc vapour is removed in a stream of gas containing carbon dioxide and carbon monoxide, while molten lead is separated from slag at the bottom of the furnace.

243.   The zinc vapour is condensed by a spray of molten lead, from which it can be separated with 98.8% purity; the lead coolant is recirculated continuously.   Some of the zinc is further purified by fractional distillation, to a purity of 99.99%.   In addition to the coke used in the blast furnace, electricity is used for mechanical power, and oil is used for sintering ore and fractional distillation.

244. <u>Galvanising</u> Galvanising is carried out by immersing steel in molten zinc. This is done in a continuous process, with steel strip or wire, or in a batch process, with steel shapes in various forms. The energy requirement per tonne of zinc is very high because the much larger quantity of steel substrate must be heated to the melting point of zinc. The PER in Table 35 is an average figure, assuming a ratio continuous to batch processing of approximately 2:1.

245. <u>Zinc Die-Casting</u> Zinc castings are made by injecting molten zinc into a steel die under pressure. After the casting has solidified, it is removed from the die and trimmed to remove excess bits of metal. The alloy for zinc die-casting is made from high purity zinc.

246. Most of the energy used in zinc die-casting is used in melting the metal and holding it molten. As with the other non-ferrous metal casting proceses, the melting and holding of metal is very inefficient, but much of the losses could be avoided by using highly insulated holding furnaces.

247. <u>Lead Refining</u> Lead bullion is refined by adding zinc to molten lead to form a zinc-silver alloy, wich can be separated and the zinc distilled. Most of the energy used is for melting the lead. Similarly, the recovery of lead from metallic scrap involves only a simple melting and casting. However, about 50% of lead scrap is in non-metallic form, and requires a smelting process to recover the lead.

248. <u>Lead and Zinc Semi-Fabrication</u> The production of semi-fabricated products in lead and zinc (sheet, pipe, etc) uses only modest amounts of energy. Lead is soft, and can be cold-worked easily, while zinc is hot-rolled. In both cases, the PER is about 3.0 GJ/tonne.

ENERGY CONSERVATION POTENTIAL IN THE LEAD AND ZINC INDUSTRIES

249. Most of the energy used in the lead and zinc industries is in furnaces of various types. The energy saving opportunities are mainly in the improved control and operation of furnaces already discussed in previous sections. Table 36 quantifies these savings for each of the main sub-sectors of the industry. For each sub-sector, a specific energy consumption is derived from the more detailed analysis of process energy requirements in Table 35. The energy conservation potential is expressed in relation to the SEC for each sub-sector.

250. In primary smelting of lead/zinc ores, the main opportunities for energy saving are:

- combustion of low calorific value gases for power generation (about 30% of the gases produced are currently wasted);

- recovery of heat from water-cooled launders;

- heat recovery from the sintering plant.

251. The energy saving measures which could be applied to the furnaces used for melting and holding metal are characterised as:

- good housekeeping (using existing equipment with care);

TABLE 36: Energy Conservation Potential in the Lead and Zinc Industries

Process/Measure	Average PER (GJ/tonne)	Potential Energy Saving		Cost-effectiveness (payback years)	UK Prospects
		%	GJ/tonne		
Primary Smelting	52.5				
- combustion of blast furnace gas		10	5.2	5	Likely to be installed by 1990
- heat recovery from sintering plant		10	5.2	10	c.a. 2000
- heat recovery from cooling water		10	5.2	10	c.a. 2000
Zinc Processing	17				
- good housekeeping		10	2.3	<1	
- improved furnace control		20	4.6	5	
- scheduling		5	1	5	
- waste heat recovery		20	4.6	10	
- product yield		5	1	5	
Secondary Lead Smelting	16.3				
- good housekeeping		10	1.6	<1	
- waste heat recovery		10	1.6	10	
Lead Refining and Processing	5.1				
- good housekeeping		10	0.5	1	
- improved furnace control		20	1	10	
- waste heat recovery		10	0.5	15	Not likely to be used
- product yield		5	0.3		

- improved furnace control (better instrumentation, automatic controls, insulation);

- improved scheduling;

- waste heat recovery;

- better yield of product, reducing the reprocessing of scrap.

These measures could ultimately reduce the energy use in zinc processing by 60% and in lead processing by 45%.

PROJECTIONS OF FUTURE SPECIFIC ENERGY CONSUMPTION

252. The assumptions which have been made in projecting future specific energy consumption are shown in Table 37. These relate to both the future structure of the industry and the technical improvements which are assumed to be taken up by the industry.

253. It is assumed that there will be no change in the primary smelting capacity in the UK - currently 100 kilotonnes per year. The proportion of total demand for zinc which is met by the primary smelter is expected to remain approximately constant up to 1990; smelter output should increase to full capacity as demand rises. After 1990, increases in demand are assumed to be met by increasing imports of slab zinc, resulting in a declining proportion of primary smelting in the UK.

254. The energy efficiency of the Avonmouth smelter is assumed to improve by 10% due to recovery of blast furnace gas. This is assumed to occur by 1990 in a high output scenario, but not until 2000 in a low output scenario. The specific energy consumption for zinc processing is assumed to fall from 17 GJ/tonne in 1980 to 14 GJ/tonne by 2000 in a high output scenario, and 15 GJ/tonne by 2000 in a low output scenario.

255. In the lead industry, output of refined lead in 1980 was somewhat higher than domestic consumption of refined lead in semi-fabrication processes, the surplus being exported. This situation has existed for a number of years, and there is no reson to believe that it will change in the future; it is partly a result of the role played by the London Metal Exchange in world markets. Consequently, it is assumed that there will be no change in the ratio of lead refining to domestic consumption.

256. The specific energy consumption for lead refining and processing is assumed to fall by 10% by 2000 in both scenarios. For secondary smelting, however, it is assumed to fall by 20% by 2000 in the high output case, and by 14% in the low case.

257. The overall specific energy consumption for the zinc and lead industries is determined by assuming that the output of zinc remains at 44% of the combined out of the two industries.

258. The specific energy consumption for the zinc and lead production is thus projected to fall from 21.6 GJ/tonne in 1980 to 17 GJ/tonne by 2000 in a high output scenario (an improvement of 21.3%) and to 18.3 GJ/tonne by 2000 in a low output scenario (an improvement of 15.3%). Note, however, that the change

in energy efficiency is sensitive to assumptions on the structure of the industry. For example, if there were no change in the proportion of primary zinc output (36%), the overall SEC by 2000 in the high output case would be 18.5 GJ/tonne. Hence, approximately half of the apparent improvement is due to an assumption of structural change; the alternative assumption, however, would imply the construction of additional smelter capacity in the UK, which is not thought to be likely.

TABLE 37:   Projections of Future Specific Energy Consumption in the Lead and Zinc Industries

	1980	1990 High	1990 Low	2000 High	2000 Low
Zinc:					
Primary production output (% of total output)	36	36	36	29	32
SEC for primary production (GJ/tonne)	52.5	47	52	47	47
SEC for processing	17	15	16	14	15
Total SEC for zinc (GJ/tonne)	35.9	31.9	34.7	27.6	30.0
Lead:					
Refining output (% of domestic lead consumption)	70	70	70	70	70
SEC for refining	2.1	2.0	2.0	1.9	1.9
Secondary smelting output (% of domestic lead consumption)	36	36	36	36	36
SEC for secondary smelting (GJ/tonne)	16.3	14	15	13	14
SEC for lead processing (GJ/tonne)	3	2.9	2.9	2.7	2.7
Total SEC for lead (GJ/tonne)	10.3	9.1	9.7	8.7	9.7
Overall SEC (zinc and lead) (GJ/tonne)	21.6	19.1	20.7	17.0	18.3

REFERENCES

1.   'Energy Projections 1982'. Proof of Evidence for the Sizewell 'B'
      Public Inquiry. Department of Energy, 1982.

2.   'Annual Statistics 1981', UK Iron and Steel Statistics Bureau, 1982.

3.   'Energy Audit Series No 16. The Iron and Steel Industry', HMSO, 1982.

4.   F Feltoe, 'Energy Management in the British Steel Corporation',
      AIEI/UNIDO Conference, New Delhi, 1983.

5.   Report of the Sub-Committee on Energy of the IISI Committee on
      Technology.

6.   Energy Audit Series No 6. The Aluminium Industry. British Non-Ferrous
      Metals Technology Centre. Issued jointly by the Departments of Energy
      and Industry, 1979.

7.   Report on the Censuses of Production 1974/75 Aluminium and Aluminium
      Alloys. Business Monitor PA 321. HMSO 1978.

8.   'Census of Production and Purchases Inquiry, 1979'. Business Monitor
      PA 1002.1. HMSO 1983.

9.   'Inquiry into Private Generation of Electricity in Great Britain, 1977'.
      Department of Energy, 1979.

10.   W J Roscow, 'Influence of Energy Costs on Non-Ferrous Melting Practice',
      The Metallurgist and Materials Technologist, p 554, December 1982.

11.   Metal Statistics 1970-1980. 68th Edition. Metallgesellschaft AG,
      Frankfurt-am-Main, 1981.

12.   World Copper Survey 1980. Published by Metal Bulletin.

13.   Energy Audit Series No 12. The Copper Industry. British Non-Ferrous
      Metals Technology Centre. Issued jointly by the Departments of Energy
      and Industry, 1981.

14.   Energy Use in the Copper Sector of the Non-Ferrous Metals Industry.
      IETS Report No 11. Department of Industry 1979.

15.   Report of the Censuses of Production 1974/75. Copper, Brass and Other
      Copper Alloys. Business Monitor PA 322. HMSO 1978.

16.   Digest of UK Energy Statistics, 1980. Department of Energy. HMSO
      1981.

17.   Energy Audit Series No 10. The Zinc and Lead Industries. British
      Non-Ferrous Metals Technology Centre. Issued jointly by the Departments
      of Energy and Industry, 1980.

18.   Report of the Censuses of Production 1974/75.   Miscellaneous Base
      Metals.   Business Monitor PA 323.   HMSO 1978.

19.   Annual Abstract of Statistics 1983, Central Statistics Office.   HMSO.
      1983.

ENERGY USE AND ENERGY EFFICIENCY IN UK MANUFACTURING INDUSTRY

UP TO THE YEAR 2000

SECTOR 2. CERAMIC MATERIALS

D H Buckley-Golder

Subsector	Paragraph
Introduction	1
Assumptions on future levels of output	9
2.1 Cement	32
The structure of the cement industry	34
Cement manufacturing processes	37
Energy use in the cement industry	44
Energy conservation potential	51
Projected levels of energy consumption	78
Conclusions	88
2.2 Bricks	92
The structure of the brick industry	95
Energy use in the brick industry	105
Energy conservation potential	116
Projected levels of energy consumption	162
Conclusions	176
2.3 Refractories	181
The structure of the refractories industry	185
Refractories manufacture	193
Energy use in the refractories industry	198
Energy conservation potential	209
Projected levels of energy consumption	238
Conclusions	246
2.4 Potteries	249
The structure of the pottery industry	252
Pottery manufacture	258
Energy use in the pottery industry	263
Energy conservation potential	277
Projected levels of energy consumption	297
Conclusions	306

Subsector   Paragraph

2.5   Glass   308

       The structure of the glass industry   312
       Glass manufacturing processes   317
       Energy use in the glass industry   324
       Energy conservation potential   334
       Projected levels of energy consumption   373
       Conclusions   379

2.6   Miscellaneous building materials   382

       Energy use   384
       Energy conservation potential   394
       Projected levels of energy consumption   397
       Conclusions   400

References

SECTOR 2.   CERAMIC MATERIALS

INTRODUCTION

1.   In 1980 this group of industries consumed approximately 214 x 10^6 GJ
of energy, on a heat supplied basis. Of this ~ 37% was used in cement
manufacture, ~ 20% in glass manufacture, ~ 9% by the brick industry, ~ 6% by
the pottery industry and ~ 5% in the manufacture of refractory materials. The
remaining 23% was used in the manufacture of miscellaneous building materials.

2.   This report contains a detailed analysis of energy consumption within
each of these industries, and the potential for energy conservation.
Projections are made of the future levels of specific energy consumption by
each industry, taking account of the various energy conservation technologies
which may be applied and the time required for these to be taken up by the
industry.

3.   Over 80% of the energy consumed by this sector is used in kilns or
furnaces for melting, firing or drying products, and it has been estimated
that by the year 2000 the mean specific energy consumption of the sector as
whole could be reduced by around 26% from ~ 34 GJ per £ thousand gross output,
to ~ 25 GJ per £ thousand. Energy conservation measures are aimed primarily
at reducing the requirements for energy in melting, firing and drying through
improved heat recovery, insulation and process control. In certain
industries, eg cement, significant savings can be achieved through process
modification or change.

4.   This study focuses mainly on energy conservation technologies which may
reduce the energy requirements of this sector over the period 1980 to 2000.
However, in the brick and pottery industries certain technologies have been
identified which may offer further savings beyond 2000. For completeness such
technologies and the associated savings have been described, but will not be
discussed in detail.

5.   Certain industries within this sector are characterised by a
particularly diverse product mix, for example pottery and glass, where certain
products are aimed at the market for luxury goods such as bone china and cut
crystal, while other products may be classed as necessities, such as window
glass and sanitaryware. Such characteristics make it difficult to anticipate
how the balance of production may alter between the product types in the
future, and the effect this may have on the mean specific energy consumption
of the industry.

6.   Where adequate information is available, account has been taken of any
effects of structural changes within an industry, on its future energy
requirements. However, where insufficient information exists it was
considered inappropriate to introduce arbitrary changes in energy
requirements, related to structural changes within the industry.

7.   Two scenarios of future industrial output have been adopted. These are
based on the central projections (YU and YL) of the Department of Energy's
1982 Energy Projections (1), and account has been taken of the specific
characteristics of each industry within the sector. Both scenarios relate to
a 1½% growth rate in GDP, but differ in the contribution made to GDP growth
by manufacturing industry. The overall rates of change of output of the

sector as a whole, under the two scenarios designated 'higher' and 'lower', are those specified under the DEn scenarios for the building materials industries. The scenarios developed for each industry are described in some detail in the following section.

8.   Figure 1 shows a graphical representation of the reduction in the projected energy requirements of the sector, by the introduction of the various energy conservation technologies. The broken line represents the total energy requirement of the sector if no conservation technologies were implemented, and the continuous line represents the projected energy requirement if the technologies are implemented as described in this report.

ASSUMPTIONS ON FUTURE LEVELS OF OUTPUT FROM THE CERAMIC MATERIALS SECTOR

1. Cement

9.   Over the period 1964-1980 the output of the cement industry has, in general, changed in line with that of the building materials sector as a whole, and both showed peak output in 1973. However, for the period 1964-73 output from the building materials sector increased by an average of 3.4% per year, while that for cement increased by only 1.8% per year. On the other hand, between 1973 and 1980 output from the building materials sector fell by an average 3.2% per year while that from cement fell by 4.2% per year.

10.   In the Department of Energy's 1982 Energy Projections, output from the building materials sector is assumed to increase continuously between 1980 and 2000 under all scenarios of the future. For the purposes of this report the two central scenarios have been adopted and it has been assumed that the rate of increase in output from the cement industry will be lower than that for the sector as a whole by the same proportion as between 1964-1973, at least for the first 10 years of the period under review. For the higher scenario considered here, an average increase of 0.8% per year has been taken for the period 1980-1990 compared with 1.5% per year for the sector as a whole under the DEn higher output scenario, but the rate of increase taken from 1990 onwards is in line with the rate projected by DEn for the rest of the building materials sector at 2.2% per year. This results in an overall increase in output between 1980 and 2000 of ∼ 1.5% per year, which is in line with that anticipated by the industry itself.

11.   A similar approach has been adopted for the lower output scenario, with output increasing by 0.5% per year up to 1990 and by 1.5% per year thereafter. This gives an overall increase in output of 1% per year between 1980 and 2000. Future levels of output from the cement industry under two such scenarios of the future may be as shown in Table 1.

Table 1. Projected Levels of Clinker Output from the Cement Industry
10^6 tonnes per year

Output \ Year	1980	1985	1990	1995	2000
Higher Scenario	13.9	14.4	15.0	16.8	18.7
Lower Scenario	13.9	14.2	14.6	15.7	16.9

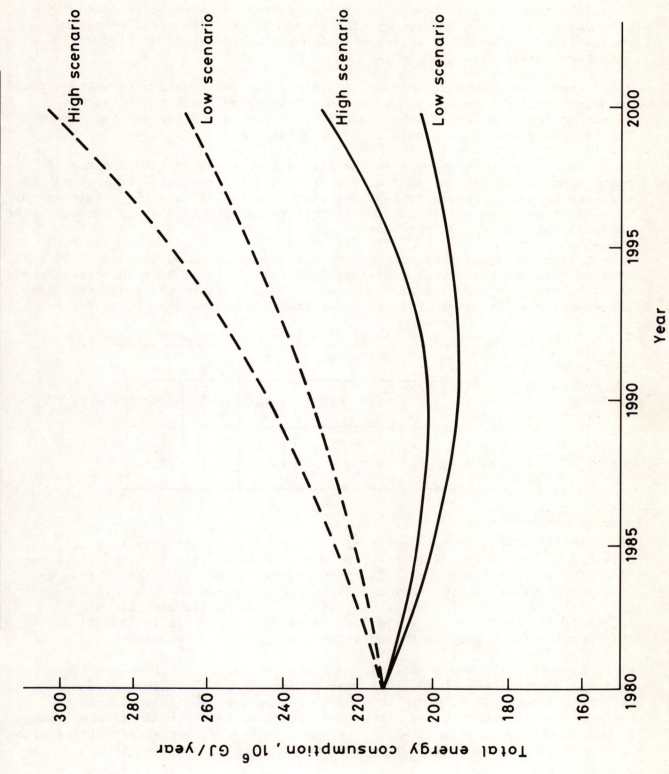

Figure 1. Projected energy requirements of the ceramic materials sector

2. Bricks

12.   Over recent years bricks have been replaced in many situations, either by lightweight concrete blocks or by timber frame construction. Such building techniques reduce the requirement for common bricks, although facing and engineering bricks are still required for their special decorative and structural qualities.

13.   Continued growth in the use of alternative construction techniques could lead to structural change within the brick industry, with the balance of output moving more towards the production of facing and engineering bricks. This is described in more detail in paragraphs 169 to 171 of section 2.2.

14.   The projected change in the overall output of the brick industry is based upon the Department of Energy's two central scenarios. Under the higher scenario output increases by 1.5% per year between 1980 and 1990, followed by 2.2% per year between 1990 and 2000. This represents an overall increase over the whole period of 1.9% per year.

15.   Under the lower scenario the overall change in output between 1980 and 2000 is 1.2% per year, with an increase of 1.0% per year over the first decade and 1.5% per year over the second decade. Table 2 shows the projected levels of output from the brick industry under these two scenarios of the future.

Table 2. Projected Levels of Output from the Brick Industry

10^6 tonnes per year

Output \ Year	1980	1985	1990	1995	2000
Higher scenario	9.2	9.9	10.6	11.9	13.2
Lower scenario	9.2	9.6	10.1	10.9	11.7

3. Refractories

16.   Around 60-65% of the output of the refractories industry is used directly in the UK iron and steel industry. Between 1974 and 1980 the index of production for the refractories industry has moved with that of iron and steel production. In the period 1973/74 there appears to have been a lag of one year between the iron and steel industry peak output and that of refractories manufacture. A similar lag may have occurred in 1980/81 due to the overall uncertainty of the durability of the increased market.

17.   However, over short periods, such as five years, it is reasonable to say that the output index of the refractories industry follows that of the iron and steel industry. In order to project future levels of output from the refractories industry it has, therefore, been assumed that the rate of change in output will be that applied in the Department of Energy's central scenarios for the future production of iron and steel.

18.   These indices suggest that even under the higher scenario the output from the refractories industry is unlikely to regain even its 1974 level by 2000.

19. In accordance with the projected levels of output used in section 1.1 of this report, on iron and steel, under the higher scenario, production of refractories will increase by 3.9% per year between 1980 and 1990, and by 1.9% per year between 1990 and 2000. This represents an overall increase of 2.6% per year over the whole period. Under the lower scenario output increases by 1.4% per year between 1980 and 1990, but decreases by 1% per year between 1990 and 2000, which represents an overall increase of 0.4% per year.

20. Future levels of production from the refractories industry under two such scenarios of the future may be as shown in Table 3.

Table 3. Projected Levels of Output from the Refractories Industry
10^6 tonnes per year

Output \ Year	1980	1985	1990	1995	2000
Higher scenario	0.82	0.99	1.20	1.28	1.37
Lower scenario	0.82	0.90	0.98	0.94	0.89

4. Potteries

21. The index of overall output for the building materials group of industries in 1980 was 90.6 (1975 = 100) although those for the individual industries varied somewhat, for example, that for the brick industry was 85.4, for cement 85.6 and for potteries 96.5.

22. However, the output from the pottery industry slumped after 1980, and in 1981 and 1982 the index was only 85.5 and 77.4 respectively. It is very unlikely that the 1980 level could be regained by 1985, as this would represent an increase in output of 7.7% per year. Even to regain it by 1990 would represent an increase of 2.8% per year.

23. The B Ceram RA feel that the most pessimistic scenario of the future could be a continuing decline in the industry, and even the most optimistic scenario might have only a modest increase in output over the next 20 years. For the purposes of this study, the percentage annual increase in output used in the DEn central projections has been applied, taking the 1982 index of production as a starting point. Therefore, for the higher scenario output increases by 1.5% per year up to 1990 and 2.2% per year from 1990 to 2000. For the lower scenario the corresponding figures are 1.0% and 1.5%.

24. The overall increase in output between 1980 and 2000, under the higher scenario is, therefore, 0.6% per year, and under the lower scenario 0.04% per year. The output of the pottery industry is usually measured in terms of the value of the product, rather than the weight. Table 4 therefore shows the projected indices of production for the pottery industry under two such scenarios of the future, and the corresponding value (in 1980 money values).

Table 4. Projected Indices of Production and Value of Output from
the Pottery Industry

Year / Index	1980	1985	1990	1995	2000
Higher scenario	96.5	80.9	87.2	97.2	108.4
Lower scenario	96.5	79.7	83.8	90.3	97.3

Year / Value £ million	1980	1985	1990	1995	2000
Higher scenario	629	527	568	634	707
Lower scenario	629	519	546	589	634

5. Glass

25. Output from the glass industry over the past decade has remained fairly steady at around 3 million tonnes of glass per year, although in 1981 it fell to only 2.6 million tonnes, due to the general recession in British industry.

26. In 1980 the index of production for the building materials group of industries was 90.6 (1975 = 100), while, according to the British Glass Industry Research Association, that for glass alone was actually higher than in 1975, the relative index being 105.2.

27. The projected levels of output from this industry have been based upon the central scenarios of the Department of Energy's 1982 Projections, with an overall increase in output between 1980 and 2000 of 1.9% per year under the higher scenario and 1.2% per year under the lower scenario.

28. Table 5 shows the projected output figures for the glass industry over the period 1980-2000, calculated on the basis of such scenarios of the future.

Table 5. Projected Levels of Output from the Glass Industry
10^6 tonnes per year

Year / Output	1980	1985	1990	1995	2000
Higher scenario	2.90	3.12	3.37	3.76	4.19
Lower scenario	2.90	3.05	3.20	3.45	3.71

6. Miscellaneous Building Materials

29. Future levels of production of miscellaneous building materials are difficult to estimate, in view of the diversity of products to be considered.

Output from this sector may grow at a slightly higher rate than the building materials group of industries as a whole, as certain products within this sector may be used as alternatives to other ceramic materials; for example concrete building materials as a replacement for bricks.

30.   The output assumptions used for this sector of the ceramic materials industries are shown in Table 6. Under the higher scenario output increases at a rate of 1.9% per year between 1980 and 1990, followed by 2.3% per year up to 2000. Under the lower scenario the rate of increase between 1980 and 1990 is 1.5% per year with 1.6% per year over the following decade.

31.   These values combine with the projected levels of output developed for the other ceramic materials industries, to give the overall rates of change in gross industrial output used in the two central scenarios of the Department of Energy's 1982 Energy Projections, for the building materials industries. These are shown in Table 7.

Table 6. Projected Levels of Output for Miscellaneous Building Materials.
£ million per year

Year / Output £ million	1980	1990	2000
Higher scenario	3,148	3,794	4,785
Lower scenario	3,148	3,643	4,270

Table 7. Projected Average Annual Rates of Change of Industrial Output for the Building Materials Industries. DEn 1982 Energy Projections. Scenarios YU and YL (1)

Period / Rate of change	1980-1990	1990-2000
Higher scenario	1.5% per year	2.2% per year
Lower scenario	1.0% per year	1.5% per year

ENERGY USE AND ENERGY EFFICIENCY IN UK MANUFACTURING INDUSTRY

UP TO THE YEAR 2000

2.1 THE CEMENT INDUSTRY

2.1   THE CEMENT INDUSTRY

INTRODUCTION

32.   According to the Cement Makers' Federation, in 1980 the cement industry used 79 million GJ of energy (heat supplied) to produce 13.9 million tonnes of cement clinker. The energy consumed by this industry amounted to 3.9% of the total UK industrial energy consumption in that year.

33.   An Energy Audit of this industry was published in July 1980 [2] and was based on data applying mainly to 1979. The Cement Makers' Federation carry out an energy survey each year of all firms within the industry; this gives information on the type and output of all kilns, the total energy consumption and types of fuel used. This information is not published, but has been supplied on a confidential basis for the purposes of this study. In this report projections of the future potential for energy conservation within the cement industry have been based on the Energy Audit and the work of the Energy Conservation Branch of ETSU.

THE STRUCTURE OF THE CEMENT INDUSTRY

34.   In 1980 there were six companies manufacturing and marketing Portland cement in the UK, and these are listed in Table 7, together with the output of each in 1980. (There are now only three companies as Ketton, Ribblesdale and Tunnel now comprise RTZ Cement Ltd and Aberthaw are now part of Blue Circle Industries PLC.) ICI produces a small amount of cement to make profitable us of limestone slurry, which is a residue from that used in some of its other production processes. However, this cement is marketed by Blue Circle Industries PLC.

35.   Figure 2 shows the output of the UK cement industry over the period 1970 to 1981. The data given in the Central Statistics Office Annual Abstract of Statistics [3] are somewhat higher than those supplied by the Cement Makers' Federation (and for the purposes of this report CMF data has been used throughout), as CMF output figures relate to cement clinker and the Abstract of Statistics data relate to finished cement. In this report, quantities expressed as tonnes of cement refer to the clinker equivalent of finished cement production. Clinker output is used as the basic data of the report.

36.   Output peaked at 20 million tonnes in 1973 and declined steadily until 1977. It rallied slightly between 1977 and 1979 but has continued to decline sharply. Output of Portland cement has fallen by 11% between 1979 and 1980, and by a further 16% between 1980 and 1981, although in 1982 output showed an increase of about 4% over the previous year.

CEMENT MANUFACTURING PROCESSES

37.   The main raw materials used in cement production are chalk, clay, limestone, shale and marl. These are ground and mixed to produce a raw feedstock of uniform chemical composition. The mixture is first heated to drive off any moisture and then heated to around 800°C for calcining or decarbonation, when carbon dioxide is driven off.

38.   As the temperature rises to around 1,400°C the oxides of calcium, silicon, aluminium and iron react together to form calcium silicates,

Table 7.   The Structure of the Cement Industry in 1980

Company	1980 Output x 10^6 tonnes	% of Total	Number of Kilns
Aberthaw Cement PLC+	0.74	5.3%	3
Blue Circle Industries PLC	8.35	60.1%	37
Ketton Portland Cement Ltd*	0.80	5.8%	7
Ribblesdale Cement Ltd*	0.86	6.2%	6
The Rugby Portland Cement PLC	2.19	15.8%	21
Tunnel Holdings Ltd*	0.95	6.8%	5

* Now RTZ Cement Ltd
+ Now part of Blue Circle Industries PLC.

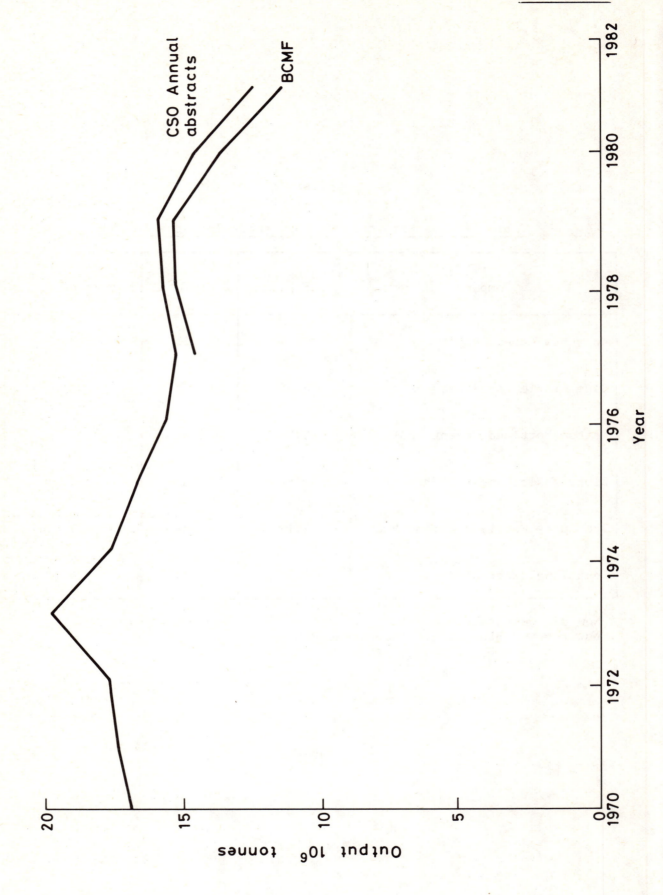

Figure 2. Output from the Cement Industry 1970–1981

aluminate and aluminoferrite which are the principal active compounds of Portland cement. The product is referred to as clinker.

39.   Finally clinker is ground to the specified fineness with the addition of a small proportion of gypsum to control the setting time of the finished cement. It is then stored in silos before despatch either in bag or bulk.

40.   The manufacturing methods may be divided into two broad categories, the wet and dry processes, which differ in the way materials are dealt with up to the calcining stage. In the wet process the raw materials are ground in water, blended and fed to the kiln as a slurry. The water typically amounts to 30-45% by weight of the slurry and is driven off in the initial stage of processing.

41.   In the dry process the raw materials are, wherever possible, dried during their initial grinding - generally using exhaust gases from the kiln system. Once the dry materials are blended they are passed through a preheater system to begin calcining.

42.   Between these two major processes are the semi-wet and semi-dry processes. In both the raw feedstock, prepared either by the wet or dry methods according to the nature of the raw materials, is formed into pellets or granules with a minimum moisture content. These can be fed into the kiln either directly or via a grate preheater on which they are dried and brought up to calcining temperature by heat from the kiln. Alternatively, in the semi-wet process, a disintegrator-dryer and preheater may be used.

43.   Hitherto, the choice of process has primarily been determined by the nature of the raw materials. The wet process was mainly used to prepare chalk and clay, while the dry process was used for limestone and shale. However, the former process is now being replaced by the semi wet process or the dry process, which are less energy intensive. In the UK in 1980 there were nine dry process kilns, 57 wet process kilns and 13 semi wet/dry kilns. However, most of the wet process kilns, largely as a result of their earlier design and construction, have a much lower capacity than any of the other processes, and the output figures for each process are shown in Table 8. (By the end of 1982 the number of wet process kilns had decreased to only 37, while there were 14 semi wet/dry kilns and nine dry process kilns.)

ENERGY USE IN THE CEMENT INDUSTRY

44.   The manufacture of cement is necessarily energy intensive, in that the chemical and physical reactions involved in the production of cement clinker take place at high temperatures. In 1980 approximately 92% of the energy used by the industry (in heat supplied terms) was used in drying and calcining. The remaining 8% was used in the form of electricity, mainly for grinding the raw materials and clinker.

45.   In 1980 only one kiln was oil-fired producing only 39,000 tonnes of clinker per year, by 1981 this kiln was no longer in use and all kilns were coal-fired. In 1980 some oil was used for kiln start-up, and gas and oil were used to some extent for drying of raw materials, but 90% of the energy used in the industry was coal. Fuel use by type and cement manufacturing process is detailed in Tables 9 and 10.

Table 8.   Output of Cement from Various Process Plant

Process	Number of Kilns	1980 Output, 10^6 tonnes	% of Total
Dry	9	2.99	21.5%
Wet	57	7.71	55.5%
Semi dry	11	2.81	20.2%
Semi wet	2	0.37	2.7%
TOTAL	79	13.88	

Table 9.   Total Energy Consumption in the Cement Industry (heat supplied)

Process	1980 Fuel Use, 10^6 GJ/year				
	Coal	Oil	Gas	Electricity*	Total
Dry	10.31	0.38	0.77[+]	1.66	13.12
Wet	49.37	0.38	0.02	2.92	52.69
Semi dry	9.50	0.67	0	1.33	11.50
Semi wet	1.76	0	0	0.17	1.93
TOTAL	70.94	1.43	0.79	6.08	79.24

* Electricity consumption in raw materials and clinker grinding for
Dry process = 0.47 GJ/tonne,
Wet = 0.32 GJ/tonne,
Semi wet/dry = 0.40 GJ/tonne.
Total measured consumption of 6.08×10^6 GJ has been divided proportionately among the processes.

+ Gas was used at only one works and does not relate to the dry process as a whole.

46. The greatest use of energy occurs in the manufacture of clinker in the cement kiln, and within the industry major efforts have been made for many years to economise on this input. Figure 3 and Table 11 show the energy consumption as kiln fuel per tonne of clinker produced for the period 1970 to 1981 [2].

47. Overall, specific energy consumption for the wet process has shown a steady decrease since 1973, mainly due to the closure of small and inefficient plant and to concerted efforts by cement manufacturers to improve the efficiency of old wet process plant still in use. For the semi-wet and semi-dry processes there has been an overall slight increase in specific energy consumption since 1976, probably due to an increased use of the semi-wet process in that period, which uses mechanical dewatering and has a higher energy requirement than the semi-dry (semi-wet 4.76 GJ/tonne, semi-dry 3.62 GJ/tonne kiln fuel requirement).

48. There has also been a slight increase in the specific energy consumption for the dry process since 1974. This is due to several factors:

- recently installed dry process plant uses less favourable raw materials than the earlier plant;

- a move towards higher quality product has involved higher energy consumption in firing, which also applies to other processes.

However, due to a continuing replacement of wet process plant with installations operating on a drier process the mean specific energy consumption of the industry has shown a steady decline since 1973, amounting to an average decrease of 2.3% per year over the period from 1973 to 1980.

49. In the wet process a great amount of energy is required to evaporate the water, and this accounts for around 40% of the total energy consumption for that process. In 1980 the average energy consumption of wet process kilns was 6.45 GJ/tonne compared with only 3.81 GJ/tonne for the dry process and 3.75 GJ/tonne for the semi wet/dry processes.

50. On the whole, electrical energy consumption per tonne of cement has risen over recent years for three main reasons:

- harder raw materials have been introduced, which require more grinding energy;

- stricter environmental regulations have required the more rigorous use of electrostatic precipitators and dust extraction plant;

- new plant may save manpower, but has a greater power consumption;

- market requirements for higher quality cement.

ENERGY CONSERVATION POTENTIAL

51. Clearly, a large amount of energy could be saved if all cement was produced by the dry or semi wet/dry processes. However, this would involve massive capital investment and conversion could only occur slowly.

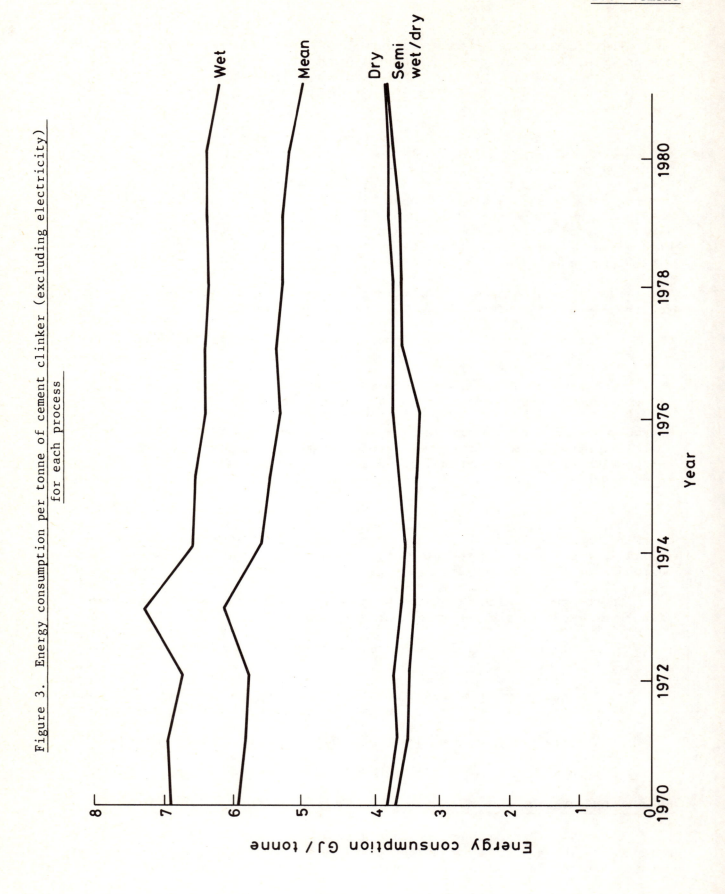

Figure 3. Energy consumption per tonne of cement clinker (excluding electricity) for each process

2.1 Cement

- 16 -

Table 10.   Energy Consumption per Tonne of Product

Process	1980 Fuel Use, GJ/tonne				
	Coal	Oil	Gas	Electricity*	Total
Dry	3.45	0.13	0.26	0.56	4.40
Wet	6.40	0.05	0.003	0.38	6.83
Semi dry	3.38	0.24	0	0.47	4.09
Semi wet	4.76	0	0	0.47	5.23
Mean	5.11	0.10	0.06	0.44	5.71

*   Electricity consumption in raw materials and clinker
    grinding for
    Dry process = 0.47 GJ/tonne,
    Wet   = 0.32 GJ/tonne,
    Semi wet/dry = 0.40 GJ/tonne.
    Total measured consumption of 6.08×10^6 GJ has
    been divided proportionately among the processes.

Table 11.   Kiln Fuel Requirements, GJ/tonne Clinker Produced

Process	Year											
	1970	1971	1972	1973	1974	1975	1976	1977	1978	1979	1980	1981
Dry	3.85	3.70	3.74	3.63	3.58	3.67	3.77	3.76	3.76	3.81	3.81	3.86
Wet	6.99	7.02	6.81	7.36	6.67	6.61	6.48	6.48	6.44	6.45	6.45	6.28
Semi wet/dry	3.72	3.53	3.51	3.47	3.45	3.42	3.39	3.61	3.64	3.66	3.75	3.84
Mean	5.99	5.91	5.84	6.20	5.68	5.54	5.39	5.45	5.34	5.35	5.27	5.05

52.   At first sight it might appear that energy could be saved within each process group if energy consumption was at the level of the 'best practice' works, i.e. the works with the lowest energy consumption per tonne of product (Table 12).   However, it is unlikely that such savings could be made in practice, due to the constraints imposed by the raw materials.   The water content of the materials can vary widely, requiring more or less energy to dry them, and the varying hardness of the material determines the amount of grinding energy required.   In fact, the works which appear to the operating under 'best practice' conditions are those which have the most favourable supply of raw material, requiring least energy for drying and/or for grinding. The industry has become increasingly energy conscious over recent years and many are already operating at that level with the raw materials available to them.

Future Energy Savings in the Cement Industry

53.   Industry sources have estimated that the cost of the energy required in the production of cement amounts to around 40% of the total production cost in the UK.   In continental Europe, however, energy costs are a smaller proportion of the total as most of the cement is produced by the less energy intensive dry process.   In the UK a substantial part of the cement market lies in the south-east of England.   As distribution costs contribute a large part of total costs to the industry, there will continue to be a significant part of the industry based in the south-east, using chalk as the raw material and consequently using the more energy intensive processes.

54.   In order to resist the growing threat from cement imports, and to maintain the position of cement and concrete as competitively priced building materials, the industry must reduce its costs and has a continuing programme of energy conservation, covering a wide range of technologies.   The conservation measures which may be applied within the industry over the next 40 years are described below, with the potential savings and timescale over which they may be achieved.   These are summarised in Table 13.

(a)   Replacement of wet by dryer processes

55.   Replacement of wet by dryer processes offers a large potential energy saving.   In 1980 7.71 million tonnes of cement were produced by the wet process with an average energy consumption of 6.83 GJ/tonne.   If this was produced for example by the semi wet process, with an average energy consumption of 5.23 GJ/tonne the energy saving would amount to 12.3 million GJ per year, or 16% of the total 1980 energy consumption in the industry (heat supplied).

56.   However, the capital cost of plant replacement is high in relation to the potential annual fuel savings, and a straight replacement of the wet by the dry process, with no change in kiln capacity, is unlikely to offer sufficient cost savings over the life of the plant to justify the investment. In cases studied so far, replacement appears only justified by increasing the plant capacity, and in the present situation of around 2 million tonnes of over-capacity this is not generally attractive.

57.   Replacement of the wet process by the dry, rather than the semi-wet, could save larger amounts of energy (about 18.7 million GJ per year) but is determined by raw materials and the greater capital costs of the plant.

Table 12. Average and 'Best Practice' Energy Consumption for Each
Production Process

Process	Output x 10^6 tonnes/year	Average Practice GJ/tonne Kiln Fuel	'Best Practice' GJ/tonne Kiln Fuel
Dry	2.99	3.81	3.31
Wet	7.71	6.45	5.38
Semi Dry	2.81	3.62	3.36
Semi Wet	0.37	4.76	4.55
TOTAL	13.88		

58.   However, there appears to be a potential for half of the existing wet process plant to convert to the semi-wet process by 1990, with a saving of 6.2 x 10^6 GJ per year. The remaining plant would be replaced directly by the dry process, with suitable raw materials available, which would offer an energy saving of 9.4 x 10^6 GJ per year. This change is already underway and one third of the saving could be achieved by 1990 with the remaining two thirds by 2000.

```
┌─────────────────────────────────────────────────────┐
│       REPLACEMENT OF WET BY DRYER PROCESSES:          │
│   SAVING 17.7% OF WET PROCESS ENERGY BY 1990          │
│             A FURTHER 11.9% BY 2000                   │
└─────────────────────────────────────────────────────┘
```

(b)   Improved kiln insulation

59.   The insulating bricks used by the cement industry have lower mechanical strength and therefore shorter life than normal grade, higher conductivity refractories. They are subject to greater chemical attack and stress from shell deformation than ordinary refractories. Developments are proceeding with stronger fired bricks of superior mechanical and chemical performance, which would permit further insulation of the hotter zones of the kiln. However, the potential for improved kiln insulation is limited by the requirement for liquid phase solidification on the inner surface of the kiln. A layer of clinker is built up on this surface and affords some protection against abrasive wear of the refractory bricks lining the kiln. If the temperature of the inner surface is too high this solidification will not occur.

60.   On the basis of the cement output and energy consumption in 1980 and assuming reasonable success of current trials and associated research, 1.6 x 10^6 GJ of energy could be saved by improved insulation, which is 2% of the total energy consumption.

```
┌─────────────────────────────────────────────────────┐
│  IMPROVED KILN INSULATION: SAVING 2% BY 1990          │
└─────────────────────────────────────────────────────┘
```

(c)   Blended cements

61.   Waste materials from other industries, such as active pulverised fuel ash (pfa) and suitable granulated blast furnace slag, may be interground or blended with Portland cement as an energy saving measure. Certain European countries use blended cements extensively although these cements may not conform with existing British Standards. In the UK consideration is being given to more extensive use of blending materials which will decrease energy requirements in concrete production.

62.   British Standards for Portland Blastfurnace Slag Cement have been in existence for many years, which permit the use of up to 65% granulated blast furnace slag in the cement according to one standard, and 90% according to a second which is very little used. A standard for Portland Pulverised Fuel Ash Cement, specifying the pfa content to lie in the 15-35% range, has been drafted.

63.   The BSI is considering a proposal to permit the incorporation of up to 5% of active pulverised fuel ash or granulated blast furnace slag to Ordinary Portland cement complying with BS 12. A cooperative testing programme, led by the Building Research Station, is examining this proposal to determine whether concrete properties are affected. The outcome of this programme will be known in 1983. Within many countries in western Europe, standards permit the incorporation of up to 5% of inert fillers in Portland cement.

64.   It is conceivable that a proportion of Portland cement might be replaced by blended cements containing some 20-30% by weight of suitable pulverised fuel ash or granulated blast furnace slag. The extent of the replacement cannot be predicted with confidence, and any estimate of aggregate energy savings would be speculative. However, as an example, for each 10% of total Portland cement production replaced by cement containing 25% by weight of blending material, 1.89×10^6 GJ of energy could be saved each year (taking account of the extra energy used for slag grinding). This will be offset to some extent by the energy required to transport blending materials to the cement works.

```
┌────────────────────────────────────────────────┐
│                                                  │
│   BLENDED CEMENTS:  SAVING 2.4% BY 2000          │
│                                                  │
└────────────────────────────────────────────────┘
```

(d)   Waste Heat Recovery

65.   On the basis of 1980 output and energy consumption in the industry, approximately 5.5×10^6 GJ per year is available in the form of waste heat from wet process works, and 3.2×10^6 GJ per year from dry and semi wet/dry works. However, this heat is often of relatively low grade at only 100-200°C.

66.   There is little opportunity for using this heat within the works, although it may be possible, with process modifications, to use it for process power generation when the economics of retrofitting are more clearly defined.

67.   The cement industry has studied various methods of 'over-the-fence' use, such as greenhouse heating, grass drying, office heating and fish farming. However, such opportunities are limited by economic viability, and offer only a method of cost-cutting for the industry rather than savings in specific energy consumption.

68.   It has been assumed that only one third of the wet process plant will exist beyond 1990 and this will be replaced by 2000. It is therefore reasonable to assume that no investment in heat recovery plant will be made at wet process works. If half of the potential saving in heat recovery from dry and semi wet/dry processes may be realised, the energy available will amount to 6.1% of the total energy consumption within the industry. Half of this may be used to generate electricity with an efficiency of about 30% and the other half sold to external customers. The latter will offer savings in national fuel supplies rather than savings directly to the cement industry.

```
┌────────────────────────────────────────────────────────────────┐
│                                                                  │
│        WASTE HEAT RECOVERY: SAVING 0.9% BY 2000                  │
│   Plus a further 0.13 GJ/tonne of product (heat supplied) by 2000│
│                                                                  │
└────────────────────────────────────────────────────────────────┘
```

(e) Increased Sulphate Levels

69.   Sulphate is added at the grinding stage of cement production as gypsum ($CaSo_4.2H_2O$), and an increase of 0.5% in the amount added (which would require a change in the current British Standard) could save around 0.68×10^6 GJ per year in saved clinker production on current output, and as gypsum also acts as a grinding aid, 0.03×10^6 GJ of grinding energy could also be saved. The use of low sulphur fuels in the production of cement could also offer opportunities for the addition of gypsum as an energy saving measure.

```
INCREASED SULPHATE LEVELS: SAVING 0.9% BY 1990
```

(f) Improved Grinding Techniques

70.   Cement and raw feed grinding is the largest user of electrical energy in a cement plant, and the amount of energy required may be reduced by the addition of grinding aids to the clinker, and the use of ring roller mills rather than the existing ball mills. However, a disadvantage is that roller mills give a particle size distribution leading to an inferior quality cement.

71.   Approximately 0.16×10^6 GJ/year of energy could be saved on current output by the introduction of ring roller mills. An additional 0.09×10^6 GJ could be saved by the addition of grinding aids.

```
IMPROVED GRINDING TECHNIQUES: SAVING 0.3% BY 2000
```

(g) Lower Temperature Cement Production

72.   Mineralisers may be added to cement raw materials to reduce the firing temperature required and potentially increase the activity of the cement. Cost savings would be partly offset by the cost and transportation of mineralisers, but an energy saving of around 2.7×10^6 GJ/year could be economically available, based on 1980 production levels.

```
LOW TEMPERATURE CEMENT: SAVING 3.3% BY 2000
```

(h) Improved Combustion Control

73.   A range of opportunities exist for improving combustion control, including a process currently in the R&D stage which could enable information concerning kiln exhaust gas composition to be related to the product quality, and hence closer control of the combustion conditions within the kiln, to match fluctuations in material quality and feed rate.

74.   The energy savings which could be made if such controls were applied could be around 4×10^6 GJ/year, based on current levels of output.

```
┌─────────────────────────────────────────────────────────┐
│  IMPROVED COMBUSTION CONTROL: SAVING 5% BY 1990           │
└─────────────────────────────────────────────────────────┘
```

(i)   Waste Materials as a Kiln Fuel

75.   The evidence of a demonstration project currently underway suggests that waste materials such as domestic refuse, waste oil and tyres could be used as a kiln fuel, substituting for around 10% of the coal currently burned. The industry itself regards this as an opportunity for reducing costs, by using a cheap, low grade fuel, rather than primarily as an energy saving measure. However, wastes could be available in suitable quantities and economically attractive to approximately half of the existing cement makers, and offering energy savings of 3.5×10^6 GJ in 1980.

```
┌─────────────────────────────────────────────────────────┐
│  WASTE MATERIALS AS KILN FUEL: SAVING 4.4% BY 2000        │
└─────────────────────────────────────────────────────────┘
```

Slurry Moisture Additives

76.   In the wet process slurry moisture content can be as high as 45% to maintain a low slurry viscosity. Chemical additives may be used to reduce the viscosity while also reducing moisture content, and hence the amount of energy needed to dry the material. However, the additives are expensive, their performance depends on the raw materials and water quality, and their use may have detrimental environmental effects. Moisture reducing agents are already in use where this is practicable and economically viable.

77.   Although this technique was discussed in the Energy Audit of this industry as a possible method for reducing the energy requirements of the industry in view of the overall move towards dryer processes, the technique is only relevant while wet process plant remains.

PROJECTED LEVELS OF ENERGY CONSUMPTION IN THE CEMENT INDUSTRY

78.   Each of the energy conservation measures described above will be taken up by the industry over a period of time. It is anticipated that all will be making their maximum contribution to energy savings by 2000. A simple diffusion model has been used to calculate how the various technologies may affect energy consumption within the industry over the period 1980 to 2000, and Figure 4 shows the simple S-curve used to model the uptake of technologies by the industry over this period.

79.   In order to avoid double-counting savings are applied sequentially, taking account of the fact that when one technology has been applied, the savings available using other technologies are reduced proportionately. Table 7 shows the percentage reduction in the energy requirements of the industry (heat supplied) due to each conservation measure, in the years 1985, 1990, 1995 and 2000. After 2000 the savings will continue at 17.8% as all the technologies are making their maximum contribution.

80.   The potential energy savings may be expressed as a reduction in the specific energy requirements of the industry. The energy required to produce

Table 13. Energy Conservation Measures Which May be Applied in the Cement Industry, and the Associated Reduction in Energy Requirements

Conservation Measure*	1985	1990	1995	2000	Fuel Saved
1. Improved kiln insulation	1.0%	2.0%	2.0%	2.0%	Coal
2. Increased sulphate levels	0.45%	0.9%	0.9%	0.9%	All
3. Improved combustion control	2.5%	5.0%	5.0%	5.0%	Coal
4. Blended cements	0.38%	1.2%	2.0%	2.4%	All
5. Waste heat recovery	0.14%	0.45%	0.76%	0.9%	Electricity
6. Improved grinding techniques	0.05%	0.15%	0.25%	0.3%	Electricity
7. Low temperature cements	0.53%	1.7%	2.8%	3.3%	Coal
8. Refuse as a kiln fuel	0.7%	2.2%	3.7%	4.4%	Coal
TOTAL SAVING	5.6%	12.9%	16.2%	17.8%	
National Fuel Savings 9. Waste heat recovery (GJ/tonne)	0.02	0.07	0.11	0.13	All

* It has been assumed that wet to dryer process conversion will take precedence over all other measures, and that the resulting combination of dry and semi-wet/dry process capacity will form the basis for the implementation of the conservation measures listed here.

Note: Percentage savings are applied sequentially to avoid double-counting.

Table 14.   Future Specific Energy Consumption, By Process

Year Process	Energy Consumption GJ/tonne				
	1980	1985	1990	1995	2000
Dry	4.40	4.15	3.83	3.69	3.62
Wet	6.83	6.45	5.95	5.72	5.61
Semi-Dry	4.09	3.86	3.56	3.43	3.36
Semi-Wet	5.23	4.94	4.56	4.38	4.30

Figure 4.   Diffusion curve for technologies taking 20 years to reach
maximum penetration of the industry

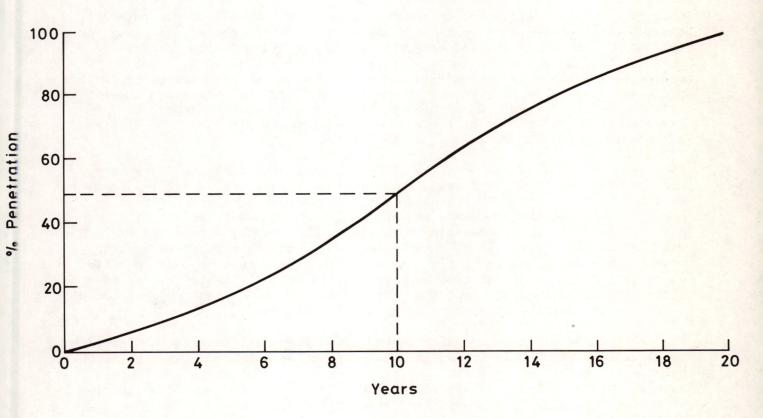

a tonne of cement differs from one process to another, and will continue to do so in the future. Table 14 shows the future energy consumption by each process when the various energy conservation measures have been applied.

81.   The total energy requirement of the cement industry as a whole will be determined by the future levels of production, and the proportions of this manufactured by each production process.

82.   In paragraph 58 above, the conversion of the wet to dryer processes was described. By 2000 no wet process plant will be operated, the capacity being taken up by the semi-wet and dry processes. It is also anticipated that the semi-dry process will have ceased production almost entirely by 1995, with that capacity being taken up wholly by dry process plant, as the semi-dry reaches the end of the plant life.

83.   By the turn of the century approximately 70% of UK cement will be produced by the dry process, with the remaining 30% produced by the semi-wet process. However, over the period 1980 to 2000 the proportions of total output produced by each process will be influenced by any overall increase, or decrease in the level of total production.

84.   The rate of retirement of semi-wet process plant will be determined by the age of existing plant, and is unlikely to be determined by any change in the overall output of the industry, or any future change in the relative fuel prices. The replacement of wet by dryer process plant is already well underway, and there is no indication that this would be accelerated by further increases in fuel prices, or how it would be affected by changes in the output of the cement industry. Consequently, it is assumed that the rate of conversion or replacement of wet and semi-dry process plant will be little influenced by varying scenarios of the future.

85.   However, any increase in overall production from the cement industry will be supplied by dry and semi-wet processes. Therefore, under any scenario of the future involving increased output, the proportion of that output supplied by these two processes between 1980 and 2000 would be slightly higher than if 1980 levels of production prevailed. This is illustrated in Table 15, which shows the percentage of total output which might be supplied by each process for a high and low output scenario, and the resulting mean specific energy requirement. Under the low scenario the SEC is slightly higher in 1985, 1990 and 1995 than under the higher scenario, as a higher proportion of output is produced by the energy intensive wet process.

86.   The change in mean specific energy consumption is, in fact, dominated by the move away from the wet process, rather than the slight differences in the balance of production from the four processes. Figure 5 shows a graphical representation of the mean specific energy requirement for cement production up to the year 2000. For comparison, the energy requirement if no conservation measures were applied is shown on the graph as a broken line, and assumes that only process conversion or replacement occurs over this period.

87.   Finally, in Table 16 the future mean specific energy requirements of this industy are broken down into the types of energy carrier, taking account of the nature of the fuel saved by each conservation measure which may be applied. The requirement for electricity increases due to the movement away from the wet process, which consumes less electricity than the dry process,

Figure 5. Mean specific energy requirements in cement production

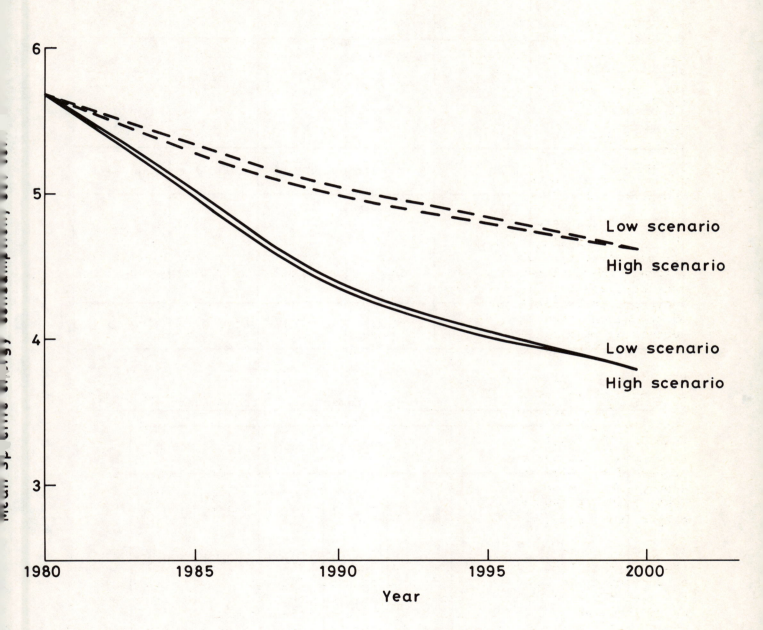

Table 15. Contribution to Total Output from Each Production Process
for a High and Low Future Output Scenario

| | Year | Percentage of Total Production Supplied By Each Process | | | | |
Process		1980	1985	1990	1995	2000
High Scenario						
Dry		21%	28%	43%	62%	70%
Wet		56%	35%	16%	7%	0
Semi–Dry		20%	19%	9%	0	0
Semi–Wet		3%	18%	32%	31%	30%
Mean SEC		5.71 GJ/t	5.04 GJ/t	4.38 GJ/t	4.05 GJ/t	3.82 GJ/t
Low Scenario						
Dry		21%	26%	41%	60%	70%
Wet		56%	37%	19%	9%	0
Semi–Dry		20%	20%	10%	0	0
Semi–Wet		3%	17%	30%	31%	30%
Mean SEC		5.71 GJ/t	5.08 GJ/t	4.42 GJ/t	4.09 GJ/t	3.82 GJ/t

Table 16. Mean Specific Energy Requirements. GJ/tonne (heat supplied)

Fuel		1980	1985	1990	1995	2000
Coal:	High scenario	)	4.43	3.84	3.51	)
	Low scenario	) 5.11	4.47	3.89	3.55	) 3.26
Oil:	High scenario	)	)	)	)	)
	Low scenario	) 0.10	) 0.10	) 0.08	) 0.08	) 0.09
Gas:	High scenario	)	)	)	)	)
	Low scenario	) 0.06	) 0.06	) 0	) 0	) 0
Elec:	High scenario	)	)	0.46	)	)
	Low scenario	) 0.44	) 0.45	0.45	) 0.46	) 0.47
Total:	High scenario	)	5.04	4.38	4.05	)
	Low scenario	) 5.71	5.08	4.42	4.09	) 3.82

and the use of gas in the dry production process will have ceased by ∼ 1990, being replaced by coal. The overall requirement for coal decreases, as the dry process requires just over half as much coal per tonne of output than does the wet process.

CONCLUSIONS

88.   The cement industry is particularly energy-conscious, and has made significant energy savings over many years. However, this report shows that further, significant savings could be made by first converting wet process plant to the semi-wet process, or replacing it with dry process plant, and secondly by implementing energy conservation technologies which are currently in the research, development or demonstration stage.

89.   These savings could amount to a reduction of 33% in the specific energy consumption of the industry (heat supplied), over a period of 20 years. The mean specific energy requirement could decrease from 5.71 GJ/tonne in 1980 to 3.82 GJ/tonne by 2000, which amounts to an average decrease of 2% per year; compared with the average decrease observed between 1973 and 1980 of 2.3% per year.

90.   Most of the energy saved would be in the form of coal, which is used for firing the kilns.

91.   Additional energy savings could be made available by the implementation of waste heat recovery, the recovered energy being supplied to neighbouring firms to reduce their conventional energy requirements. This saving, which has been termed a 'national fuel saving' as it does not accrue to the cement industry, could amount to 0.13 GJ per tonne of cement produced, by the year 2000.

ENERGY USE AND ENERGY EFFICIENCY IN UK MANUFACTURING INDUSTRY

UP TO THE YEAR 2000

2.2 THE BRICK INDUSTRY

2.2 THE BRICK INDUSTRY

INTRODUCTION

92.   In 1980 the industries manufacturing bricks, tiles, fireclay and general building materials consumed 48 x 10^6 GJ of energy on a heat supplied basis [4].   Of this 19 x 10^6 GJ were used by the brick industry [5].

93.   This industry has been the subject of an Energy Audit [6] and, together with the other industries in the group described above, has been studied as part of the Industrial Energy Thrift Scheme [7].

94.   The British Ceramic Research Association Ltd. carried out surveys of its member firms in 1974, 1975, 1977 and 1981, which were funded by their clay brick manufacturing members, to collect information on energy use within the industry.   The 1981 survey [5], which looked at energy use in 1980, covered 90% of the total brick production in that year (all of Fletton bricks, 48% of non-Fletton common bricks and 81% of non-Fletton facing and engineering). Data from these surveys have been used as a basis for this report, but as some variation in attitudes exists within the industry on the potential for energy conservation, the conclusions drawn have been based on ETSU's knowledge of the industry, gained through the Energy Conservation Demonstration Projects Scheme.

THE STRUCTURE OF THE BRICK INDUSTRY

Output

95.   The total output of the brick industry in 1980 was 4,107 million bricks (9 million tonnes).   This represents a reduction of 13% on the output recorded in 1976/7 (10.4 x 10^6 tonnes) and a steep decline on the average output over the period between 1953 and 1973 of 18 x 10^6 tonnes.

96.   During the period 1954-1968 the number of companies within the industry more than halved, while the output per plant almost doubled.   In 1972 there were 336 works and in 1975 only 274.   There are now five main manufacturers who in 1980 accounted for over 95% of the UK brick output:

- the London Brick Company manufactures 'Fletton' bricks which now account for around 46% of the total UK output by weight, or 53% by number;

- Steetley Bricks operating in the north of England;

- Butterley Building Materials and the Ibstock Group in central England and Wales;

- Redland Building Materials in south-east England.

97.   Producers have responded to the reduced market over recent years by improving the quality and widening their product range.   However, in terms of energy, this has led to a wide range in specific energy consumption which is closely linked to the wide range of products.

Types of Brick Produced

98.   The three main types of brick which are produced are as follows:

(i)   Facing bricks. These are used for the outer walls of houses and have a consistently high aesthetic standard.

(ii)   Common bricks. These are used where appearance is not important, but their other properties can be as good as those of facing bricks.

(iii) Engineering bricks. These are used where severe environmental conditions are encountered, and have mechanical and corrosion properties superior to other types of brick.

99.   In recent years common bricks have been widely replaced by lightweight concrete and other types of block for the construction of the load bearing inner walls of houses. Facing bricks of all types are now the major product

100.   These brick types may be further divided into 'Fletton' and 'non-Fletton' bricks. Fletton bricks are made exclusively by the London Brick Company of Oxford clay, which has a very high inherent carbon content (2.6 GJ per tonne of bricks produced) which acts as a fuel during the firing process. Fletton bricks can be facings and commons. Non-Fletton bricks are made from a variety of clays ranging from carboniferous shale, which also contains inherent carbon, to Weald which has none. The total output of bricks in 1980 was divided as shown in Table 17.

Forming Processes

101.   Four processes are used for forming bricks, and to some extent the process employed determines the amount of energy required to dry and fire the bricks. These processes are as follows:

- Stiff plastic process: the clay has a relatively low moisture content (13-19%) and may be dried out efficiently in the kiln without preliminary drying. This process is not associated with tunnel kilns in any of the works reported in the B.Ceram R.A. census [5]. The proportion of non-Fletton bricks reported as made by this process fell from 23% to 6% by weight between 1976 and 1980;

- Extrusion process: the bricks made by this process are dried before firing in the kiln. 81% by weight of non-Fletton bricks were reported as made by this process in 1980, as compared with 68% in 1976/7.

- Soft-mud process: these bricks have a very high moisture content (23-40%) and a large amount of energy is required for drying. 13% by weight of non-Fletton bricks were reported as made by this process in 1980, while only 9% were reported as made this way in 1976;

- Semi-dry process: this process is only used for producing Fletton bricks, and all Fletton bricks are made in this way. The moisture content of the clay is between 19% and 25%.

Table 17. Total Output of the UK Brick Industry in 1980

Brick Type	Brick Output		Change since 1976/7 (by number)
	Millions	10^3 tonnes	
Flettons (Commons and Facings)	2,178	4,200 (46%)	– 10%
Commons (non-Fletton)	148	402 (4%)	– 66%
Facings and Engineering (non-Fletton)	1,781	4,560 (50%)	– 4%
TOTAL	4,107	9,162	– 13%

Source: Extrapolated from outputs reported to B. Ceram R.A. [5].

Brick Firing

102.   Most common and facing bricks are fired in continuous kilns which have two basic designs, annular (called chamber and Hoffman kilns) and tunnel. In annular kilns the bricks remain stationary and the fire is moved slowly around the kilns by operation of a sequence of dampers, whereas in tunnel kilns the bricks move slowly through the kiln and the fire remains stationary. Both types of kiln are designed to recover most of the sensible heat in the combustion gases by using them to pre-heat the dried bricks.   All Fletton bricks are fired in annular (chamber) kilns.   However, the number of tunnel kilns reported in the B.Ceram R.A. survey [5] increased by 38% between 1976/7 and 1980 to 51%.   The proportion by weight of non-Fletton facing and engineering bricks reported as fired in continuous tunnel kilns increased in that period from 60% to 72%, and while only 3% by weight of reported non-Fletton common bricks were fired in tunnel kilns in 1976/7, in 1980 all were fired in annular kilns.

103.   A few bricks are produced in intermittently fired kilns.   These are engineering and facing bricks with special qualities, and those reported only amount to about 4% of the total output by weight.

104.   Kiln temperatures reported range from 970°C to 1,230°C for continuous kilns, and 1,089°C to 1,167°C for intermittent kilns.   The mean temperature employed for each type of brick is as follows:

-   Fletton bricks 1,025°C

-   Non-Fletton commons 1,022°C

-   Non-Fletton facing and engineering 1,058°C (continuous kilns)
                                      1,134°C (intermittent kilns).

ENERGY USE IN THE BRICK INDUSTRY

105.   Energy consumption in brick production depends upon the type of kiln used, either continuous or intermittent, on the type of brick produced, the methods of manufacture and the raw materials.

106.   Certain clays have inherent carbonaceous material, for example Oxford clay, or may have it added before brick forming.   This material generally contains volatile hydrocarbons, some of which distil off at low temperatures in the kiln and are lost in the exhaust gases.   However, the remainder of the material burns while the brick is being fired, thus reducing the requirement for the conventional energy supply.

107.   The energy supplied for all purposes from conventional sources is termed the net energy consumption; the total energy use, including that from inherent and added carbonaceous material in the clay, is termed the gross energy.

108.   Table 18 shows the energy consumption per tonne of bricks for all those reported in the B.Ceram R.A. 1980 survey [5], and the percentage change in both net and gross energy requirements since 1976.   The change shown in columns 6 and 7 has been based on all brick production reported in the surveys.

109. Although the overall increase in net specific energy consumption for Fletton and non-Fletton common bricks could be assumed to be related to the overall reduction in output of these bricks (Table 17), this would be an over simplification. This may be so for Fletton bricks, which are made only by one company, but non-Fletton common bricks are made by a variety of firms, and some firms using carboniferous shales have, in fact, reduced their specific energy requirement.

110. On the whole, improvements in energy use have arisen from better management and housekeeping practices, although in some cases it may also be due to the reorganisation and closing of older, obsolete plant.

111. Brick manufacture may be divided into six process stages:

- raw material excavation

- transport

- body preparation (mixing and grinding)

- shaping (extruding or pressing)

- drying in driers

- drying and firing in kilns.

112. In 1980 92% of the conventional energy supplied to brickworks was used for firing and drying. Electricity, mainly for materials preparation and brick shaping, amounted to 6% of the total on a heat supplied basis; while energy for internal transport amounted to around 2%.

113. At the works reporting [5] the fuels used for firing and drying amounted to nearly all the solid fuel and gas consumed by the industry, and around 49% of the oil consumed. The remainder of the oil was used mainly for internal transport. Table 19 shows the total fuel consumption for the industry in 1980, on a heat supplied basis.

114. Table 19 shows that there has been a marked reduction in the amount of solid (-14%) and liquid (-36%) fuels used by this industry since 1976, but an increase of 23% in the amount of gaseous fuels consumed. The reported consumption of natural gas increased by 66% and of mine drainage gas by 157%, while butane decreased by 7%.

115. Fletton bricks are all produced in coal-fired chamber kilns. Table 20 shows the types of kiln used to fire non-Fletton bricks and the fuels used to fire them.

ENERGY CONSERVATION POTENTIAL

116. In response to the reduction in the size of the market for bricks over recent years, manufacturers have widened their product range. However, the outcome of this is that each individual brick type is effectively satisfying one specific area of the market, and to suggest a greater unification of the output for the sake of energy conservation would conflict with the manufacturer's overall objectives.

Table 18. Summary of Specific Energy Consumption in 1980

1. Brick Type (continuous kilns)	2. Clay	3. % by weight	4. Average Net GJ/tonne	5. Energy Gross GJ/tonne	6. % Change Net GJ/tonne	7. Since 1976 Gross GJ/tonne
Fletton	Oxford	100%	0.91	3.46	+ 14	+ 1
Common- Non-Fletton	Carboniferous	72%	1.17	4.00	+ 8	+ 32
	Boulder	28%	2.64	3.32	- 3	+ 2
	Average		1.60	3.81	+ 9	+ 24
Facing and Engineering (Continuous kilns)	Carboniferous*	42%	3.07	3.52	- 13	- 14
	Keuper Marl#	18%	3.03	3.22	- 7	- 1
	Etruria Marl	14%	3.05	3.05	- 3	- 3
	Weald	6%	3.77	3.77	0	0
	Boulder#	4%	2.24	2.76	- 8	- 20
	Various	8%	3.23	3.58	- 3	+ 1
Stock°	Various	8%	3.24	5.10	- 8	0
	Average		3.09	3.51	- 8	- 8
(Intermittent kilns) Stock°	Various	42%	5.25	5.60	- 4	- 4
	Various	46%	2.97	5.35	- 14	- 4
Blue and Brindled×	Etruria Marl	12%	6.61	6.61	- 14	- 14
	Average		4.38	5.60	- 20	- 11

Source: B. Ceram R.A. [5].

Notes: * The fuel consumption for firing these bricks could often be lowered, but for problems associated with a high carbon content which requires a slow firing cycle to avoid discolouration of the brick surface.

Coal slurry is sometimes added to these clays to increase the carbon content.

° Stock bricks are generally made by hand moulding or the soft-mud process. High proportions of coal slurry, coke breeze, town refuse etc. and sometimes chalk slurry are added to give the type and colour of brick required.

x Blue and brindled bricks require a reducing atmosphere during the final stages of firing to give colour and required engineering properties.

Table 19. Total Energy Consumption of the Brick Industry in 1980
(Heat Supplied)

Output x 10^6 Bricks	Fuel	Conventional 10^3GJ yr^{-1}	Additives 10^3GJ yr^{-1}	Inherent 10^3GJ yr^{-1}	% Change Since 1976	Total 10^3GJ yr^{-1}
3,690 Reported	Coal	3,590) 3,651			− 14%	
	Coke	61)				
	Oil:					
	Heavy	160)				
	Medium	208)				
	Gas	404) 776			− 36%	
	Kerosene	2)				
	Petrol	2)				
	Gas:					
	Natural	5,137)				
	Butane	4,102)				
	Propane	493)10,519			+ 23%	
	Mine	447)				
	Mixed	340)				
	Elect.*	961			+ 14%	
	Additives		1,042		+ 84%	
	Inherent			11,940	− 19%	
417 Unreported		3,087	287	124		
Total 4,107		18,994	1,329	12,064	− 17%	32,387

* Electricity consumption is reported in B.Ceram R.A. data in terms of the primary fuel consumption. In 1976 an efficiency of 23.85% was taken for the generation plant, in 1980 this was 29.6%.
Source: B.Ceram R.A. [5].

Table 20. Numbers of Kilns Used For Firing All Reported Non-Fletton Bricks

Kiln Type	Solid	Oil	Nat. Gas	Butane	Propane	Methane	TOTAL
Tunnel			30	15	2	4	51
Chamber	4		7	12	1		24
Hoffman	3	2	2	6	2		15
Down-Draught (intermittent)	6	13		20			39
Shuttle (intermittent)			4	12			16
Clamp (intermittent)	3	2	4	10			19
TOTAL	16	17	47	75	5	4	164

Source: B.Ceram R.A. [5].

117. For example, within the group of non-Fletton facing and engineering bricks fired in continuous kilns and which were reported to the B.Ceram R.A., there are 25 different combinations of clay/brick forming process/kiln type and kiln fuel. Each one of these has a different specific energy requirement and it could be claimed that each essentially produces a different brick for a specific area of the market. The average net energy requirement for this group was 3.09 GJ per tonne of bricks with a range from 1.84 to 13.10 GJ/tonne. Each of these 25 different combinations is also associated with a specific amount of inherent or added carbonaceous material in the clay, which itself influences the specific energy requirements and type of kiln which may be used.

118. It cannot be assumed that if all the bricks within this group were manufactured under best practice conditions the specific net energy consumption would be 1.84 GJ/tonne for all bricks.

119. A similar argument holds for non-Fletton common bricks where the average specific energy consumption is 1.60 GJ/tonne with a range of 0.21-3.74 GJ/tonne and non-Fletton facing and engineering bricks fired in intermittent kilns, with an average specific energy consumption of 4.37 GJ/tonne and range of 2.13-12.61 GJ/tonne.

120. In order to estimate the amount of energy which could be saved if all bricks were produced under best practice conditions the total of 37 different combinations of clay, forming process, kiln type and kiln fuel in the Fletton and non-Fletton sectors, were first divided into 25 groups, as shown in Table 21. The best practice energy consumption was identified for each of these groups, and the conservation potential was the difference between the average net energy consumption and the best practice consumption for that group.

121. The brick production reported in the B.Ceram R.A. survey covers all of Fletton brick production, approximately 48% of the non-Fletton common bricks and 81% of the facing and engineering bricks. When account is taken of the unreported production the total best practice conservation potential is as shown in Table 22.

122. Table 22 shows that the energy which may be saved by all firms operating at best practice, as described earlier, amounts to 4.81×10^6 GJ per year or 25.3% of the total conventional energy consumed by the industry (15% of gross energy consumption). Of this 38% could be saved on Fletton bricks; 51% on non-Fletton facing and engineering bricks fired in continuous kilns; 8% on intermittent kilns and the remaining 3% on non-Fletton common bricks.

123. This saving may also be expressed in terms of the saving per tonne of bricks produced, and is shown in Table 23.

124. If the more simplistic approach had been adopted, and it was assumed that all bricks within each group could be produced with a net energy consumption equal to the lowest in the group, irrespective of clay or kiln type etc., the potential savings would amount to 44% of the total conventional energy consumption of the industry, or 26% of the total gross energy consumption.

125. The energy savings described above would mainly consist of fuels used for drying and firing, i.e. coal, oil and gas, and amount to 50.4% of the

Table 21. Data Used in Calculating Best Practice Conservation Potential

	Brick Type / Clay	Process	Kiln	Fuel	Output 10³tonne yr⁻¹	Net Best Practice GJ/tonne	Net Average Practice GJ/tonne	Total Saving x 10³GJ yr⁻¹
1.	Fletton	S.D.P.	Ch.	Coal	4,200	0.48	0.91	1,806
	Common							
2.	Carboniferous shale	S.P.	Hoff.	Coal	88.3	0.20	0.26	5.3
3.	" " "	E.	Hoff.	Gas	48.3	1.91	2.83	44.4
4.	Boulder	E.	Hoff/Ch	Gas/Coal	54.4	2.45	2.64	10.3
	Facing & Eng.							
5.	Carboniferous	S.P.	Ch.	Coal/Gas	49.4	3.19	3.49	14.8
6.	"	E.	Hoff.	Oil	39.2	2.99	2.99	0
7.	"	E.	Hoff.	Gas	19.5	1.84	1.84	0
8.	"	E.	Tk.	Gas	1,300.7	2.09	3.05	1,248.7
9.	"	S.M.	Hoff.	Gas	9.1	6.73	6.73	0
10.	Keuper	E.	Hoff.	Gas	109.5	3.11	3.11	0
11.	"	E.	Ch.	Gas	135.4	3.63	3.82	25.7
12.	"	E.	Tk.	Gas	329.6	2.19	2.55	118.7
13.	"	S.M.	Ch.	Gas	22.9	4.92	4.92	0
14.	Etruria Marl	E.	Tk.	Gas	465.1	2.47	3.06	274.4
15.	Weald	SP/E	Ch.	Gas	64.1	4.15	4.32	10.9
16.	"	SP/E	H/Ch/Tk	Gas	136.8	3.52	3.52	0
17.	Boulder	E.	Ch.	Gas/Coal	150.3	2.02	2.25	34.6
18.	Alluvial	S.M.	Ch.	Gas	60.6	5.86	5.86	0
19.	Various	E.	Tk.	Gas	210.3	2.42	2.48	12.6
20.	Stocks	E.	Tk.	Gas	64.5	2.20	2.20	0
21.	"	S.M.	H/Ch/Tk	Gas	217.3	2.22	3.56	291.2
	Intermittent Kilns							
22.	Facing & Eng.	E/S.M.	D.D.	Coal/Oil/Gas	114.8	3.52	4.82	149.2
23.	" "	E.	Sh.	Gas	14.0	7.12	8.83	23.9
24.	Stocks	S.M.	Cl.	Coal/Oil/Gas	138.4	2.15	2.97	113.5
25.	Blue & Brindled	E.	Sh.	Gas	36.2	5.61	6.61	36.2
TOTAL					8,079 x 10³ tonne			4,220 x 10³GJ yr⁻¹

KEY:

S.D.P.	Semi-dry process	Ch.	Chamber kiln	
S.P.	Stiff plastic	Hoff.	Hoffman kiln	
E.	Extrusion	Tk.	Tunnel kiln	
S.M.	Soft-mud	D.D.	Down draught	
		Sh.	Shuttle	
		Cl.	Clamp	

Table 22. Summary of Data Used in Calculation of Best Practice
Conservation Potential

Brick Type	Output tonnes/year (a)	Mean Best Practice GJ/tonne (b)	Mean Average Practice GJ/tonne (c)	Conservation Potential GJ yr^{-1} (c-b) x (a)
Flettons	4,200 x 10^3	0.48	0.91	1,806 x 10^3
Non-Fletton Common	402 x 10^3	1.27	1.60	133 x 10^3
Non-Fletton Facing and Engineering				
Continuous Kilns	4,185 x 10^3	2.50	3.09	2,469 x 10^3
Intermittent Kilns	375 x 10^3	3.31	4.38	401 x 10^3
TOTAL	9,162 x 10^3			4,809 x 10^3

Table 23. Best Practice Conservation Potential

Brick Type	Average Energy GJ/tonne Net	Gross	Potential Savings GJ/tonne	Savings as % of net
Flettons	0.91	3.46	0.43	47%
Non-Flettons Common	1.60	3.82	0.33	21%
Non-Flettons Facing and Engineering				
Continuous Kilns	3.09	3.51	0.59	19%
Intermittent Kilns	4.38	5.60	1.07	24%

Table 24. Potential Fuel Savings by Plant Operating at Best Practice

Fuel	Potential Savings, 10^6 GJ
Solid Fuels	1.05
Oil	0.13
Gases	3.63
TOTAL	4.81

energy used for drying and firing Fletton and common bricks in 1980, and 21.1% of that used for firing facing and engineering bricks. If the savings are divided among these fuels in proportions equal to their contribution to the total energy consumption, then the potential fuel savings would be as shown in Table 24.

126. However, even within the industrial groups described in paragraph 32 above, the best practice potential may be further constrained by the product range and variations in raw materials. Savings may also be limited by the scale of operations at individual works; kilns of larger capacity having lower specific energy requirements than those of smaller capacity.

127. In view of these further limitations it is unlikely that the full 25.3% savings could be achieved, particularly as many brick producers already regard their fuel consumption as being low, within the constraints of the raw materials available and requirement of their specific product.

128. A more realistic target for the best practice energy savings within this industry would be about half of the savings indicated above, i.e. 13% of the total conventional energy consumption, which could be achieved by about 1990. This would amount to 28.3% of the fuel used for drying and firing Fletton, 12% of that used for common bricks, and 10.8% of that used for non-Fletton facing and engineering bricks.

129. Such savings may be achieved primarily by improved management and control of energy use, and could include such measures as improved plant scheduling; reduction in the duration of firing times where possible; control of clay moisture content; closer control of fuel feed and more rigorous plant and equipment maintenance.

Future Energy Savings Within the Brick Industry

130. The processes involved in brick production are shown schematically in the diagram below. In 1980 the industry consumed 19.0×10^6 GJ of energy on a heat supplied basis. Of this 92% was used in firing and drying (3.28×10^6 GJ Fletton, 0.57×10^6 GJ common and 13.60×10^6 GJ non-Fletton); 6% was used as electricity (310×10^3 GJ Fletton, 58×10^3 GJ common and 793×10^3 GJ non-Fletton) and 2% was used for other miscellaneous purposes (218×10^3 GJ Fletton, 17×10^3 GJ common and 150×10^3 GJ non-Fletton).

131. A variation in attitudes exists within the industry on the potential for energy conservation and the various conservation technologies which may be applied. However the technologies, potential savings and timescale in which they may be achieved, which are described below, represent an overall view of the industry, gained through the work of the Energy Conservation Branch in ETSU.

(a) Carbonaceous additives

132. Certain clays contain carbonaceous material which, as described in paragraph 15 above, burns while the brick is being fired, thus contributing to the energy requirement for this process. Other clays, which do not contain such inherent material, may have carbonaceous additives incorporated into them before brick forming and firing, with a consequential reduction in the amount of conventional energy (coal, gas, oil) required during firing. At present

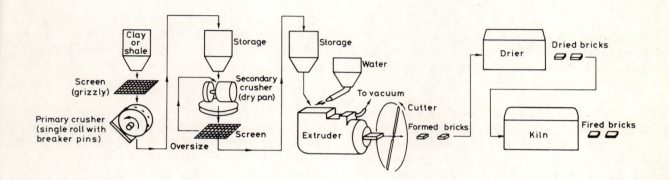

Figure 6. Process flow diagram of a brickworks

this practice is not carried out for reasons of energy saving, but is used to produce various colour effects in the bricks.

133. Although it is financially attractive to extend the practice of adding such combustible materials as coal slurry and coke breeze to the clay, there is a cost associated with purchasing these materials and the additional mixing plant required. Other factors which must be considered include alterations to the firing cycle and increased overall heating times which are likely to be required to oxidise the carbon and the maintenance of production rates and brick quality.

134. Studies are being carried out by the B.Ceram R.A. to assess the problems associated with this practice, and it is currently assumed that inherent or added carbon in bricks can save up to 2 GJ/tonne of bricks. This is based on the amount of energy known to be released by the inherent carbon in Oxford clays during the firing cycle.

135. If carbonaceous material was added to all clays to bring the total energy available up to 2 GJ/tonne then, taking account of the reduced output of bricks in 1980 and the existing trend towards higher levels of additives in the clay, 7.2×10^6 GJ/year could be saved in this way. This is compared with an estimate of 10.6 GJ/yr given in the Energy Audit of the Brick Industry.

136. The total saving amounts to 40% of the kiln fuel used in the non-Fletton common brick sector of the industry and 51% of that used in the facing and engineering sector. Approximately half could be achieved economically by 1995, and the further savings would occur very slowly indeed, due to resistance to change within the industry, and any inability to achieve the same aesthetic appearance of the bricks.

```
CARBONACEOUS ADDITIVES: SAVING 40% COMMON BRICK KILN FUEL
        20% BY 1995 + A FURTHER 20% AFTER 2000.
     51% FACING AND ENGINEERING BRICK KILN FUEL.
     25.5% BY 1995 + A FURTHER 25.5% AFTER 2000.
```

(b)   Increased brick perforations

137. Bricks which are formed by the extrusion process may have their mass reduced by the incorporation of cavities or 'perforations' into their structure during forming. These perforations are holes up to 3,000 mm^2 in area, passing through the thickness of the brick.

138. The practice can save energy in various ways:

- by increasing the surface area of the brick, thus enabling the heat to penetrate the brick more rapidly;

- by reducing the volume of the brick such that the heat may penetrate to the centre of the brick more quickly, thus reducing the firing time;

- reducing the thermal capacity of kiln cars in tunnel kilns because of the lower mass of the bricks they have to carry;

- reducing the mass of brick material needed to be fired and also the amount of water in each brick to be evaporated during drying.

139. Studies by B.Ceram R.A. have shown that the average level of perforations adopted in the UK is about 17%. Increasing the level of perforations can lead to problems with extrusion, and may meet with customer resistance. However, there appears to be general support within the industry for a move to 25% perforations, which could save 3.7% of the kiln fuel use in the non-Fletton sector by 2000. Perforations up to about 35% should be possible in the future, offering energy savings of 1.9% of kiln fuel by 2000 and a further 5.6% beyond that date.

```
INCREASED PERFORATIONS: SAVING 11.2% NON-FLETTON KILN FUEL.
5.6% BY 2000 + A FURTHER 5.6% AFTER 2000.
```

(c) Kiln heat recovery

140. Energy losses from kilns, in the form of sensible and latent heat in exhaust gases, can amount to 40-80% of the total energy input to brickworks. It is unusual to recover any of this energy, which must not be confused with the energy in the brick cooling air which is recovered and used as the main energy input to driers.

141. Kiln exhaust gas is dirty, containing moisture, sulphur and fluorine, and the acid dew point could be as high as 130°C. This leads to acid condensation on the inside of ducting and corrosion, necessitating the replacement of metalwork as frequently as three times each year in certain cases.

142. One solution is to reduce the acid dew point temperature by mixing exhaust gas with clean cooling air, or alternatively to replace ducting with more expensive material to resist corrosion.

143. The temperature of the exhaust gases leaving continuous kilns is 100-250°C. However, many brickworks have no ready economic use for the thermal energy once it is recovered as the investment in heat recovery equipment is high compared with the value of the low grade heat recovered, and brickworks are generally situated in isolated geographical positions where the energy cannot readily be sold to nearby customers.

144. Although heat recovery from continuous kilns in the non-Fletton sector offers an energy saving potential of 0.43 GJ/tonne of the product it is unlikely to be adopted on a widespread basis until low-cost recuperator systems capable of withstanding heavy acid contamination are available, and a ready market for the energy exists.

145. By around 2000 such savings may be technically achievable, and an economic use for energy savings of 0.10 GJ/tonne may exist. However, the further savings of 0.33 GJ/tonne will only become available as new kiln types

come into use, or local markets for the energy develop near to the brickworks.

```
HEAT RECOVERY FROM CONTINUOUS KILNS: SAVING 0.43 GJ/TONNE OF OUTPUT
                          (NON-FLETTON)
    0.10 GJ/TONNE BY 2000 + A FURTHER 0.33 GJ/TONNE AFTER 2000.
```

146. Intermittent kilns offer more attractive opportunities for heat recovery as exhaust gas temperatures are higher and a market could exist for the heat, for example in preheating the combustion air.

```
HEAT RECOVERY - INTERMITTENT KILNS: SAVING 0.05 GJ/TONNE
 OF NON-FLETTON FACING AND ENGINEERING OUTPUT BY 2000.
```

(d)   Heat recovery from driers

147. Approximately 80% of the non-Fletton brick industry uses supplementary heat for drying, in addition to the heat recovered from kilns. This amounts to about 0.3 GJ/tonne of bricks. Energy recovery from dryer exhausts using run-around coils or heat pumps is a relatively low-cost option and could totally eliminate the need for supplementary heating.

148. The energy savings available amount to 8.4% of the energy used in kilns and driers in the non-Fletton sector of the industry in 1980.

```
HEAT RECOVERY FROM DRIERS: SAVING 0.3 GJ/TONNE NON-FLETTON OUTPUT
                  KILN + DRIER FUEL BY 2000.
```

(e)   Improved process control

149. A range of measures can be applied which come under the heading of process control. These include the automatic control of kiln temperatures, improved drier programming and clay moisture control. The overall energy savings which could be available using these measures amount to 10.9% of the kiln energy used throughout the industry, and this could be achieved both technically and economically by the year 2000.

```
IMPROVED PROCESS CONTROL: SAVING 10.9% OF ALL KILN FUEL BY 2000.
```

(f)   Alternative gas supplies - colliery methane and landfill gas

150. In 1980 the brick industry used approximately 0.5×10^6 GJ of mine drainage gas, over double the amount used in 1976.

151. The NCB produces a surplus of this gas, as new sources become available when new pits are opened and old pits can continue to produce gas for many years after mining has ceased. Although the NCB have encouraged the use of

this gas by the brick industry, and increasing resources may be available in the future, the amount which can be used by the industry is limited by the suitable location of brickworks within about five miles of a coal mine.

152. It is, therefore, impossible to predict how much of the conventional energy supply could, in the future, be replaced by colliery methane.

153. Methane, produced by the decomposition of domestic refuse in landfill sites, is already being used as a fuel by the brick industry. These sites are often exhausted clay pits near to the brickworks, which are filled with refuse and reclaimed for agricultural purposes.

154. ETSU have estimated that landfill gas could contribute about 3×10^6 GJ of energy to the total UK energy requirement around 2000. As this gas is, in many cases, available close to existing brickworks, and London Brick Landfill Ltd. have pioneered the use of landfill gas in the UK, it seems reasonable to assume that a large proportion of this three million GJ may be used by the brick industry.

155. As a conservative estimate, it is considered that landfill gas will replace 1.5×10^6 GJ of conventional energy in the brick industry, and the actual proportion of the energy replaced will, of course, depend on the total requirement of the industry in the future. In terms of the 1980 consumption it amounts to 8.6% of the total fuel consumption in kilns.

> ALTERNATIVE GAS SUPPLIES: SAVING 1.5×10^6 GJ BY 2000.

156. Most of these energy saving measures would be applied predominantly in the production of non-Fletton bricks. Only improved process control and alternative gas suplies would reduce the energy consumption in the production of Fletton bricks.

157. Heat recovered from kiln exhaust gases may not find a ready market within the brick industry itself, and therefore may not save energy in the production of bricks, although the heat recovered from intermittent kilns may be more readily used within the industry than that recovered from continuous kilns. Heat recovery from kilns has, therefore, been considered as a method of saving national energy resources rather than simply saving energy within the brick industry. This leaves opportunity for the recovered heat to be used by either the brick industry, or a nearby customer, or both.

158. In Tables 25 and 26 the energy saving measures are listed according to the dates by which they might be fully adopted by the industry. Table 25 shows the measures which may be applied in the non-Fletton sector, and Table 3.10 shows those which are applicable to the industry as a whole.

159. Savings could be made predominantly in fuels for drying and firing (in 1980 22% coal, 75% gas and 3% oil). A small amount of fuel for space heating may be replaced by heat recovered from kiln exhaust gases, and improved process controls may save a small amount of electricity in the brick preparation stage.

Table 25. Energy Savings in the Non-Fletton Sector

Conservation Measure	Timescale	Energy Savings
Drying and Firing Fuels		
1. Increased carbonaceous additives:		
– common bricks	by 1995	20.0%
– facing and engineering	by 1995	25.5%
2. Increased brick perforations.	by 2000	5.6%
3. Heat recovery in driers.	by 2000	0.3 GJ/tonne
4. Increased brick perforations.	beyond 2000	5.6%
5. Increased carbonaceous additives:		
– common bricks	beyond 2000	20.0%
– facing and engineering	beyond 2000	25.5%
Heat Recovery in Kilns		
6. Heat recovery in intermittent kilns: facing and engineering bricks.	by 2000	0.05 GJ/tonne
7. Heat recovery in continuous kilns.	by 2000	0.10 GJ/tonne
8. Heat recovery in continuous kilns.	beyond 2000	0.33 GJ/tonne

Table 26. Energy Savings Applying to the Whole of the Brick Industry

Conservation Measure	Timescale	Energy Savings
Drying and Firing Fuels		
9. Improved process controls.	by 2000	10.9%
10. Alternative gas supplies.	by 2000	1.5M GJ/year *

* Regardless of total consumption.

160. However, as 92% of the total energy consumed within the industry is used
in drying and firing, any savings which apply to the other 8% will be very
small in comparison.

161. In this study it has, therefore, been assumed that the energy
requirements per tonne of product, in the form of electricity or fuels used
for miscellaneous purposes around the site, will be unaffected by the energy
conservation technologies described above.

PROJECTED LEVELS OF ENERGY CONSUMPTION

162. Each of the energy conservation measures described above and listed in
Tables 25 and 26 will be taken up by the industry over a period of time. A
diffusion model has been used to calculate how the various conservation
measures may enter the industry, and the associated reduction in specific
energy consumption which will occur.

163. Figure 7 shows the typical S-curves which were used to model diffusion
of technologies over periods of 15 and 20 years and Table 27 gives the
reduction in the fuel requirement due to each conservation measure, in each
fifth year between 1980 and 2020. It has been assumed that all technologies
are making their maximum contribution by 2020.

164. The potential saving which could be made if all plant operated at the
current best practice has been treated as another conservation technology,
diffusing into the industry over a period of 10 years.

165. The projected specific energy consumption in the production of each
brick type is shown in Table 28. The SEC in the production of Fletton bricks
is reduced by 32% between 1980 and 2000 (1.9% per year on average); that for
common brick production by 54% over the same period (3.8% per year) and that
for facing and engineering brick production by 51% (3.5% per year).

166. However, although carbonaceous material added to clay during the
brick-forming process has been considered as a measure to conserve
conventional energy resources, it may also be considered as an alternative
energy source for brick firing. Carbonaceous additives may contribute
~ 0.27 GJ/tonne to the total requirement for the production of common bricks
by the year 2000, and ~ 0.74 GJ/tonne for the production of facing and
engineering bricks. In this case the reduction in the total SEC in the
production of common bricks is only 37% between 1980 and 2000 (2.3% per year)
and in the production of facing and engineering bricks 28% over the same
period (1.6% per year).

167. By 2020 it is anticipated that carbonaceous additives could contribute
around 0.37 GJ/tonne to the energy required to produce common bricks and 1.06
GJ/tonne in the production of facing and engineering bricks. Table 29 shows
the contribution of such additives to energy consumption in brick production
over the period 1980 to 2020.

168. Further savings of conventional energy resources may be achieved by the
use of alternative gas supplies, expressed as measure C in Table 27. However,
the use of such resources does not reduce the total energy requirement of the
brick manufacturing process, and therefore cannot be expressed as a reduction

Figure 7.   Diffusion curve for technologies taking 15 years to reach
100% penetration of the industry

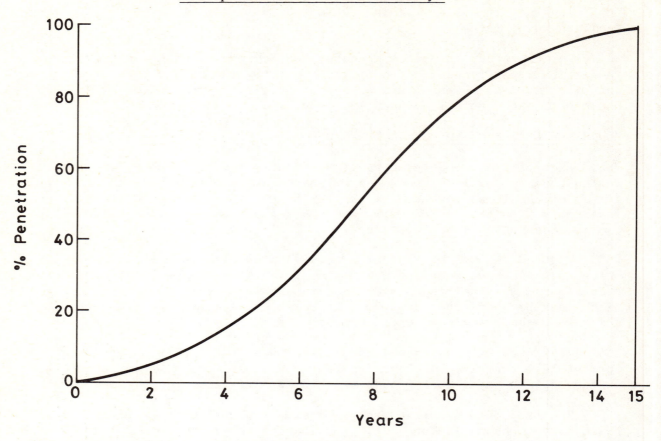

Diffusion curve for technologies taking 20 years to reach
100% penetration of the industry

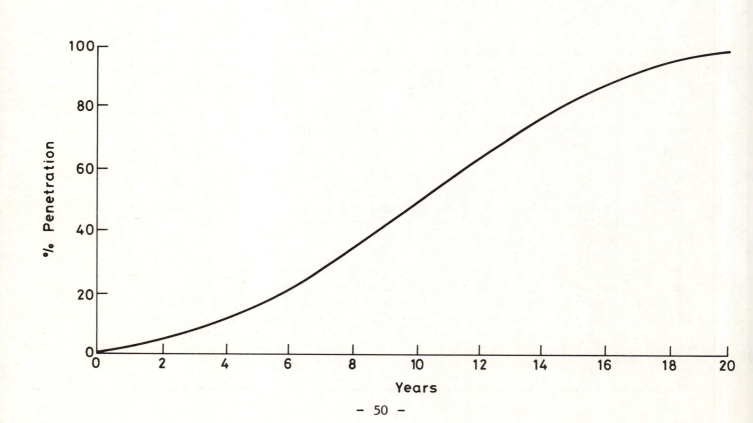

Table 27. Reduction in Energy Consumption due to Various Conservation Measures

Conservation Measure	Year 1985	1990	1995	2000	2005	2010	2015	2020
A. Non-Fletton Production (Fuel for Drying and Firing)								
1. Best practice – Facing + Engineering	5.4%	10.8%	10.8%	10.8%	10.8%	10.8%	10.8%	10.8%
– (Common)	(6.0%)	(12.0%)	(12.0%)	(12.0%)	(12.0%)	(12.0%)	(12.0%)	(12.0%)
2. Increased carbonaceous additives								
– Facing + Engineering	5.9%	19.6%	25.5%	25.5%	25.5%	25.5%	25.5%	25.5%
– (Common)	(4.6%)	(15.4%)	(20.0%)	(20.0%)	(20.0%)	(20.0%)	(20.0%)	(20.0%)
3. Increased perforations	0.9%	2.8%	4.7%	5.6%	5.6%	5.6%	5.6%	5.6%
4. Heat recovery in driers, (MJ/tonne)	48	150	252	300	300	300	300	300
5. Increased perforations	0	0	0	0	0.9%	2.8%	4.7%	5.6%
6. Increased carbonaceous additives								
– Facing + Engineering	0	0	0	4.1%	12.8%	21.4%	25.5%	25.5%
– (Common)	(0)	(0)	(0)	(3.2%)	(10.0%)	(16.8%)	(20.0%)	(20.0%)
7. Improved process control	1.7%	5.5%	9.2%	10.9%	10.9%	10.9%	10.9%	10.9%
(National Fuel Savings)								
8. Heat recovery – intermittent kilns, MJ/tonne	8	25	42	50	50	50	50	50
Facing + Engineering bricks								
9. Heat recovery – continuous kilns, MJ/tonne	16	50	84	100	153	265	377	430
B. Fletton Production (Fuel for Drying and Firing)								
1. Best practice	14.2%	28.3%	28.3%	28.3%	28.3%	28.3%	28.3%	28.3%
2. Improved process control	1.7%	5.5%	9.2%	10.9%	10.9%	10.9%	10.9%	10.9%
C. All Production Alternative gas supplies, (10³ GJ)	240	750	1260	1500	1500	1500	1500	1500

Note: Percentage savings are applied sequentially to avoid double-counting.

- 51 -

Table 28. Specific Energy Consumption

YEAR	SPECIFIC ENERGY CONSUMPTION, GJ/TONNE								
	1980	1985	1990	1995	2000	2005	2010	2015	2020
Fletton Bricks	0.91	0.78	0.66	0.63	0.62	0.62	0.62	0.62	0.62
Non-Fletton Common	1.60	1.38	1.01	0.82	0.74	0.71	0.65	0.63	0.62
Non-Fletton Facing & Engineering	3.19	2.70	2.03	1.70	1.55	1.41	1.28	1.20	1.19

Table 29. Specific Energy Consumption

YEAR	SPECIFIC ENERGY CONSUMPTION, GJ/TONNE								
	1980	1985	1990	1995	2000	2005	2010	2015	2020
Fletton Bricks	0.91	0.78	0.66	0.63	0.62	0.62	0.62	0.62	0.62
Non-Fletton Common:									
Conventional fuels	1.60	1.38	1.01	0.82	0.78	0.71	0.65	0.63	0.62
Carbonaceous additives	0	0.06	0.19	0.25	0.27	0.31	0.35	0.37	0.37
Non-Fletton Facing & Engineering:									
Conventional fuels	3.19	2.70	2.03	1.70	1.55	1.41	1.28	1.20	1.19
Carbonaceous additives	0	0.17	0.52	0.68	0.74	0.88	1.01	1.06	1.06

in SEC. However, alternative gas supplies do reduce the demand on conventional energy resources.

169. The total energy consumption of the brick industry in the future will depend upon the level of output, and the balance of production between the different types of brick. Over the past years bricks have been replaced in many situations, either by lightweight concrete blocks or by timber frame construction. An expansion in the use of these alternative building materials could represent a significant reduction in the use of bricks. In the extreme bricks could become a luxury decorative item, and could be reduced to tile-like facing materials which would be applied over other, cheaper building materials.

170. Under a high output scenario of the future it is likely that the proportion of total output contributed by Fletton brick production will decrease slowly, with a comparable increase in the proportion of production made up of facing and engineering bricks.

171. Under a low output scenario these changes in the balance of output are likely to happen more rapidly, with Fletton production falling away at a greater rate and the balance being made up by facing and engineering brick production. Table 30 shows the possible balance of production under two such scenarios of the future, and the resulting mean specific energy consumption for the industry.

172. Under a high output scenario the mean SEC of the brick industry might be reduced by 44% between 1980 and 2000 (2.9% per year) and by a further 16% between 2000 and 2020 (0.8% per year).

173. Under a low output scenario of the future the SEC might fall by 40% between 1980 and 2000 (2.5% per year) and by a further 12% between 2000 and 2020 (0.6% per year).

174. These figures are based on the assumption that carbonaceous additives reduce the requirement for conventional energy supplies. If such additives are considered simply as alternative sources of energy in the process of brick production then the changes in SEC will be less marked, and between 2000 and 2020 the SEC increases under both scenarios, due to the changing balance of output of each brick type. Table 31 summarises the changes which may be observed.

175. Finally, Table 32 shows the future mean specific energy requirements of the brick industry in terms of the individual energy carriers. It has been assumed that the balance of fossil fuels remains unchanged to 2000 but that petroleum products will be replaced by coal between 2000 and 2020, and that natural gas consumption will be supplemented by the use of mines drainage and landfill gas.

CONCLUSIONS

176. Despite the fact that the period from 1974 has been a time of recession and declining output in the brick industry, manufacturers have become more energy conscious and, in fact, have already achieved reductions in specific energy requirements.

Table 30. Projected Balance of Output from the Brick Industry

Year	1980	1985	1990	1995	2000	2005	2010	2015	2020
	Balance of Output % of Total								
High Scenario:									
Fletton	46%	45%	43%	40%	38%	35%	34%	34%	34%
Common	4%	4%	4%	4%	4%	3%	3%	3%	3%
Facing & Engineering	50%	51%	53%	56%	58%	62%	63%	63%	63%
Mean SEC GJ/tonne	2.08	1.78	1.40	1.24	1.16	1.11	1.04	0.99	0.98
Low Scenario:									
Fletton	46%	44%	41%	36%	30%	22%	15%	15%	15%
Common	4%	4%	4%	4%	3%	2%	1%	1%	1%
Facing & Engineering	50%	52%	55%	60%	67%	76%	84%	84%	84%
Mean SEC GJ/tonne	2.08	1.80	1.43	1.28	1.25	1.22	1.17	1.11	1.10

Table 31. Mean Specific Energy Consumption

Year	Carbonaceous Additives as a Conservation Measure			Including Contribution from Carbonaceous Additives		
	SEC GJ/tonne	% Change	Average Annual Change	SEC GJ/tonne	% Change	Average Annual Change
High Output Scenario						
1980	2.08			2.08		
2000	1.16	- 44%	- 2.9%	1.60	- 23%	- 1.3%
2020	0.98	- 16%	- 0.8%	1.66	+ 4%	+ 0.2%
Low Output Scenario						
1980	2.08			2.08		
2000	1.25	- 40%	- 2.5%	1.75	- 16%	- 0.8%
2020	1.10	- 12%	- 0.6%	1.99	+ 14%	+ 0.6%

Table 32. Specific Energy Consumption by Fuel, GJ/tonne

	Specific Energy Consumption, GJ/tonne								
Year	1980	1985	1990	1995	2000	2005	2010	2015	2020
High Scenario									
Coal	0.42	0.36	0.27	0.24	0.22	0.27	0.36	0.45	0.50
Petroleum products	0.72	0.61	0.47	0.41	0.38	0.30	0.17	0.05	0
Natural and alternative gases	0.81	0.68	0.53	0.46	0.43	0.41	0.38	0.36	0.35
Electricity	0.13	0.13	0.13	0.13	0.13	0.13	0.13	0.13	0.13
Carbonaceous additives	0	0.09	0.28	0.39	0.44	0.55	0.65	0.68	0.68
Total*	2.08	1.87	1.68	1.63	1.60	1.66	1.69	1.67	1.66
Low Scenario									
Coal	0.42	0.36	0.28	0.25	0.24	0.30	0.41	0.51	0.57
Petroleum products	0.72	0.62	0.48	0.43	0.41	0.34	0.20	0.06	0
Natural and alternative gases	0.81	0.69	0.54	0.48	0.47	0.45	0.43	0.41	0.40
Electricity	0.13	0.13	0.13	0.13	0.13	0.13	0.13	0.13	0.13
Carbonaceous additives	0	0.09	0.29	0.42	0.50	0.68	0.85	0.89	0.89
Total*	2.08	1.89	1.72	1.70	1.75	1.90	2.02	2.00	1.99

* Including carbonaceous additives.

177. However, this report shows that further savings could be made by implementing technologies which are currently in the research, development or demonstration stage. Such technologies could reduce the mean specific energy consumption of the industry by up to 44% between 1980 and 2000, under a high output scenario of the future, which is equivalent to an average decrease of $\sim$2.9% per year. Under a low output scenario the overall savings might be $\sim$40% or $\sim$2.5% per year.

178. The greatest energy savings may be made by the addition of carbonaceous material to clay during the brick-forming process. If this material is regarded as an alternative source of energy, rather than a conservation measure, then the reduction in SEC due to other technologies could be up to 23% between 1980 and 2000 ($\sim$1.3% per year) under a high output scenario of 16% (0.8% per year) under a low output scenario of the future.

179. By the year 2000 up to 1.5×10^6 GJ of the energy consumed by the brick industry could be supplied as gas from landfill sites or mines drainage schemes. The magnitude of this resource is independent of the level of output from the brick industry itself.

180. Most of the energy savings in this industry may be made in fuels used for drying and firing bricks. Additional savings may be made by recovery of waste heat from kilns; these have been regarded as savings of the national fuel supplies, as the recovered energy would be sold to neighbouring consumers with no savings accruing to the brick industry. By the year 2000 this could amount to 150 MJ per tonne of non-Fletton bricks produced.

ENERGY USE AND ENERGY EFFICIENCY IN UK MANUFACTURING INDUSTRY

UP TO THE YEAR 2000

2.3   THE REFRACTORIES INDUSTRY

2.3 THE REFRACTORIES INDUSTRY

Introduction

181. Detailed energy analyses of the refractories industry are not available, primarily because several of the major firms regard information on fuel and power consumption as commercially sensitive.

182. However, the industry was studied as part of the Industrial Energy Thrift Scheme with a report being published in December 1978 [7], only a month after the publication of a report on the refractories industry in the Energy Audit Series [8]. Both of these reports are based on data relating, primarily, to 1976.

183. In the Digest of Energy Statistics [4] refractories are grouped with bricks, tiles and other building materials, making it difficult to identify how much energy is used by the refractories industry itself.

184. In the absence of any more up-to-date information this report has been based on information given in the Audit and Thrift reports, and on information in reference 9.

THE STRUCTURE OF THE REFRACTORIES INDUSTRY

185. Refractory materials are those which can withstand high temperatures while remaining undeformed and chemically stable. Almost any substance with a melting point above 1,500°C is a possible refractory, but the main products are those based on oxides of aluminium, silicon, magnesium, calcium and chromium.

186. The industry produces shaped products (bricks or furnace lining shapes) and powders which may be used as cement, ramming materials, mouldables or castables. Shapes are usually fired in kilns before use, although some may be chemically bonded and fired in-place. Castables are typically based on pre-fired granular mixtures bonded by calcium aluminate cements. Mouldings, rammings and castings are formed at the point-of-use and fired in situ.

187. Refractory materials are not generally expected to survive for the lifetime of the equipment and are replaced or repaired at regular intervals.

188. There are about 30 companies of differing sizes within this industry, but it is dominated by three - Steetley Refractories, J&J Dyson Ltd and G R-Stein Refractories. The smaller companies tend to specialise in particular types of refractories such as pottery kiln furniture, although all works tend to make a range of products. Steetley also produce refractory magnesia, which may be used by other companies to manufacture basic refractories.

189. The greater part of the refractories output is used by the metals manufacturing industries and the large cut-back in production by these industries has had a significant effect on the refractories industry. Current production levels are low and rationalisation of plant and products has occurred within the larger companies. Some plants are operating down to 50% and less of their output in the early and mid 1970s, as may be seen from the

indices of production shown in Table 33, and the output figures shown for the period 1972 to 1981 in Table 34.

190.   This industry suffers competition from overseas producers both in the UK and export markets, unlike the brick industry. However, it is more technically aware than the brick industry, and has always been aware of the need for energy cost savings.

191.   Output from the industry has declined since 1974 for several reasons:

- prolonged depression within the iron and steel industry;

- the operational lifetime of refractories is being improved by manufacturers;

- new or modified processes installed by users require less refractory material per tonne of output;

- refinements to existing user practice increase refractory life.

192.   It may be seen from Table 34 that production of firebricks and shapes has declined from around 39% to 26% of total output, over the period 1972-1981, while that of bricks and shapes containing magnesia and chrome has declined from 14% to 10%. However, an increase has been observed in the percentage of total output made up of high alumina bricks with more than 60% alumina content (6.5% increased to 11%) and castables, mouldables and ramming materials (19-28%). The percentage of output of the other refractory materials has remained relatively steady over the period, and the six groups together contribute around 25% of total production.

REFRACTORIES MANUFACTURE

193.   Figure 8 shows the main production route for refractories.

194.   Certain raw materials may require heat treatment to stabilise their structure before mixing. For example, dolomite ($MgCO_3.CaCO_3$) is calcined at 1,400°-1,500°C to provide an input for the manufacture of synthetic magnesia (MgO); magnesium hydroxide and dolomite are dead-burned at 1,700°-2,000°C. After mixing and grinding the material may be sold as an unshaped product or shaped, dried and fired for sale. Many products are shaped in hydraulic presses from material containing $\sim$ 4-5% moisture, so the energy used in drying is far less than that used in the brick industry, where moisture content may be $\sim$ 13-40%.

195.   Continuous tunnel and annular kilns may be used for firing shaped products, although tunnel kilns are generally preferred because of their better temperature control, and throughput can be varied to a greater extent to match market conditions. The use of annular kilns has diminished considerably over the past decade. Rotary kilns are used in this industry solely for firing raw materials.

196.   Although intermittent kilns are thermally less efficient than tunnel kilns, they are used for firing items which require very long firing times, and are often preferred because they are very flexible in the quality and quantity of material fired. There is an increasing number of intermittent

Table 33. Indices of Production for the Refractories Industry

Year	1972	1973	1974	1975	1976	1977	1978	1979	1980	1981
Index	100	111	118	100	102	98	90	93	62	55

Table 34. Output from the Refractories Industry 1972-1981, as a percentage of total production

	1972	1973	1974	1975	1976	1977	1978	1979	1980	1981
Firebricks and shapes	39.3	38.6	37.7	36.3	34.7	32.1	29.8	31.2	29.4	26.0
High alumina bricks (44-60% alumina)	2.6	2.9	2.8	2.9	2.6	3.1	2.4	2.6	2.4	3.7
High alumina bricks (> 60% alumina)	6.5	6.5	7.3	8.4	8.6	9.6	8.8	9.5	11.5	10.6
Bricks and shapes containing magnesia and chrome	13.8	13.7	12.8	11.8	11.1	11.9	11.5	10.4	9.8	10.1
Refractory insulating bricks	1.4	1.4	1.5	1.7	1.6	1.7	1.7	1.5	2.0	1.6
Refractory holloware for casting pits	8.4	8.5	7.9	8.5	8.5	7.9	8.5	9.1	9.5	12.6
Blocks and crucibles of graphite and other materials	1.7	1.7	1.8	2.1	2.0	2.2	2.7	2.4	3.1	3.2
Refractory jointing cement	4.6	5.0	4.6	4.7	4.7	4.8	4.6	5.6	4.2	4.4
Castables, mouldables and ramming materials	19.0	19.0	21.4	21.3	24.1	25.3	27.9	26.4	28.2	27.7
Silica bricks and shapes	2.8	2.6	2.3	2.2	2.2	1.5	2.2	1.3	U	U
Total production, 10^6 tonnes	1.307	1.449	1.537	1.306	1.331	1.285	1.179	1.210	0.815	0.712
Index of production. 1975 = 100	100	111	118	100	102	98	90	93	62	55

U = Unavailable

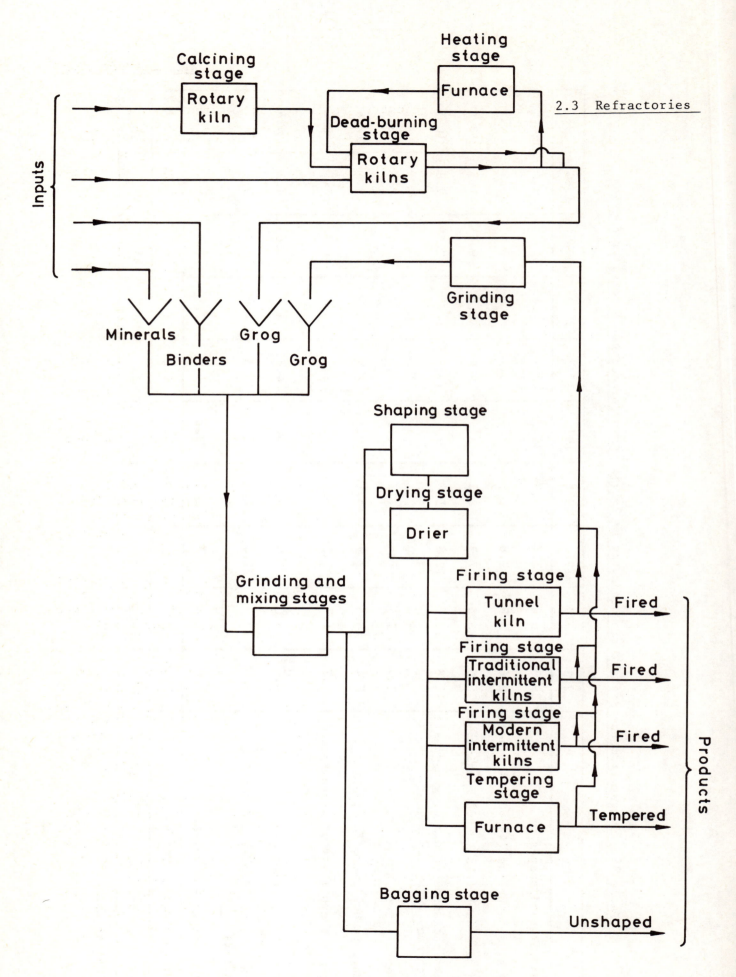

Figure 8. Bulk Refractories

kilns used in the refractories industry, as the effect of the contracting market has forced producers to manufacture a wider range of products in relatively small batches, and the intermittent kiln is more appropriate for firing these than are continuous kilns. For modern, intermittent kilns of low thermal mass, the energy cost is roughly proportional to the weight of product fired. In a tunnel kiln the energy cost per unit of time is approximately constant, so that for low throughputs, energy cost per tonne can be very high.

197. Dryers are frequently heated with waste warm air from the kilns; although heat recovery from intermittent kilns can be difficult, hot air may be extracted during the cooling part of the operating cycle for use in dryers.

ENERGY USE IN THE REFRACTORIES INDUSTRY

198. Approximately 90% of the energy used within this industry is used for drying and firing. Electricity supplies the motive power for crushing, grinding and shaping and also for the fans in dryers and kilns. Only a very small proportion of the energy is used for lighting and space heating.

199. There is little, reliable information on the amount of energy consumed by this industry, or the nature of the energy carriers.

200. The Energy Audit estimates that 16×10^6 GJ of energy were used in 1976, to produce 1.33×10^6 tonnes of product. This gives a mean specific energy consumption of 12.0 GJ/tonne primary energy [8]. With electricity contributing around 15.4% of this, the specific energy requirement in heat supplied terms amounts to around 10.8 GJ/tonne.

201. Table 35 shows the proportions of each energy carrier or fuel estimated in the Energy Audit of 24 establishments in 1976, and the estimate produced for this industry in the 1979 Purchases Enquiry [10]. The latter shows a much higher proportion of coal, mostly used in raw materials manufacture, and a price-induced move away from LPG towards Natural Gas.

202. The specific energy requirement for refractories manufacture can vary widely due to differences in raw materials, the type of product and the equipment in use. If calcining or dead-burning of raw materials is necessary, substantial amounts of energy (up to 18 GJ/tonne) may be consumed; on the other hand, although the production of fireclay bricks requires little more than the energy to fire the kiln, this can be fairly high due to the energy required to drive off the high levels of water associated with fireclay.

203. Refractories made from synthetic materials, such as magnesia and alumina, can have widely differing energy requirements. These synthetics are made both in the UK and overseas, and since large amounts of energy are used in their manufacture, variations in the energy requirement of the refractory products are caused by changing the proportions of UK to imported materials. Taking two extreme cases, magnesia, unfired products will have energy requirements of about 1-40 GJ/tonne (primary) if they are made completely of imported or home produced magnesia respectively; for fired magnesia bricks the values are about 10-50 GJ/tonne (primary).

Table 35. Estimates of Energy Consumption in the Refractories
Industry [8 and 10] (heat supplied)

	Coal	Oil	LPG	Natural Gas	Electricity
Energy Audit 1976	1.5%	32.9%	32.4%	27.5%	5.7%
Purchases Enquiry 1979	26.5%	22.5%	5.1%	39.8%	6.1%

204. Table 36 shows the energy requirements in 1976 of the various products of the industry, and Table 37 shows the amounts of material heated to different temperatures in that year. (Tables taken from the Energy Audit [8].)

205. Since 1976 the mean specific energy consumption in the refractories industry has increased, mainly because of a shift by users to higher performance products. The increase may also have been influenced by the dramatic decline in output, with the result that much of the equipment is operating below capacity. As approximately half of the fuel input to the kilns is used simply to maintain the temperature of the kiln, regardless of the throughput of ware, a reduction in throughput will cause an increase in energy consumption per tonne of ware.

206. It has been estimated (by the Energy Conservation Branch in ETSU) that the specific energy consumption, in primary fuel, may have increased by some 17% between 1976 and 1980, to around 14.0 GJ/tonne. As electricity still supplies about the same proportion of energy consumed as it did in 1976, the specific energy consumption in 1980, on a heat supplied basis, would be 12.60 GJ/tonne (compared with 12.1 GJ/tonne in 1979 estimated from the Purchases Enquiry [10]). Output in that year amounted to 0.815×10^6 tonnes, requiring an estimated total energy consumption of 10.27×10^6 GJ (heat supplied).

207. However, savings have been made within the industry over recent years; for example, the replacement of fired magnesia products by unfired, carbon-bonded magnesia bricks. These offer a saving of some 5.5 GJ/tonne of product or ~ 75% of the energy previously used for firing the bricks.

208. Assuming that the proportions of the energy requirement supplied by the different fuels has not changed since the 1979 Purchases Enquiry, then the estimated amounts of each fuel consumed in 1980 are shown in Table 38. Natural gas supplies almost 40% of the total requirement, coal and petroleum products each supply around one quarter, and LPG and electricity supply similar proportions at 5% and 6% respectively. The move away from LPG since 1976 has been caused mainly by its price relative to natural gas, and the large amounts of coal are used in rotary kilns for raw materials manufacture.

ENERGY CONSERVATION POTENTIAL

209. In considering other industries, as part of this study, an attempt has been made to identify 'best practice' within the industry, which is defined as the lowest energy consumption per tonne of product. An estimate has then been made of the amount of energy which could be saved, if all firms within the industry operated at 'best practice' levels, with existing technologies.

210. In order to do this it is necessary to have some idea of the range in specific energy consumption for the industry, and the factors which may influence, or hinder, a move towards best practice by all firms.

211. There is insufficient information available on energy consumption in the refractories industry for such an analysis to be carried out. It is not even certain how much energy the whole industry used in 1980, let alone the consumption and output of individual firms.

Table 36. Energy Requirements

Type of Kiln	Product	Energy Requirement GJ of required product
	Unfired fireclay Unfired high alumina Unfired basic cement (mainly magnesia)	0.8 – 1.6 3.6 31.0
Continuous	Fireclay ladle bricks Fireclay shapes for steel industry Fireclay saggars for pottery industry Fireclay shapes made from china clay for pottery industry Magnesite-chrome bricks (includes some alumina) Pitch impregnated magnesia bricks	7.8 9.0 17.2 19.6 25.0 28.0
Intermittent	Fireclay shapes for steel industry Fireclay shapes for pottery industry Fireclay saggars for pottery industry	19.3 21.2 16.1

(Source: B Ceram RA Energy Audits)

Table 37. Approximate Amounts of Materials Heated to Different Temperatures in 1976

Maximum Temperature of Process	Amount of Material, Thousand Tonnes	
	Finished Product	Synthetic Raw Material
Not heated (unfired)	400	–
Less than 500°C (tempered)	90	–
Fired at 1100–1600°C	690	300
Fired at 1600–1800°C	150	250

Table 38. Estimated Energy Consumption in the Refractories Industry, 1980 (heat supplied). 10^6 GJ/year

Output 10^6 tonnes	Coal	Oil	LPG	Natural Gas	Electricity	TOTAL
0.815	2.72	2.31	0.52	4.09	0.63	10.27

212. The only estimate of this type which can be made is on the amount of energy which might be saved by implementing simple, good housekeeping measures throughout the industry. The Energy Audit of the industry estimated that 5% of the energy consumed by the industry could be saved in this way, and it is probable that the opportunity still remains within the industry for such savings to be made.

213. On the basis of a total energy consumption in 1980 of 10.27×10^6 GJ, this saving would amount to some 0.51×10^6 GJ per year, and may take around 10 years to achieve.

Future Energy Savings Within the Refractories Industry

214. Energy savings through good housekeeping can be made at relatively little cost to the industry, and with very little time delay. However, greater amounts of energy could be saved by implementing technologies which are currently in the research, development or demonstration stages, and which have higher capital costs associated with them.

215. The rate of up-take of such technologies by the industry depends on a variety of factors, such as the payback period required on any investment; the state of technical development of the technology; the attitude of firms towards energy use and their expectation of future rises in the real price of fuel.

216. The technologies which could be applied, the energy savings available and timescale over which they could be achieved are described below, and summarised in Table 39.

(a)   Heat Recovery in Rotary Kilns

217. These kilns are used in the refractories industry solely for raw materials preparation, but in 1980 consumed approximately 35% of the energy used by the industry. This amounted to around 3.59×10^6 GJ.

218. A rotary kiln is a long cylinder which rotates slowly and is inclined to the horizontal, so that the feed of raw material enters at the upper end and travels slowly to the lower end. It is heated by burning fuel at the lower end and the combustion gases flow countercurrent to the charge. The hot material is cooled in a current of air after discharge from the cylinder, and resultant hot air is used to pre-heat combustion air. The kiln operates at temperatures between 1,400° and 2,000°C.

219. The temperature of the exhaust gas leaving the kiln can be as high as 800°-900°C, and the energy contained in it at least 6 GJ/tonne of material being processed. However, the gas is dusty and contains oxides of nitrogen and sulphur.

220. Approximately 10% of the energy used in kilns is recoverable ($\sim 0.36 \times 10^6$ GJ in 1980) as high temperature heat, with suitable heat exchangers capable of handling the dirty gases. This energy could be used for generating electricity or as process heat for raw materials drying.

221. A further 10% of the input energy is recoverable as lower temperature heat, but there is no ready use for this within the industry.

222. Energy savings through high temperature heat recovery could reach their maximum contribution in around 20 years.

```
┌─────────────────────────────────────────────────────────┐
│              Heat Recovery in Rotary Kilns:              │
│     Saving 3.5% of total energy consumption by 2000      │
└─────────────────────────────────────────────────────────┘
```

(b)   Heat Recovery in Continuous Tunnel Kilns

223. In 1980 these kilns used approximately 35% of the energy consumed by the refractories industry (i.e. $\sim$ 3.59 x 10^6 GJ), and are used for firing shaped refractory products.

224. Tunnel kilns consist of a long tunnel through which the ware and gases move in opposite directions, thus making them self-recuperative. The ware is fired by burning fuel in the firing zone in the middle of the kiln; hot ware moves slowly away from the flames in one direction and the combustion gases move in the other direction. A counter-current of cold air progressively cools the fired ware, and is itself heated as it moves towards the firing zone. It therefore acts as preheated combustion air, or may be ducted away before reaching the firing zone for use in driers. The combustion gases are cooled by moving against the flow of unfired ware, which is preheated before firing.

225. Approximately 40% of the kiln firing energy is lost in exhaust gases (not including the cooling air) at temperatures of 200-400°C. Around 50% of this energy is recoverable for use either to preheat combustion air, or for space and water heating. However, the equipment required for combustion air preheat may be difficult to retrofit, because of the large volumes of gas which must be handled.

226. Heat exchangers must be used as exhaust gases contain contaminants such as sulphur, and precautions must be taken to avoid corrosion of ducts etc.

227. In 1980 the recoverable energy amounts to around 0.72 x 10^6 GJ, 7% of the energy used in the refractories industry. This may take 20 years to achieve its maximum contribution to energy savings, and if the industry remains at the current level of vastly reduced output and a wide range of products, there may be a significant move towards the use of intermittent kilns.

```
┌─────────────────────────────────────────────────────────┐
│              Heat Recovery in Tunnel Kilns:              │
│      Saving 7% of total energy consumption by 2000       │
└─────────────────────────────────────────────────────────┘
```

(c)   Heat Recovery in Intermittent Kilns

228. Intermittent kilns are less thermally efficient than continuous kilns, but have a number of advantages which make them attractive to producers in the current situation of greatly reduced output, with a particularly diverse nature. Intermittent kilns are used for items requiring extended firing cycles, as tunnel kilns would have to be unacceptably long, and when the amount of a particular product requiring a certain firing temperature

profile is too small to enable a tunnel kiln to operate efficiently. Intermittent kilns also offer product flexibility which is not available with continuous kilns.

229. They are thermally inefficient as the whole kiln structure has to be reheated and cooled for each new batch of ware fired, and only a limited amount of energy from the combustion gases is recovered due to the variability of its availability.

230. The kilns used around 20% of the energy used in the industry in 1980 ($\sim 2.05 \times 10^6$ GJ) of which some 30% could be recovered from combustion gases and cooling air ($\sim 0.62 \times 10^6$ GJ in 1980). Although some recovered energy is already used in combustion air preheating the use is limited by the intermittent nature of the supply, and depending upon the products being fired, by the formation of local high temperatures leading to material failure.

231. On the limited amount of information available it seems that a further 20% of the recoverable heat ($\sim 0.12 \times 10^6$ GJ in 1980) could find a use, either in dryers or as combustion air preheat, over the next 15 years or so.

> Heat Recovery in Intermittent Kilns:
> Saving 1.2% of total energy consumption by 1995

(d)   Heat Recovery from Afterburned Exhausts

232. Exhaust after-burn is practised extensively in the pitch and tar-bonded refractories and carbon electrode industries. Four fairly large operators in the UK refractories industry and six smaller operators together use an estimated 0.15×10^6 GJ of energy per year. This operation is carried out to clean up emissions from the factories, but heat recovery from these high temperature exhausts could offer significant opportunities for fuel saving. The energy available could be used for steam raising or for combustion air preheat, and in 1980 potential savings amounted to some 0.08×10^6 GJ, or 0.8% of the fuel used in that year. This level of savings could be achieved by the industry in about 15 years.

> Heat Recovery from Afterburned Exhausts:
> Saving 0.8% of total energy consumption by 1995

(e)   Improved Insulation for Rotary Kilns Operating at > 1,400°C

233. The energy used in rotary kilns in 1980 amounted to around 3.59×10^6 GJ; these kilns operate at high temperatures and radiation and convection losses are high.

234. Insulating materials are under development which could reduce these losses, and could save about 10% of the fuel consumed in rotary kilns. Such savings could be achieved by around 2000.

```
┌─────────────────────────────────────────────────────────┐
│              Improved Insulation for Rotary Kilns:        │
│      Saving 3.5% of total energy consumption by 2000      │
└─────────────────────────────────────────────────────────┘
```

(f)   Improved Process Control

235. A range of opportunities exist within the refractories industry for improving process control, such as improved kiln temperature control, reduction of excess air and dryer programming. Savings of around 5% of the energy consumed in the industry could be achieved by implementing such measures, but could take some 15 years to achieve.

```
┌─────────────────────────────────────────────────────────┐
│              Improved Process Control:                    │
│      Saving 5% of total energy consumption by 1995        │
└─────────────────────────────────────────────────────────┘
```

236.   In the Energy Audit of this industry [8] opportunities were identified for energy savings through air circulation in tunnel kilns, and reduction of heat losses from the outside of tunnel kilns. None of these has proved to be commercially viable, and, if the refractories industry remains at the current level of vastly reduced output and a wide range of products is manufactured, there may be a significant move towards intermittent kilns. In which case, any change in the future viability of such measures would be irrelevant.

237.   It has also been suggested that improved drying techniques could offer energy savings, where intermittent kilns are used for firing shaped products. These techniques would involve heat recovery from dryer exhausts using run-around coils, or heat pumps, and would be aimed at reducing the need for supplementary drying energy, in excess of the energy supplied by kiln cooling air. However, in view of the low moisture content of most refractories, and the abundance of waste heat available in most works, these measures are no longer considered to be viable.

PROJECTED LEVELS OF ENERGY CONSUMPTION IN THE REFRACTORIES INDUSTRY

238. The energy conservation measures described above, and summarised in Table 39, will be taken up by the refractories industry over a period of time. The rate of uptake will be determined by a number of factors, including the state of technical development of a technology, the rate of return on capital investment, the ease with which it may be applied to existing plant or equipment, etc.

239. A simple diffusion model has been used to determine how the technologies may enter the industry, and the two S-curves used to model diffusion over a period of 15 and 20 years are shown in Figure 9.

240. The contribution made by each technology to a reduction in energy consumption, over the next 30 years, is shown in Table 40. The savings are applied sequentially to avoid any double-counting, and as they are applied as percentage reductions on energy consumption, can be applied in any order to achieve the same total saving.

Table 39. Energy Conservation Measures which may be applied in the
Refractories Industry

Conservation Measure	Timescale	Energy Saving
1. Good housekeeping	1990	5%
2. Heat recovery – intermittent kilns	1995	1.2%
3. Heat recovery – afterburned exhausts	1995	0.8%
4. Improved process control	1995	5%
5. Heat recovery – rotary kilns	2000	3.5%
6. Heat recovery – tunnel kilns	2000	7%
7. Improved insulation – rotary kilns	2000	3.5%
TOTAL SAVING:		23.4%

Note: Percentage savings are applied sequentially to avoid
double-counting.

Table 40. Reduction in Energy Consumption due to Various
Conservation Measures

Conservation Measure	1985	1990	1995	2000
1. Good housekeeping	2.5%	5%	5%	5%
2. Heat recovery – intermittent kilns	0.3%	0.9%	1.2%	1.2%
3. Heat recovery – afterburned exhausts	0.2%	0.6%	0.8%	0.8%
4. Improved process control	1.2%	3.8%	5%	5%
5. Heat recovery – rotary kilns	0.6%	1.8%	2.9%	3.5%
6. Heat recovery – tunnel kilns	1.1%	3.5%	5.9%	7%
7. Improved insulation – rotary kilns	0.6%	1.8%	2.9%	3.5%
TOTAL SAVING	6.3%	16.2%	21.5%	23.4%

Note: Percentage savings are applied sequentially to avoid
double-counting.

241. Table 41 shows the projected specific energy consumption for the refractories industry, up to the turn of the century. This shows a fall of 23% between 1980 and 2000, which is equivalent to an average decrease of 1.3% per year.

242. No account has been taken of possible changes in the structure of the refractories industry, towards more, or less energy intensive products under different scenarios of future output from the industry. This is because the paucity of information on the energy intensity of each product in 1980 makes the projection of changes in specific energy consumption, due to structural changes in the industry, almost impossible and to introduce any difference in the specific energy consumption of the industry for different scenarios of the future would be purely arbitrary.

243. The nature of the output from the refractories industry has been largely determined by the requirements of the UK iron and steel industry. Around 60-65% of the output of the refractories industry is used directly in the iron and steel industry, and Figure 10 shows the indices of production of these two industries together with those of the castings and general ferrous metal manufacturing industries, for the period 1971-1981. It is clear that between 1974 and 1980 the index for the refractories industry moved with that of ferrous metal manufacture, and there was never a difference of more than 12% between them.

244. Such factors will be important in determining the future level of output from the refractories industry, the nature of the product mix and, therefore, the future energy requirements of the industry.

245. If the balance of fuel use in the industry remains as shown in Table 38, and energy savings are made primarily in kiln and dryer fuel, then the future levels of specific energy consumption may be divided among the various fuels as shown in Table 42. However, it is expected that the industry will convert most of its oil-burning plant to gas, by about 2010, and on this basis the consumption of each fuel might be as shown in Table 43. There may also be some change in the proportion of coal use by the industry, but in the absence of any information on this it has been assumed that it remains unchanged.

CONCLUSIONS

246. There is little up-to-date information on energy use in the refractories industry, but it is probable that specific energy consumption has increased by around 17% between 1976 and 1980 to 12.6 GJ/tonne, while output has fallen by nearly 40% to only 0.82×10^6 tonnes.

247. The mean specific energy consumption of the industry could be reduced by implementing a range of energy conservation technologies, most of which are aimed at reducing kiln fuel consumption. The specific energy consumption could be reduced by 23.4% by the year 2000, from 12.60 GJ/tonne in 1980 to only 9.65 GJ/tonne in 2000, on a heat supplied basis. This amounts to an average decrease of 1.3% per year, compared with an estimated average increase of 4% per year over the past four years. However, this has been influenced largely by the production of higher quality materials and a dramatic decline in output over that period.

248.   It has not been possible to take account of changes in the structure of the industry, which may occur in the future, and could be of significance to the future energy requirements of the industry.   There is currently considerable over-capacity within the industry and further rationalisation could lead to significant energy savings.   Those tunnel kilns remaining will have to operate at, or near to maximum throughput, for most efficient energy use, with intermittent kilns being used for firing the smaller quantities of material.

Figure 9.   S-curves used to model the diffusion of conservation
technologies into the industry

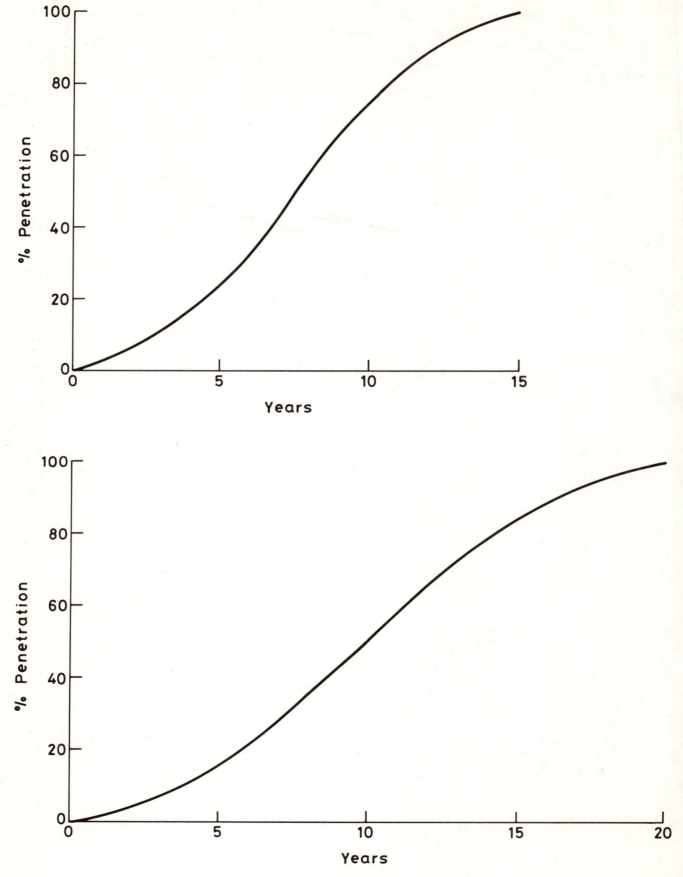

Figure 10. Indices of production 1971–1981

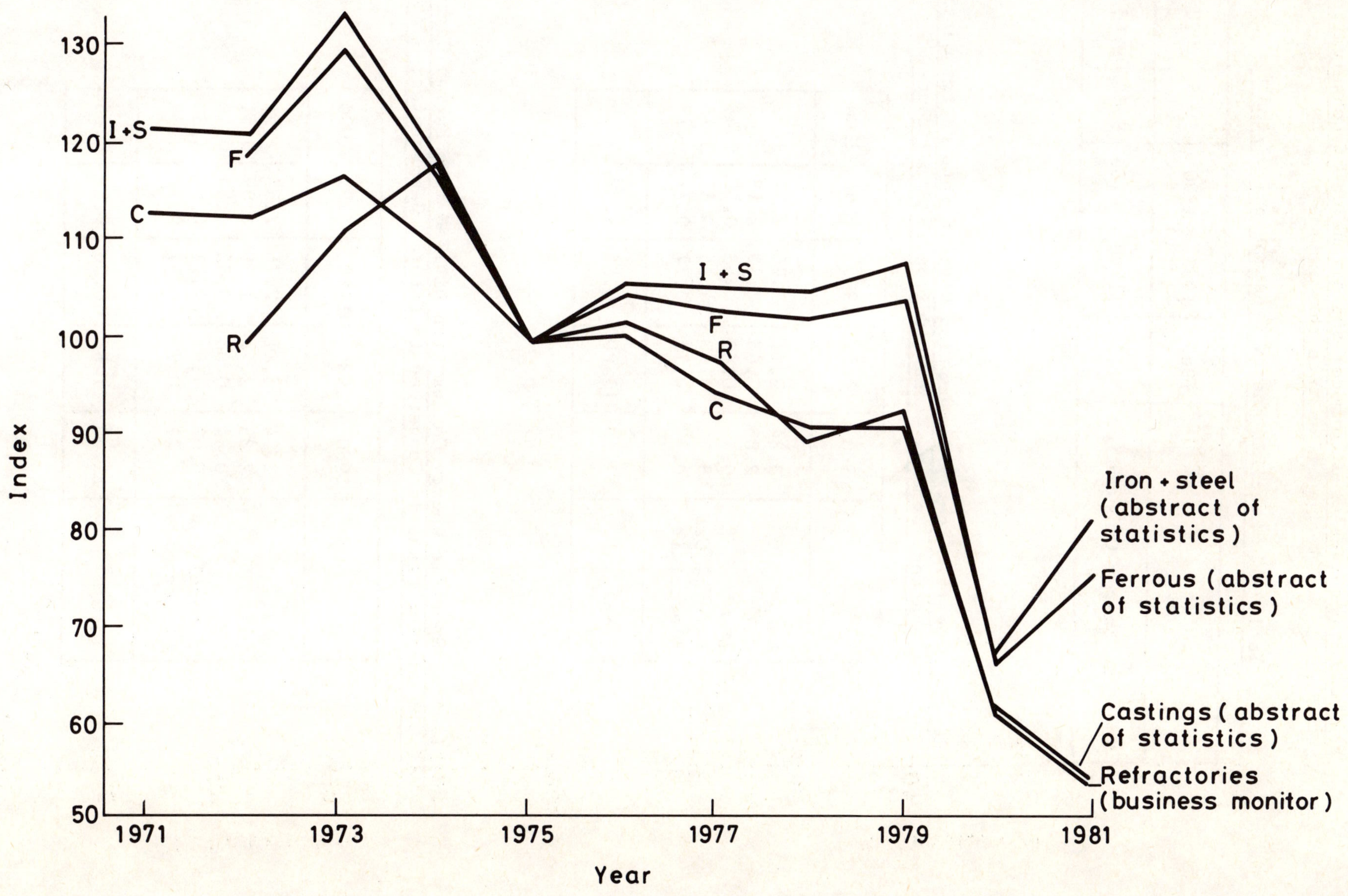

Table 41. Specific Energy Consumption, GJ/tonne

	1980	1985	1990	1995	2000
Specific Energy Consumption	12.60	11.80	10.56	9.89	9.65

Table 42. Specific Energy Consumption Divided Into Energy Carriers if Balance Remains Unchanged from 1979, GJ/tonne

	1980	1985	1990	1995	2000
Coal	3.34	3.11	2.76	2.57	2.51
Oil	2.83	2.65	2.35	2.18	2.13
LPG	0.64	0.60	0.53	0.50	0.48
Natural Gas	5.02	4.67	4.15	3.87	3.76
Electricity	0.77	0.77	0.77	0.77	0.77
Total	12.60	11.80	10.56	9.89	9.65

Table 43. Specific Energy Consumption Divided Into Energy Carriers if Oil-Burning Plant is Converted to Natural Gas-Burning, GJ/tonne

	1980	1985	1990	1995	2000
Coal	3.34	3.11	2.76	2.57	2.51
Oil	2.83	2.41	1.79	1.09	0.51
LPG	0.64	0.60	0.53	0.50	0.48
Natural Gas	5.02	4.91	4.71	4.96	5.38
Electricity	0.77	0.77	0.77	0.77	0.77
Total	12.60	11.80	10.56	9.89	9.65

ENERGY USE AND ENERGY EFFICIENCY IN UK MANUFACTURING INDUSTRY

UP TO THE YEAR 2000

2.4 THE POTTERY INDUSTRY

2.4 THE POTTERY INDUSTRY

INTRODUCTION

249. Although several sources of information on energy use in the pottery industry exist, none of them gives a complete description of the industry, and the different methods of data collection and presentation inhibit the formation of a consistent, overall picture.

250. In the Digest of UK Energy Statistics [4] the pottery industry is included under the heading 'China, Earthenware and Glass' and energy use is not subdivided further. Although the British Ceramic Manufacturers' Federation (BCMF) carry out an annual fuel and power census [11] this covers a variable proportion of the industry, usually around 75% of the total energy consumed.

251. A report on the pottery industry, prepared for the Industrial Energy Thrift Scheme [12], was published in 1978, and an Energy Audit [13] was published in 1979. Both of these reports were based on data relating mainly to 1975. However, the BCMF feel that the balance of energy use within the different sectors of the industry will not have changed significantly sinca 1975.

THE STRUCTURE OF THE POTTERY INDUSTRY

252. The principal products of the pottery industry are china and earthenware tableware, sanitaryware, electrical ware, glazed wall tiles and unglazed floor tiles.

253. The tableware industry is dominated by Royal Doulton and the Wedgwood Group, who together account for over 65% of sales turnover. Royal Worcester/Spode, Federated Potteries and Staffordshire Potteries together account for a further 20-25% of turnover, the remainder coming from a large collection of small potteries.

254. White clay tile manufacturing is carried out by two established UK manufacturers; over 65% of the output is produced by H & R Johnson Tiles, and around 15% is produced by Pilkington Tiles.

255. Sanitaryware production is dominated by three companies; Armitage-Shanks, Twyfords Bathrooms and Ideal Standard, who together produce over 80% of the UK output. The market remains fairly buoyant as a result of a strong export potential.

256. Energy costs have not been of major concern to the pottery industry as a whole, which is mainly concerned with productivity and efficiency. However, in the face of increasing competition from imports the industry will have to adopt new, automatic production techniques which should also lead to energy savings.

257. The industry has recently undergone a large programme of reorganisation and rationalisation in which several potteries have been closed, some within the larger company groups. The future will depend on the industry's ability to compete with overseas producers both in the UK market and abroad.

POTTERY MANUFACTURE

258. Clay is the basic raw material used in pottery manufacture. Silica may be added as a cheap filler and fluxes are used to promote vitrification at firing temperatures of 1,100°-1,250°C. Bone china also contains about 50% of ground animal bone which acts mainly as a filler.

259. Silica and bone are calcined and with the fluxes are then crushed and finely ground before being mixed with the clay slurry to form a suspension, called slip. Impurities are removed by sieving and magnetic separation before the water content is reduced by filter pressing to produce a plastic body. The filter cake is then processed in a pug-mill to remove air and consolidate the body.

260. Articles are then shaped in or on plaster-of-Paris moulds and scrap material is recycled. The articles and moulds are then dried before firing. However, hollow items of complex shape are made by slip casting in plaster moulds, in which case the slip does not require filter pressing or pug-milling. Tiles are usually prepared from dusts of suitable moisture content which are made by spray drying slip and are also dried before firing.

261. Ware may undergo three firing processes. Tableware and wall tiles are fired at 1,150°-1,250°C to form a hard, strong, sometimes non-porous biscuit-ware and a suspension of powdered glaze in water is applied which may be dried before firing at 1,050°-1,100°C. Decoration may be applied either before or after glazing. If decoration is carried out before glazing a further firing may be necessary before the glaze is applied. On-glaze decoration may be applied in a variety of ways, in the case of the more expensive tableware several methods of application may be used on a single article, with a separate firing after each application. On-glaze decorated ware is fired at about 800°C to weld the decoration to the glaze.

262. Electrical ware, unglazed floor tiles and sanitaryware are fired only once at about 1,200°C.

ENERGY USE IN THE POTTERY INDUSTRY

263. Table 44 shows the levels of energy consumption in the pottery industry reported in the BCMF census from 1971 to 1980 inclusive, although it is acknowledged that the census covers a varying proportion of the industry and includes between 65% and 82% of the total energy consumption for each year.

264. However, the British Gas Corporation's Annual Report and Accounts for the year ending 31 March 1981 lists the gas sales to the whole of the pottery industry over the period 1971 to 1980. In accordance with the approach adopted in the Energy Audit of the industry it is assumed that the ratio of gas consumption recorded by BGC for the whole industry to that recorded by the BCMF also applies to all other forms of energy. The estimates of total energy consumption by the pottery industry shown in Table 45 and Figure 11 are, therefore, based on this assumption.

265. It is, therefore, assumed that in 1980 a total of 14.13×10^6 GJ was used by the industry, of which 63.5% was natural gas and 16% LPG. Consumption of fuel oil has declined from 30.7% in 1971 to only 10.1% in 1980, although it increased between 1976 and 1978 from 13.4% to 18.3% before declining again.

Table 44. Energy consumption by the pottery industry reported in the
British Ceramic Manufacturers' Federation Census. 10^6 GJ/year.

Fuel	1971	1972	1973	1974	1975	1976	1977	1978	1979	1980
Electricity	0.93	0.80	0.97	0.86	0.86	0.89	1.03	0.96	1.09	1.02
Fuel Oil	3.41	2.39	2.35	2.15	1.71	1.46	1.78	2.39	1.77	1.18
Coal	1.36	0.65	0.45	0.23	0.19	0.30	0.26	0.26	0.27	0.19
Coke	0.007	0.003	0.001	0.003	0	0.002	0	0	0	0
Natural Gas	4.22	5.20	6.54	6.96	6.65	6.74	7.32	7.49	7.72	7.36
LPG	0.74*	1.47*	2.12*	1.74	1.64	1.52	1.92	1.92	2.05	1.85
Town Gas	0.42	0.22	0.01	0	0	0	0	0	0	0
TOTAL	11.09	10.73	12.44	11.94	11.05	10.91	12.31	13.02	12.90	11.60

* By deduction.

Table 45. Estimated total energy consumption by the pottery industry
10^6 GJ/year

Fuel	1971	1972	1973	1974	1975	1976	1977	1978	1979	1980
Electricity	1.37	1.22	1.29	1.13	1.21	1.26	1.42	1.34	1.38	1.24
Fuel Oil	5.02	3.63	3.11	2.84	2.41	2.06	2.46	3.32	2.25	1.43
Coal	2.01	0.99	0.60	0.31	0.27	0.42	0.36	0.36	0.34	0.23
Coke	0.01	0.005	0	0.005	0	0	0	0	0	0
Natural Gas	6.22	7.91	8.65	9.18	9.39	9.50	10.13	10.44	9.81	8.97
LPG	1.09	2.24	2.80	2.30	2.32	2.15	2.65	2.68	2.60	2.26
Town Gas	0.62	0.34	0.01	0	0	0	0	0	0	0
TOTAL	16.34	16.33	16.46	15.77	15.60	15.39	17.02	18.14	16.38	14.13

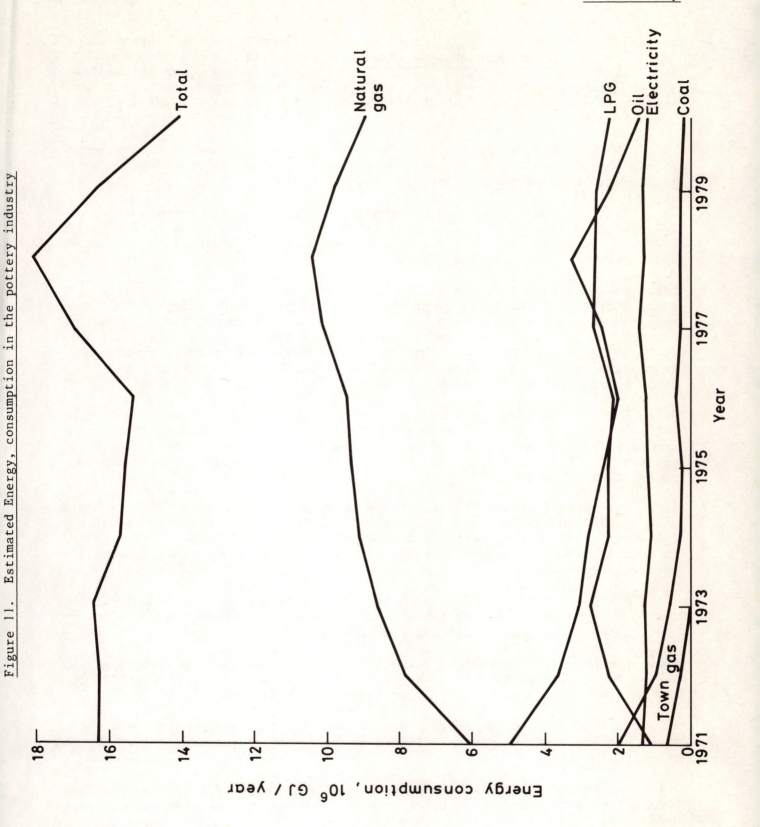

Figure 11. Estimated Energy, consumption in the pottery industry

Electricity consumption has remained relatively steady over the period at 7.2% to 8.8% of the total consumption, while coal consumption has fallen from 12.3% in 1971 to only 1.6% in 1980. The pottery industry is now committed to the use of clean, particularly gaseous fuels, as the kilns now being built are designed to use gas, and the move to direct firing of ware necessitates a clean fuel.

266. The amount of energy used by the pottery industry should be considered within the context of the changing output of the industry. However, very little detailed information on this is available.

267. Table 46 shows the index of production for the whole of the pottery industry from 1971 to 1980 [3]. The table also shows the normalised specific energy consumption for the industry in terms of each individual fuel type and the total energy used in each year.

268. The uncertainties involved in estimating the total fuel consumption of the industry, and the index of production, lead to figures which show little reduction in energy consumption per unit of output, between 1976 and 1980. However, the authors of the Energy Audit [13] feel that significant savings of up to 10% have been made over that period, predominantly through a move away from steam raising in boilers to directly fired equipment. The figures in Table 6.3 suggest savings of ~ 28% between 1971 and 1980.

269. Less coal and fuel oil is now used per unit of output than in either 1971 or 1975, and although use of electricity is now less than in 1971 it is around 6% higher than in 1975. Gas consumption, in the form of both natural gas and LPG, has increased since 1971, but is approximately the same now as consumption in 1975.

270. The overall output of the pottery industry has fallen by 17.6% from its peak in 1978 and in 1980 was just below its 1974 level. It has continued to fall and in 1982 output was a further 19.8% lower than in 1980. Until 1977 increasing levels of output were clearly associated with decreasing specific energy consumption.

271. A large proportion of the energy supplied to the pottery industry is used in the kilns for firing ware. Table 47 shows how the various fuels were used by the firms reported in the 1980 BCMF census.

272. Energy use in 1980 shows very little difference from the figures given in the Thrift Report for 1975, when 61% of energy was used in firing; 33% in drying and space heating, and 6% for power and light.

273. It is difficult to separate energy used for drying from that used for space heating, as there is very little sub-metering of fuels used for purposes other than firing.

274. Within the pottery industry there is a wide range of products from fine bone china to unglazed floor tiles, and the various products within this range undergo different production processes. The estimated specific energy consumption for each product type is shown in Table 48, and ranges from 212.5 GJ/tonne for bone china to only 9.7 GJ/tonne for unglazed floor tiles.

Table 46. Production and Specific Energy Consumption by The Pottery Industry

Year	Index of Production	Index of Specific Energy Consumption					
		All Fuels	Electricity	Fuel Oil	Coal	Nat Gas	LPG
	(1975 = 100)						
1971	79.8	131.3	141.9	261.0	932.9	83.0	58.9
1972	85.6	122.3	117.8	176.0	428.3	98.4	112.8
1973	93.7	112.6	113.8	137.7	237.2	98.3	128.8
1974	98.9	102.2	94.4	119.2	116.1	98.9	100.2
1975	100.0	100.0	100.0	100.0	100.0	100.0	100.0
1976	104.1	94.8	100.0	82.1	149.4	97.2	89.0
1977	113.7	96.0	103.2	89.8	117.3	94.9	100.5
1978	117.1	99.3	94.6	117.6	113.9	94.9	98.6
1979	109.2	96.2	104.4	85.5	115.3	95.7	102.6
1980	96.5	93.9	106.2	61.5	88.2	99.0	100.9

Table 47. Energy Use Within the Pottery Industry in 1980

Fuel Use and Fuel Type			Total
Energy for firing:	Natural Gas	45.4%	
	LPG	13.4%	
	Fuel Oil	2.0%	
	Electricity	3.0%	63.8%
Energy for drying and space heating:	Natural Gas	18.1%	
	LPG	2.6%	
	Fuel Oil	8.1%	
	Coal	1.6%	30.4%
Energy for power and light:	Electricity	5.8%	5.8%

275. Table 49 gives the distribution of energy use within the firms surveyed as part of the Energy Audit [13] Study. Bearing in mind that this information relates primarily to 1975 and that a very limited number of firms were surveyed, there may be some differences between the information given in the Audit and the corresponding data relating to 1980 (which is not currently available in this level of detail).

276. The estimated total energy consumption in 1980 by each product group within the industry is shown in Table 50, and when combined with Table 49 gives a measure of the energy consumed by the industry for firing, drying, motive power, space heating and lighting, as shown in Table 51.

ENERGY CONSERVATION POTENTIAL

277. Over the past ten years the pottery industry has made significant energy savings which have been achieved through improved management and house-keeping, reorganisation and closing of older, obsolete plant. Direct energy saving projects have also been introduced, such as improved burners, retrofit heat recovery systems and attempts to increase throughput to reduce specific energy consumption. Savings have also been made through the move away from steam raising to directly fired equipment for space heating and drying.

278. However, discussions with those familiar with the pottery industry suggest that further savings in the order of 15-20% over those shown in Table 46 could be made by continued improvement in management and control of energy use, and the natural replacement of older plant by more energy efficient equipment. These savings could be achieved by around 1990.

279. In the absence of more detailed information on the range of specific energy consumption within each product group of the industry, no more refined estimate can be made of current best practice, or the energy savings which could be made if all firms operated at the level of the best practice firm. Therefore, it is assumed that overall savings of 20% in the total energy consumption of the whole industry may be made by 1990. On current levels of output this would amount to a saving of 2.83×10^6 GJ.

Future Energy Savings in the Pottery Industry

280. In 1980 14.13×10^6 GJ of energy was used in the pottery industry, of which 9.01×10^6 GJ was used for firing ware, 4.30×10^6 GJ for drying and space heating and 0.82×10^6 GJ in the form of electricity for motive power and lighting.

281. In addition to the energy savings described in paragraph 279 above, further savings could be made by implementing a range of energy conservation technologies. These technologies are described below with the potential savings and timescale over which they may be achieved, they are also summarised in Table 52.

(a) Heat recovery from tunnel kiln exhausts

282. Around 80% of the firing energy used in the pottery industry is used in continuous tunnel kilns, of this approximately one third is lost as waste heat in exhaust gases. These gases can sometimes be used directly for drying, but once passed through a heat exchanger can provide a source of clean, hot

Table 48.   Estimated Specific Energy Consumption by Sectors of the Pottery Industry in 1980* (heat supplied)

Product	Total Energy Consumption GJ/tonne saleable product
Bone China	212.5
Earthenware	42.9
Sanitaryware	44.0
Electrical Ceramics	77.1
Glazed Wall Tiles	13.3
Unglazed Floor Tiles	9.7

* Based on Tables 5 and 6 of Energy Audit [13].

Table 49.   Distribution of Energy Use as Reported in the Energy Audit [13] (heat supplied)

Product	Firing Energy		Electricity for Power and Light	Drying and Space Heating
	Fuel	Electricity		
Bone China	51.6%	1.6%	7.0%	39.8%
Earthenware	66.7%	3.6%	6.0%	23.8%
Sanitaryware	62.1%	0	3.6%	34.3%
Electrical Ceramics	39.4%	0	4.7%	55.8%
Wall Tiles	58.0%	0	13.0%	29.0%
Floor Tiles	50.7%	0	5.1%	44.1%
BCMF DATA 1980	60.8%	3.0%	5.8%	30.4%

Table 50.   Energy Consumption by each Product Group

Product	Energy Consumption 10^6 GJ	1980
Bone China	1.84	13%
Earthenware	4.04	28.6%
Sanitaryware	3.23	22.9%
Electrical Ceramics	1.40	9.9%
Tiles	3.62	25.6%
TOTAL	14.13	

Based on BCMF detailed survey 1973.

air.   The recovered heat can be used in waste heat boilers to raise steam or provide hot water and for drying ware prior to firing.

283.   The recoverable energy amounts to about half of the energy in the exhaust gases, and although it is not cost effective to recover and use this energy in all situations, about half of that available could be recovered and used within the pottery industry itself.   In 1980 this would amount to 0.60 x 10^6 GJ per year.   Approximately two thirds of the recoverable energy could be available by the turn of the century, the remaining one third being available by 2010.

```
HEAT RECOVERY FROM TUNNEL KILN EXHAUSTS.
SAVING:        2.8% of total energy consumed by 2000
A FURTHER      1.4% by 2010.
```

(b)   Heat recovery from intermittent kiln exhausts

284.   Energy recovered from the exhaust gases of intermittent kilns can be used to preheat the combustion air for the gas burners.

285.   Twenty per cent of the energy used within the pottery industry for firing ware is used in intermittent kilns and of this 75% is used in gas fired kilns where a ready use exists for recovered energy.   Approximately 40% of the energy used in firing these kilns can be recovered from the exhaust gases and in 1980 this would amount to 0.54 x 10^6 GJ.   Around half of the energy available could be recovered and used by 1995, the remainder being taken up much more slowly, with a further 25% of the available energy being recovered and used by around 2010.   It may not be technically feasible to use the final 25% of the recoverable energy due to the inappropriate location of kilns.

```
HEAT RECOVERY FROM INTERMITTENT KILN EXHAUSTS.
SAVING:        1.9% of total energy consumed by 1995
A FURTHER      1.0% by 2010.
```

(c)   Improved kiln design and firing control in tunnel kilns

286.   Closer control of firing conditions could save significant amounts of energy within this industry, together with improvements to burner configuration and kiln design to optimise heat transfer to the ware.   It is technically feasible to save 25% of the energy currently used in tunnel kilns but this will occur in increments, over a long period of time.   A large proportion will have to await plant replacement as retrofit equipment cannot often be justified on an economic basis.

287.   Of the energy saving which could be made, an overall saving of 10% of the energy used in tunnel kilns may be achieved by 2000, with the remainder taking up to 40 years to achieve as some plant will not be due for replacement for many years.   On the basis of 1980 fuel consumption the total savings would amount to 1.8 x 10^6 GJ per year.

```
IMPROVED KILN DESIGN AND FIRING CONTROL IN TUNNEL KILNS.
SAVING:      5.1% of total energy consumed by 2000
A FURTHER    7.7% by 2020.
```

(d)   Improved dryer recirculation and optimisation

288.   Dryers operate at much lower temperatures than kilns, and are consequently much easier to construct. However, this often results in little attention being given to designs for their efficient operation. In all types of dryer a recirculation system can be used to increase their thermal efficiency, and improvements to these systems and the overall design of dryers could save significant amounts of energy.

289.   Energy use for space heating within this industry is not distinguished from that used for drying. Although the energy saving measures described above apply only to drying energy, savings have to be measured as a percentage of the energy used for both drying and space heating. Major modifications or replacement of dryers could save 50% of the energy currently used in this way. One third of this saving could be achieved by 1990 through heat recovery from dryer exhausts and early replacement, while further replacement and major improvement would achieve the remaining two thirds of the savings by 2010.

290.   On the basis of energy utilisation in 1980 this would amount to a total of 2.15×10^6 GJ per year.

```
IMPROVED DRYERS
SAVING:      5.1% of total energy consumed by 1990
A FURTHER    10.1% by 2010.
```

(e)   Direct firing of sanitaryware

291.   The sanitaryware sector of the pottery industry has conventionally used muffle (indirectly fired) tunnel kilns for firing ware. However, significant energy savings can be made by changing to open flame firing in gas-fired tunnel kilns. During the transition period, while both types of kiln are in operation, there will be some difficulty with colour matching in the ware. The glaze quality and surface finish is also more difficult to control in open flame kilns and these problems must be overcome if the full energy saving potential is to be achieved.

292.   The potential savings of 1.8% of the energy consumed within the pottery industry could be achieved by 1990, and in terms of the 1980 level of consumption would amount to 0.26×10^6 GJ.

```
DIRECT FIRING OF SANITARYWARE
SAVING:   1.8% of total energy consumed by 1990.
```

(f)   Fast firing

293.   Firing cycles in pottery manufacture may often be shortened considerably by increasing the firing temperature, particularly within the tableware and tile sectors of the industry.   Some process development work is still required in this area, for example to develop improved glazes suitable for open flame firing and developments in the translucency of bone china.

294.   However, the energy savings which may be achieved in this way could amount to 22% of the energy used in firing ware if kiln furniture is reduced and the use of low thermal mass kilns is extended, which in 1980 would represent savings of 2.0×10^6 GJ.   Two thirds of the saving could be achieved by the year 2000 with the remaining one third being achieved more slowly over a further period of about 15 years.

```
FAST FIRING
SAVING:      9.4% of total energy consumed by 2000
A FURTHER    4.7% by 2015.
```

(g)   Once firing

295.   Many products of the pottery industry currently undergo multiple firing cycles, particularly those from the tableware and tile sector of the industry.   Large energy savings would be made if this could be reduced to a single firing stage, although significant technical difficulties must be overcome first.   These are factors such as the need to improve dimensional and shape stability of ware during high temperature, once firing and the need for under/in-glaze decoration systems of wider visual appeal and technically better quality.

296.   The potential energy savings which could be made if this technology was implemented within the industry amount to 25.3% of the energy used for firing, which in 1980 would be 2.28×10^6 GJ.   Of the total potential saving 68% might be achieved by the year 2000 and the further 32% by 2020.

```
ONCE FIRING
SAVING:      11% of total energy consumed by 2000
A FURTHER    5.2% by 2020.
```

PROJECTED LEVELS OF ENERGY CONSUMPTION IN THE POTTERY INDUSTRY

297.   The energy conservation measures which are described above, and listed in Table 52, will be taken up by the pottery industry over a period of time. An estimate has been made of the date by which each technology might make its maximum contribution to energy savings within the industry, these dates are shown as the timescale in Table 52.   Certain technologies show two tranches of savings, one of which is achieved in the medium-term future (to around 2000) and the other which is often smaller and achieved more slowly over the medium and long-term future (say 2000 to 2020).   In order to avoid double-counting of savings it has been assumed that technologies are applied sequentially, and

Table 51. Distribution of Energy Use Within the Pottery Industry

Product	Firing Energy 10^3 GJ		Power and Light 10^3 GJ	Drying and Space Heating 10^3 GJ
	Fuel	Electricity		
Bone China	949	29	129	732
Earthenware	2,695	145	242	962
Sanitaryware	2,006	0	116	1,108
Electrical Ceramics	552	0	66	781
Tiles*	2,026	0	391	1,203
Total	8,228	174	944	4,786
BCMF 1980 Survey	8,591	424	820	4,296

* Estimate

Table 52. Energy Conservation Measures Which May be Applied in the Pottery Industry

Conservation Measure	Timescale	Energy Saving
1. Best practice	by 1990	20%
2. Improved dryers	by 1990	5.1%
3. Direct firing	by 1990	1.8%
4. Heat recovery – intermittent kilns	by 1995	1.9%
5. Heat recovery – tunnel kilns	by 2000	2.8%
6. Improved kiln design	by 2000	5.1%
7. Fast firing	by 2000	9.4%
8. Once firing	by 2000	11.0%
9. Heat recovery – tunnel kilns	by 2010	1.4%
10. Heat recovery – intermittent kilns	by 2010	1.0%
11. Improved dryers	by 2010	10.1%
12. Fast firing	by 2015	4.7%
13. Improved kiln design	by 2020	7.7%
14. Once firing	by 2020	5.2%
TOTAL SAVING		60.2%

Note: Percentage savings are applied sequentially to avoid double-counting.

Figure 12. Diffusion curve for technologies taking 15 years to reach maximum penetration of the industry

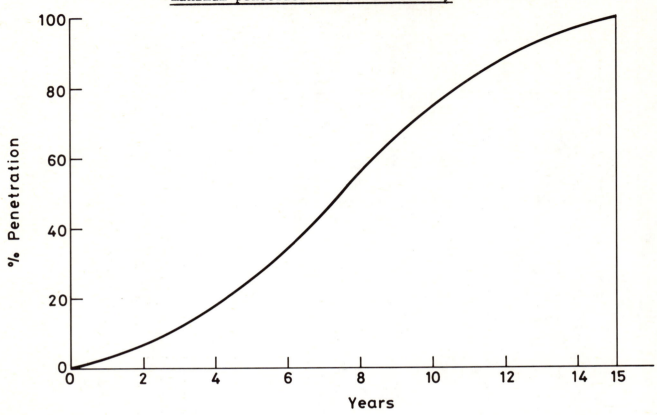

Diffusion curve for technologies taking 20 years to reach maximum penetration of the industry

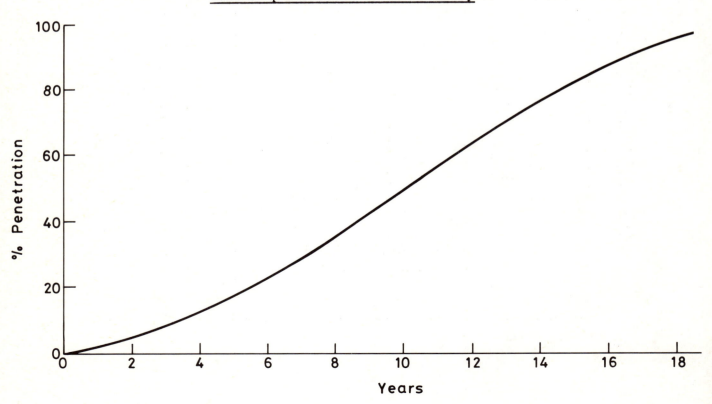

Table 53. Reduction in energy consumption due to various conservation measures

Conservation Measure	Year							
	1985	1990	1995	2000	2005	2010	2015	2020
1. Best practice	10%	20%	20%	20%	20%	20%	20%	20%
2. Improved dryers	2.5%	5.1%	5.1%	5.1%	5.1%	5.1%	5.1%	5.1%
3. Direct firing	0.9%	1.8%	1.8%	1.8%	1.8%	1.8%	1.8%	1.8%
4. Heat recovery – intermittent kilns	0.5%	1.4%	1.9%	1.9%	1.9%	1.9%	1.9%	1.9%
5. Heat recovery – tunnel kilns	0.5%	1.4%	2.3%	2.8%	2.8%	2.8%	2.8%	2.8%
6. Improved kiln design	0.9%	2.6%	4.2%	5.1%	5.1%	5.1%	5.1%	5.1%
7. Fast firing	1.7%	4.7%	7.7%	9.4%	9.4%	9.4%	9.4%	9.4%
8. Once firing	2.0%	5.5%	9.0%	11.0%	11.0%	11.0%	11.0%	11.0%
9. Heat recovery – tunnel kilns	0	0	0	0	0.7%	1.4%	1.4%	1.4%
10. Heat recovery – intermittent kilns	0	0	0	0.25%	0.7%	1.0%	1.0%	1.0%
11. Improved dryers	0	0	1.8%	5.1%	8.3%	10.1%	10.1%	10.1%
12. Fast firing	0	0	0	0	1.2%	3.5%	4.7%	4.7%
13. Improved kiln design	0	0	0	0	1.4%	3.9%	6.3%	7.7%
14. Once firing	0	0	0	0	0.9%	2.6%	4.3%	5.2%
TOTAL SAVING	17.8%	36.4%	43.5%	48.5%	52.5%	56.9%	59.2%	60.2%

Note: Percentage savings are applied sequentially to avoid double-counting.

Table 54. Index of Specific Energy Consumption for the Pottery Industry
1975 = 100

Year	1975	1980	1985	1990	1995	2000	2005	2010	2015	2020
Index of Consumption	100.0	93.9	77.2	59.7	53.0	48.3	44.6	40.5	38.3	37.4

Table 55. Estimate of the Future Specific Energy Consumption by Sector, GJ/tonne

Sector	Year								
	1980	1985	1990	1995	2000	2005	2010	2015	2020
Bone China	212.5	177.2	139.5	124.7	114.1	105.0	94.8	89.8	87.8
Earthenware	42.9	34.8	25.9	21.9	19.3	17.4	15.2	13.9	13.4
Sanitaryware	44.0	36.4	29.1	27.7	26.2	24.7	23.2	22.6	22.3
Electrical Ceramics	77.1	66.4	55.4	52.7	50.0	47.0	44.3	43.1	42.4
Glazed Wall Tiles	13.3	10.9	8.3	7.1	6.4	5.8	5.1	4.7	4.6
Unglazed Floor Tiles	9.7	8.1	6.4	5.7	5.2	4.8	4.3	4.0	4.0

Table 56. Index of Specific Energy Requirements by the Pottery Industry
1975 = 100

Fuel	1980	1990	2000	2010	2020
Natural Gas	99.0	69.4	60.5	53.7	49.6
LPG	100.9	44.1	18.4	0	0
Fuel Oil	61.5	22.9	8.1	0	0
Coal	88.2	31.1	11.1	0	0
Electricity	106.2	94.0	99.9	104.8	96.9
Total	93.9	59.7	48.3	40.5	37.4

once energy savings have been made using one technology the potential savings which may be made by the next to be applied are reduced proportionately.

298.   A simple diffusion model has been used to calculate how each technology may affect energy consumption within the industry over the period 1980 to 2020.   Figure 12 shows the simple S-curves which were used to model diffusion of technologies into the industry over a period of 15 and 20 years respectively.   Table 53 shows the estimated percentage reduction in the total energy consumption due to each conservation technology, for each fifth year between 1980 and 2020.

299.   By the year 2000 the energy required to produce a tonne of product could be 48.5% lower than in 1980, which represents an average annual decrease of 3.3%.   Further savings could be achieved in the longer term to 2020, which amount to an annual decrease of 1.3% in specific energy requirements.   Table 54 shows the index of specific energy consumption for the pottery industry, for the period 1975 to 2020.

300.   For the purposes of this report is has been assumed that the balance of output from each sector of the pottery industry remains at the 1980 level. However, it must be recognised that any change in this balance may significantly affect future levels of energy consumption.   For example, in 1980 bone china required around five times as much energy per tonne of product (212.5 GJ/tonne) as did earthenware at only 42.9 GJ/tonne, but the bone china sector consumed only 13% of the total energy requirement of the industry, while earthenware consumed almost 29%.

301.   A 1% increase in the total output of the pottery industry in 1980, if all accredited to an increase in bone china production, would have an associated increase of around 7.7% in the total industry energy consumption. However, a similar increase in output accredited solely to earthenware production would cause an increase of only 1.5% in total energy consumption in 1980.

302.   It is possible that any overall increase in the output of the pottery industry, in future years, could come predominantly from certain sectors of the industry.   This could lead to a change in the specific energy consumption (either an increase or decrease) the magnitude of which could be dependent upon the future level of output.   However, there is little information on how the balance of output has changed in the past, how it may change under different scenarios of the future, nor how this could influence the SEC.   It is therefore inappropriate to introduce any arbitrary difference in specific energy consumption for different scenarios of the future.

303.   The differences in energy requirements of each sector of the pottery industry may be reduced with the introduction of fast and once firing in the tableware and tile sectors, and direct firing in the sanitaryware sector.   By taking account of the specific processes and sectors to which energy conservation technologies apply, an estimate has been made of the possible changes in specific energy consumption for each sector of the industry.   This is shown in Table 55, where it appears that the greatest percentage savings up to the year 2000 may be made in the earthenware sector, where annual savings of ~ 3.9% may be achieved, while in the electrical ceramics sector the lowest percentage savings of only 2.1% per year appear possible.   However, over the

total period between 1980 and 2020 the earthenware sector shows annual savings of ~ 2.9% while electrical ceramics shows ~ 1.5% per year.

304. The pottery industry is becoming increasingly committed to the use of clean, predominantly gaseous fuels. Steam raising plant for space heating and drying has been replaced by direct gas fired heaters, and many of the kilns now being built are designed to be gas fired. With the price of petroleum products rising in the future it is likely that fuel oil and LPG will not be used by 2010, and the balance between natural (or coal-derived) gas and electricity consumption will be determined by price relativities.

305. If electrically fired kilns could be operated on fast firing cycles during the off-peak tariff period, significant cost savings could be achieved. Gas firing is also inherently less efficient than electric firing. There is no information available on the possible future balance of fuel use by the industry, but Table 56 shows the possible specific energy requirement for each fuel type if all drying and space heating is gas-fired and electricity use in kilns replaces half of the current LPG and fuel oil consumption.

CONCLUSIONS

306. Energy consumption per unit of output from the pottery industry has fallen by about 28% over the past decade (an average decrease of 3.7% per year or 8.0% per year between 1971 and 1974 followed by 1.4% per year between 1974 and 1980). Further energy savings may be achieved by implementing energy conservation technologies which are currently in the research, development or demonstration stage and over the next two decades these savings could amount to a further 49% of the 1980 energy consumption per unit of output (an average decrease of 3.3% per year), with an additional 12% saving by the year 2020 (1.3% per year).

307. However, any change in the balance of production from each sector of the industry could enhance or reduce this saving, as the specific energy consumption per tonne of product differs by a factor of 22 for the two extremes of bone china and unglazed floor tiles.

ENERGY USE AND ENERGY EFFICIENCY IN UK MANUFACTURING INDUSTRY

UP TO THE YEAR 2000

2.5 THE GLASS INDUSTRY

2.5 THE GLASS INDUSTRY

INTRODUCTION

308. In 1980 the glass industry consumed a total of 42.5×10^6 GJ of energy to produce 2.90×10^6 tonnes of glassware. This is approximately the same amount of energy as that consumed by the group of industries producing bricks, refractory goods and building materials, and represents 2.5% of the energy used by UK industry in 1980 (excluding Iron and Steel).

309. An Energy Audit [14] of this industry was published in June 1979, based on data relating to 1975, and a report prepared as part of the Industrial Energy Thrift Scheme [15], based on data for 1977, was published in December 1980.

310. In the Department of Energy's Digest of UK Energy Statistics the glass industry is grouped with china and earthenware, and the group as a whole was estimated to have used 66.9×10^6 GJ of energy in 1980. This is some 18% higher than the total of 56.6×10^6 GJ estimated in this study.

311. The data on energy consumption and output from the glass industry which has been used in this report, was supplied by the British Glass Industry Research Association (BGIRA). In order to avoid repetition of detail which is available in the Energy Audit of this industry, only brief summaries will be given of the production methods etc. employed by the glass industry.

THE STRUCTURE OF THE GLASS INDUSTRY

312. The British glass industry is the sixth largest in the world, with a fairly steady output of around 3 million tonnes per year, over the past decade, although this fell to 2.6×10^6 tonnes in 1981 (Table 57). In 1980 the turnover value was approximately £960 million.

313. The largest proportion of production is in the form of containers (69%) and flat glass (17%), with the remainder being made up of glass fibre, domestic and scientific glassware. Table 58 shows that there has been only a small fluctuation in the balance of output over the period 1975-1980.

314. Glassware can be found in many other industries; for example, as food and beverage containers, components in the electronics industry, in buildings and vehicles, and as fibreglass for textiles and construction.

315. Manufacturing processes range from the hand-forming of articles by an individual craftsman, making ornamental items, to the highly automated processes used in the production of containers and flat glass.

316. The industry may also be divided into glass melters, who manufacture glass from its raw materials, and glass manipulators, who manufacture artifacts using glass as their raw material. The latter make items such as double glazed unit and glass components for vehicles.

GLASS MANUFACTURING PROCESSES

317. Most of the raw materials used in glass manufacture are indigenous, in much of the glass produced these materials are sand (~ 60%), limestone (~ 18%)

Table 57. Output from the UK glass industry 1970-1981

Year	1970	1971	1972	1973	1974	1975	1976	1977	1978	1979	1980	1981
Output 10^6 tonnes	2.513	2.646	2.748	3.064	2.913	2.756	2.848	2.974	3.078	3.065	2.898	2.634

Source: BGIRA

Table 58. Balance of output from the glass industry x 10^6 tonnes

Year	1975	1977	1980
Flat Glass	0.539 (19.6%)	0.488 (16.4%)	0.494 (17.0%)
Containers	1.825 (66.2%)	2.091 (70.3%)	1.996 (68.9%)
Other (including domestic and fibre)	0.392 (14.2%)	0.396 (13.3%)	0.408 (14.1%)
TOTAL	2.756	2.974	2.898

and soda ash ($\sim 20\%$). However, in flat glass production the limestone content is reduced to around 4% and the difference made up of dolomite at 14%. In crystalware the raw materials are somewhat different, with sand ($\sim 50\%$), red lead ($\sim 30\%$), potash ($\sim 16\%$), and saltpetre ($\sim 2\%$).

318. Scrap and recycled glass is also used to improve the rate of melting of the raw materials, anything from 5-25% of the weight of glass can be made up of recycled material. Scrap produced 'in house' is preferred, as the composition is known and compatible with the melt. Bought-in scrap (e.g. from bottle banks) can cause problems as the composition is unknown and bottle caps can cause serious problems if accidentally included in the melt.

319. Once the raw materials are mixed in the appropriate quantities they are charged into the furnace and fired at a temperature of 1,450°-1,550°C to produce an homogeneous melt. Depending upon the scale of operations at the glass works, furnaces may be very large, melting up to 5,000 tonnes of glass per week, or small pots melting only about two tonnes per week.

320. The melted glass is then shaped; this induces high levels of stress into the article which could cause it to shatter on cooling, and are relieved by contolled heating and cooling of the article, called annealing. This is accomplished by slowly cooling the glass from around 600°C to ambient temperature by passing it through an annealing lehr, or tunnel.

321. The articles are then inspected and packed for despatch.

322. The methods of shaping glass obviously depend upon the nature of the article being made - containers, flat glass and glass fibre clearly require very different forming processes. These are discussed in some detail in the Energy Audit, and will simply be summarised here:

- Containers. Figure 13 shows the steps involved in container manufacture.

- Flat glass. Most is produced by the float process, in which the melted glass flows over a bath of molten tin. This ensures near perfect flatness and very uniform thickness.

- Glass fibre. Insulation glass fibre is produced as discontinuous filaments, which are combined with a binder material, formed into a mat and cured in an oven (Figure 14). Continuous fibres are produced for reinforcement and textiles, and are coated in an organic size before being wound onto spools for storage and/or transportation (Figure 15).

- Crystal glass. This is a skilled process, carried out by hand and involves rotating, blowing and moulding the glass into shape. Stems, handles and feet have to be welded onto the main article and some reheating may be required during forming. Cutting of the glass is carried out after annealing.

- Domestic and scientific glassware. This is formed in a machine which automatically carries out the necessary pressing and blowing operations. Glass tubing is made by a process shown in Figure 16. The diameter and wall thickness of the tubing are determined by the

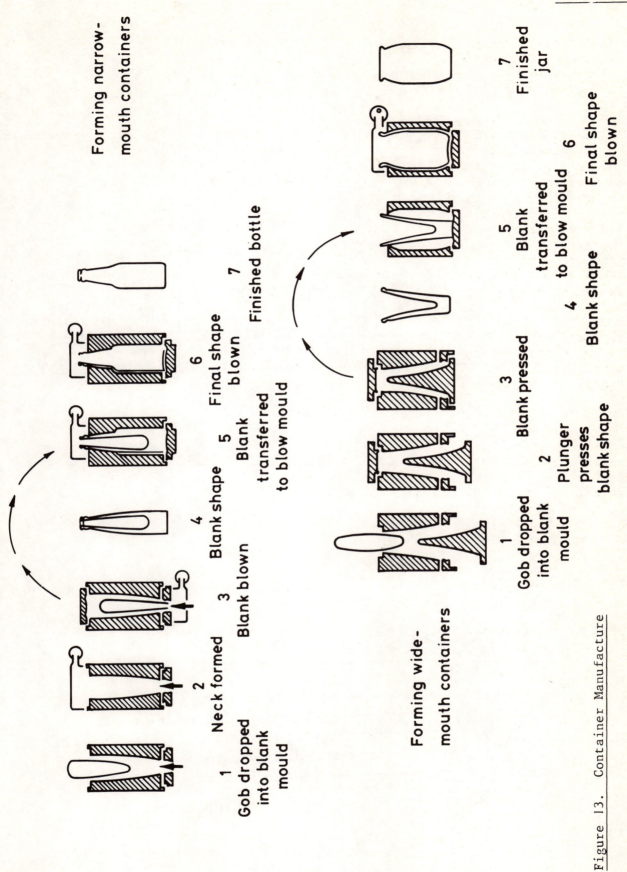

Forming narrow-mouth containers

1 Gob dropped into blank mould

2 Neck formed

3 Blank blown

4 Blank shape

5 Blank transferred to blow mould

6 Final shape blown

7 Finished bottle

Forming wide-mouth containers

1 Gob dropped into blank mould

2 Plunger presses blank shape

3 Blank pressed

4 Blank shape

5 Blank transferred to blow mould

6 Final shape blown

7 Finished jar

Figure 13. Container Manufacture

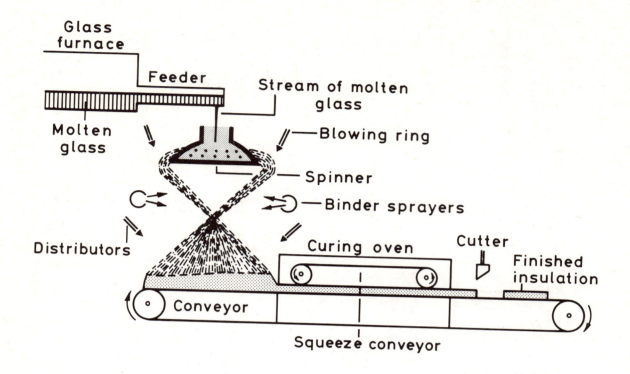

Figure 14.   Insulation Glass Fibre

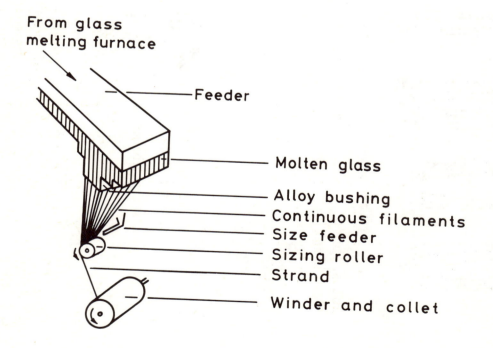

Figure 15.   Reinforcement Glass Fibre

amount of air blown through the centre of the mandrel, and the rate at which the tubing is drawn.

323. These processes and the articles produced are very different, and in some cases large amounts of glass are wasted in the process. The efficiency of a particular manufacturing process is regarded as the percentage of the glass initially melted which actually appears in the saleable product. For example, the production of glass containers is approximately 76-85% efficient [16], while the production of lead crystal is only 50% efficient.

ENERGY USE IN THE GLASS INDUSTRY

324. The British Glass Industry Research Association estimate that in 1980 42.5×10^6 GJ of energy were used by the industry, of which about 51% was oil, 39% gas and the remainder electricity. No solid fuel has been used since 1972. Table 59 shows the energy consumption of the glass industry over the period from 1970 to 1981, and Table 60 shows the normalised specific energy consumption of each fuel over that period (1975 = 100).

325. Between 1975 and 1980 the fall in electricity consumption per tonne of product appears to have followed the overall fall in specific energy consumption, while that of oil has moved more rapidly, showing an overall decrease of 52% over a period of 10 years. Gas consumption per tonne of output has increased slightly since 1975, but its strong move into the industry occurred in the late 1960s and early 1970s. Specific gas consumption increased by 156% between 1970 and 1980. In 1981 marked changes occurred, with gas consumption per tonne increasing by 15%, oil decreasing by 12% and electricity increasing by 8%, with an overall increase in specific energy consumption of 0.6%.

326. Table 61 shows the absolute values of specific energy consumption within the industry over the period 1970-1981, with the year-on-year percentage changes in consumption. The average change over the period amounts to a compound decrease of 2.8% per year, giving an overall reduction of 26.8% between 1970 and 1981.

327. As observed in paragraph 313 above, there has only been a small fluctuation in the balance of output from the glass industry, over the past decade. This suggests that any decrease in specific energy consumption cannot simply be due to a change in the balance of output, towards less energy intensive products. It can more properly be attributed to modernisation of plant and improvements in management and control, which have occurred within the industry.

328. The specific energy consumption varies widely across the product range. Figures given in the Energy Audit suggest that it can range from about 8.5 GJ/tonne of glass melted (a minimum of 10.5 GJ/tonne final product) for containers, to around 86 GJ/tonne (a maximum of 173 GJ tonne final product) for lead crystal. The values for the different product types for 1975 [14] are given in Table 62, with the associated production efficiencies where available. The values of energy consumption shown are in GJ per tonne of glass melted, not per tonne of finished product.

329. However, it should be noted that, for the Energy Audit, only a limited sample of firms within the industry were studied; six container factories, two

Figure 16. Danner Tube-drawing Process

After cutting, the lengths of tubing are warehoused ready for despatch or further processing

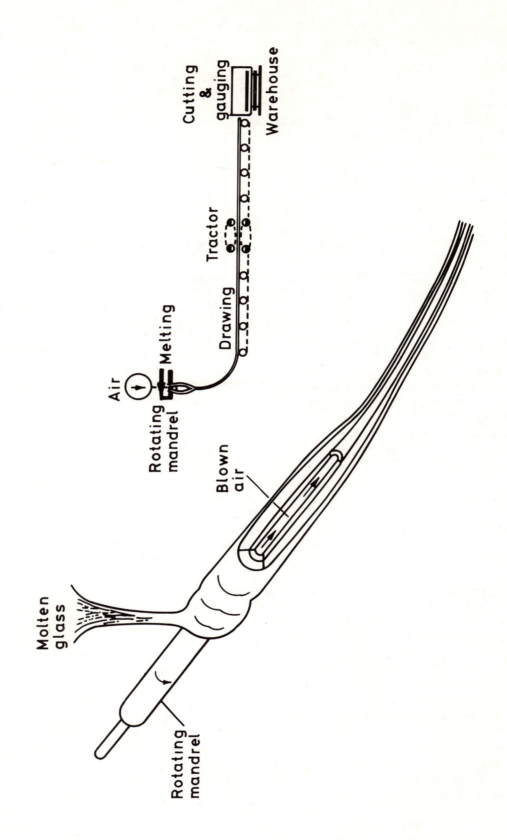

Table 59. Energy consumption in the glass industry 1970–1981 10^6 GJ/year

Year	Output 10^6 tonnes	Electricity [1]	Oil [2]	Gas [3]	Solid [4]	Total
1970	2.513	4.00	39.30	5.59	1.75	50.64
1971	2.646	4.10	42.40	7.28	1.42	55.20
1972	2.748	4.20	39.20	9.60	0.69	53.69
1973	3.064	4.25	37.44	11.18	0	52.87
1974	2.913	4.25	33.88	15.19	0	53.32
1975	2.756	4.20	25.84	15.09	0	45.13
1976	2.848	4.41	26.36	16.67	0	47.44
1977	2.974	4.55	24.56	19.83	0	48.94
1978	3.078	4.35	24.29	18.15	0	46.79
1979	3.065	4.46	23.60	19.94	0	48.00
1980	2.898	3.95	21.87	16.67	0	42.49
1981	2.634	3.88	17.55	17.41	0	38.84

Source: BGIRA from original sources as follows:

(1)   BGIRA estimate
(2)   Institute of Petroleum
(3)   Annual report of the Gas Council
(4)   DEn Digest of Energy Statistics

Table 60. Index of specific energy consumption 1970–1981 (1975 = 100)

Year	Total	Electricity	Oil	Gas	Solid (1972 = 100)
1970	123	104	167	41	277
1971	127	102	171	50	214
1972	119	100	152	64	100
1973	105	91	130	67	0
1974	112	96	124	95	0
1975	100	100	100	100	0
1976	102	102	99	107	0
1977	100	100	88	122	0
1978	93	93	84	108	0
1979	96	95	82	119	0
1980	89	89	80	105	0
1981	90	97	71	121	0

Table 61. Specific energy consumption in glass manufacture 1970–1981

Year	1970	1971	1972	1973	1974	1975	1976	1977	1978	1979	1980	1981
Consumption GJ/tonne	20.15	20.86	19.54	17.26	18.30	16.38	16.66	16.46	15.20	15.66	14.66	14.75
% Change		+3.5	−6.3	−11.7	+6.0	−10.5	+1.7	−1.2	−7.7	+3.0	−6.4	+0.6

each of lead crystal and glass fibre, one each of scientific and domestic glass, and the only factory producing float glass. Although visits were made to a total of 54 glass factories under the Industrial Energy Thrift Scheme, the report produced concentrates on energy use within each factory rather than energy use per unit of output. Consumption is expressed as TJ/year, GJ/m^2 floor area/year, TJ/employee/year and TJ/£1M gross sales with no measure of tonnage output from each factory, or what proportion of the industry it represents. Flat glass production was not represented at all in the sample.

330. However, using the figures presented in the IETS report for energy use per £ million gross sales, information available in the Business Monitor for total gross sales from the industry in the year studied (1977), and information from the BGIRA on total production for that year, an estimate of specific energy consumption in the production of glass containers has been made. This amounted to 13.3 GJ/tonne of final product, which corresponds fairly well with the figures in the Audit of 10.5-17.8 GJ/tonne final product. There was insufficient information to calculate corresponding figures for the other products.

331. For the industry as a whole, over 80% of the energy consumed is used for melting and annealing glass in furnaces and lehrs. The proportion can vary widely for the different types of product, as shown in Table 62, but as containers and flat glass dominate production the overall figure is well above 80%. The figure for insulating fibres is only 35%, but this still amounts to some 14.6 GJ/tonne, and large amounts of energy are used in reducing the glass filament to fibres (~ 17 GJ/tonne).

332. Energy use in the glass industry as a whole can be divided approximately as follows:-

Mixing	~ 0.3%
Melting and annealing	~ 88%
Forming	2-3%
Services	7-10%
Administration	~ 0.5%
Transport	~ 0.3%

For the individual product types this balance will obviously vary.

333. There is insufficient published information available on the glass industry for a more detailed description of energy use in the different sectors of the industry.

ENERGY CONSERVATION POTENTIAL

334. In the general approach to this study, an estimate has been made of the potential for energy conservation if all firms within a sector could operate at 'best practice'. This 'best practice' has been defined as the lowest level of energy consumption per tonne of product, using existing technologies. However, such an estimate requires detailed knowledge about the existing range in specific energy consumption (consumption per tonne of product) within the industry, and any factors which may influence a move towards 'best practice'. For the glass industry such an estimate would be required for each sector of

Table 62. Specific energy consumption per tonne of glass melted in various sectors of the glass industry (1975 [14])

	Energy Use GJ/tonne	Production Efficiency	Melting and Annealing
Containers	8.5 - 14.2	76 - 85%	79 - 90%
Flat glass	11.9		~ 88%
Insulating fibre	41.7		~ 35%
Reinforcement fibre	30.5		~ 78%
Lead crystal	45.8 - 86.4	50%	~ 90%
Domestic & scientific	21.8 - 23.3	50 - 60%	56 - 68%

the industry; containers, flat glass manufacture and manipulation, glass fibres, domestic and scientific ware.

335. However, such detailed information is not available. The IETS report does show ranges of energy consumption in 1977 per employee, per unit floor area, and per unit value of gross sales, but not energy consumption per tonne of product. There is insufficient information available to use the figures in the IETS report to derive some meaningful estimate of 'best practice' energy conservation potential in 1980.

336. From a consideration of other industries similar to the glass industry, in which over 80% of the energy consumed is used in kilns and furnaces, and from discussion with individuals familiar with the glass industry it is reasonable to assume that savings of around 5% could be made simply by improved housekeeping, monitoring and control of the high energy consuming operations. Further savings would require technological changes, and are described in the following section.

Future Energy Savings in the Glass Industry

337. The glass industry is already energy conscious, as energy costs amount to around 30% of their total purchases [12], and a reduction of 12% in specific energy consumption has been achieved since the Energy Audit was carried out in 1976. Almost 90% of the energy is used in melting and annealing, and efforts are already made to conserve energy in these processes.

338. The British Glass Industry Research Association were consulted on the potential for energy conservation in the industry, as the authors of the Energy Audit report [14] published in 1979. However, they were reluctant to put any figures to the energy savings which could be made by implementing the various, available conservation technologies, as they felt there was insufficient, reliable information on which to base such estimates.

339. In the absence of any other source of such information it has not been possible to assess the potential for energy conservation in this industry, with the same degree of detail as the assessment for other industries. However, the main areas in which savings could be made, and where possible estimates of the magnitude and timescale of these savings, will be discussed below.

Electric Melting

340. The BGIRA feel that the main focus for energy savings in the future will be a move to all electric melting. The melting process currently consumes around 80% of the energy used within the industry, and gas or oil fired furnaces are only around 35% efficient. However, electric furnaces can be 75-80% efficient and therefore offer significant energy savings to the consumer. Although it is perceived energy savings to the consumer, and hence cost savings which drive the implementation of such technologies, when expressed in terms of primary fuel equivalents electric melting offers no energy savings, and in fact would require a greater consumption of primary fuels.

341. In this process electrodes are immersed in the melt and energy is released within the glass itself. There are several advantages over fossil fuel firing, not least the reduction of heat losses. A cleaner melt can be produced which reduces the proportion of reject glass, and there is also a reduction in material losses during melting.

342. The BGIRA are of the opinion that the industry has long wanted to move to electric melting, but is prevented by the high price of electricity relative to other fuels. This technology will not be implemented in more than a few, isolated instances until the price of electricity is around a factor of two above that of gas. In the Department of Energy's projections [1] of future fuel prices this will only occur under the high fossil fuel price assumptions, around the year 2010, when projected electricity prices to industry are about twice the price of gas supplies. These prices are shown in Table 63.

343. Electricity may be used for melting, in conjunction with fossil fuels in mixed melters. This also offers energy savings to the consumer, but implementation is again hindered by the price of electricity.

344. Even if the price relativities were favourable for the wider use of electric melting, the useful lifetime of existing furnaces (7-12 years) results in relatively infrequent opportunities for replacing fossil-fuel fired equipment by electric furnaces. The process would therefore not be introduced on a broad basis until around 1990-1995.

345. The energy savings which could be available through the introduction of electric melting amount to around 32% of total energy consumption on a heat supplied basis. But the timescale over which this saving could be made depends on future fuel price relativities.

```
                    Electric Melting:
Saving 32% (heat supplied)
Timescale dependent on future fuel price relativities.
```

346. The BGIRA feel that the only hope for significant energy savings in the glass industry, if electric melting remains unattractive, would be a revolutionary change in the manufacturing processes, on the scale of the change to float glass production in the flat glass sector. They feel that no major breakthrough is possible within the confines of the traditional reverberatory-hearth glass melting furnace.

347. Figure 17 shows a typical large glass melting tank which could be used to melt around 2,000 tonnes of glass per week. The various conservation technologies which could be applied to such equipment will be described below.

(a) Insulation

348. Ideally all parts of the furnace structure should be insulated, but this could lead to overheating of the refractory materials, and premature failure. Insulation should be considered at the time of furnace design and cannot easily be added afterwards. BGIRA estimate that energy savings of around 3-4%

Table 63. UK energy price projections 1980-2010 [1] p/therm

Low Fossil Fuel Price Assumption	1980	1990			2000			2010		
Scenario	Actual	A	B	C	A	B	C	A	B	C
Gas	17.6	26.0	20.8	16.7	37.8	33.4	23.2	60.4	50.9	34.1
Electricity	69.4	90.3	75.7	65.3	111.1	107.3	82.6	136.1	120.5	118.0
Electricity/Gas	3.94	3.47	3.64	3.91	2.94	3.21	3.56	2.25	2.37	3.46
Higher Fossil Fuel Price Assumptions	1980	1990			2000			2010		
Scenario	Actual	X	Y	Z	X	Y	Z	X	Y	Z
Gas	17.6	33.8	30.6	25.7	59.8	48.2	44.0	84.3	68.6	63.5
Electricity	69.4	113.1	88.8	79.1	151.3	120.1	104.8	174.9	137.1	127.0
Electricity/Gas	3.94	3.35	2.90	3.08	2.53	2.49	2.38	2.07	2.00	2.00

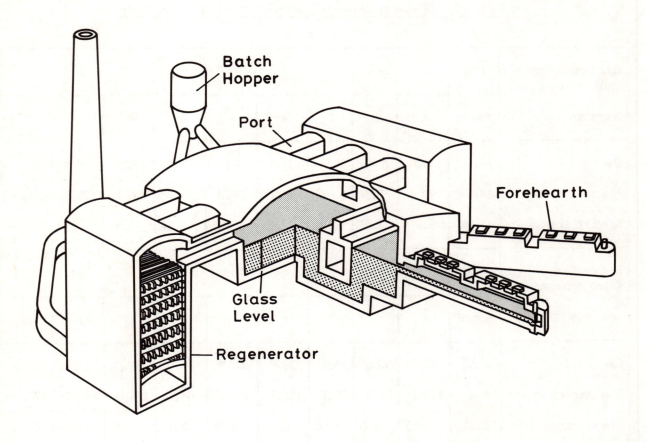

Figure 17.   Typical Large Glass Melting Tank

may have been made since 1976 through improved insulation, which has been added in a fairly random way.

349. Although in the Energy Audit estimated savings of around 6% are quoted, the BGIRA feel that as savings become progressively more difficult to achieve, it is unlikely that a further significant reduction in energy consumption could be made.

(b)   Burner Design

350. Recuperative burners are already used in the glass industry, and will continue to offer some improvements in energy efficiency. The energy Audit estimated that savings of around 2% could be achieved by improved burner design, but BGIRA now feel that there is insufficient information available to estimate the remaining potential for savings in 1980.

(c)   Regenerators

351. Glass melting furnaces are designed to recover heat from the exhaust gases before venting from the kiln. Firing is by means of burners placed on each side of the furnace, and takes place from each side alternately. The inlet air is taken from the firing side and the exhaust gases leave on the opposite side. After leaving the furnace the hot gases pass through a regenerator (Figure 17) where they heat the refractory blocks, or checker work, through which cold inlet air will be drawn when firing is reversed and is from that side.

352. Improvements in regenerator design are constantly sought, and continued improvement is expected. The Energy Audit suggests that 5% of the energy used in melting could be saved, and assuming some of this has already been achieved a further saving of 3% might be achieved by around 1990.

> Regenerator Design:
> Saving 3% of Energy Used in Melting by 1990

353. Further improvements to regenerators could involve the installation of secondary regenerators. With a single regenerator exhaust gas temperatures are around 500°C when they are vented, with a secondary regenerator this could be reduced to around 250°C, with the associated increase in energy savings. The Audit estimated that a further 5% of the energy used in melting could be saved in this way. However, this would be achieved over a much longer timescale than simple modifications to primary regenerators.

> Secondary Regenerators:
> Saving 5% of Energy Used in Melting by 2000

354. A new type of compact regenerator unit is currently under consideration. This system is designed to reduce heat losses through the large regenerator structures, and achieve around 90% heat recovery from exhaust gases. However, this development is still in the early stages and barriers to its further

development may arise - for example problems of fouling and attack by the alkaline atmosphere in the regenerator.

355. If there are no major breaks through advancing the introduction of electric melting into the glass industry, it is anticipated that such compact regenerators could be applied economically on all melting tanks saving around 20% of melting energy (~ 16% of total energy consumption) by about 1995, with the first commercial installation around 1985. However, as this is still somewhat speculative, and still in the very early stages of demonstration, this technology has not been included in projecting the potential for energy conservation in the glass industry.

356. If electric melting was introduced on a broad scale within the industry improvements to regenerators would no longer be relevant, as hot exhaust gases would not be produced.

(d)   Combustion Control

357. A small improvement in combustion control has been achieved since 1976, but further improvements using microprocessor controls should be available over the next 10 years. This is achieved by monitoring the oxygen content of the furnace exhaust gases, which indicates the amount of excess air used in combustion. This can be automatically optimised offering a saving in fuel and an improvement in furnace operation stability. Savings of at least 1% of the energy use in melting could easily be achieved using such techniques, over a period of around 10 years.

```
┌─────────────────────────────────────────────────┐
│              Combustion Control:                  │
│   Saving 1% of Energy Use in Melting by 1990     │
└─────────────────────────────────────────────────┘
```

(e)   Waste Heat Boilers

358. These have not proved attractive to the industry for a variety of reasons, although one or two have been operated successfully. The heat available from the regenerators does not occur as a steady supply, and can contain sodium sulphate carried over from the glass tank, which causes blocking of the boiler. The heat can be used for space heating and drying of raw materials, or to generate electricity. However, the availability of waste heat and the opportunities for use on-site are frequently unmatched, making the potential for energy saving using this method particularly uncertain.

359. If the industry moves strongly towards electric melting the amounts of recoverable waste heat will be limited.

360. Although an estimate of 5% savings was given in the Energy Audit, this was a simple assumption, based on little operational experience. The BGIRA and those involved with the Energy Efficiency Demonstration Projects Scheme feel that waste heat boilers are unlikely to be used in anything more than isolated instances over the next 10 years and their use beyond that period will be determined largely by any move towards electric melting.

(f)   Higher Cullet Ratios

361.   Glass manufacturers are always ready to use higher proportions of
recycled glass, provided it is available in sufficient quantity and is of good
quality.   However, energy savings made in melting are partly offset by
additional energy use in crushing, cleaning, sorting and transportation of the
cullet.   For these reasons BGIRA feel that it is not possible to quantify the
savings which could be achieved.   Although in the Energy Audit it was
estimated that for every 1% increase in the cullet ratio 0.2% of melting
energy could be saved.

362.   It is probable that environmental considerations will be most
influential in bringing about changes in the amounts of recycled glass used in
the future.

(g)   Forehearths

363.   The forehearth is a channel in which molten glass is transported to the
forming machine.   It is very important to maintain the correct temperature and
viscosity distribution in the forehearth, if the glass is to be formed into
acceptable ware.

364.   Most forehearths are gas-fired, and although electric heating has been
shown to be more efficient (savings of up to 85% have been recorded using
demonstration plant) technical problems exist which restrict the introduction
of electric heating on a broad scale.   However, modern gas-fired forehearths
could offer savings of about 30% of that currently used in forehearths
compared with the old-style equipment (3% of total energy consumption), and
are more likely to be widely adopted by the industry.   This saving could be
achieved by around 1990.

```
Improved Forehearths:
Saving 3% by 1990
```

(h)   Annealing Lehrs

365.   Some improvements have already been made in the efficiency of energy use
in lehrs, through hot gas recirculation, insulation and internal belt returns.
Electrically heated lehrs are more controllable than gas heated equipment and
may offer further energy savings.   Over the next 10 years savings of around 5%
in annealing energy may be achieved.

```
Annealing Lehrs:
Saving 5% of Annealing Energy by 1990
```

366.   Other methods of energy conservation mentioned in the Energy Audit, such
as oxygen enrichment of combustion air and submerged combustion for melting,
are not now considered to offer feasible opportunities for energy saving.

367.   Energy savings which relate to exhaust gas heat recovery and fossil fuel
combustion, would clearly not be relevant if the industry moved significantly

Figure 18. Future energy savings in the UK glass industry, expressed as a percentage of total consumption

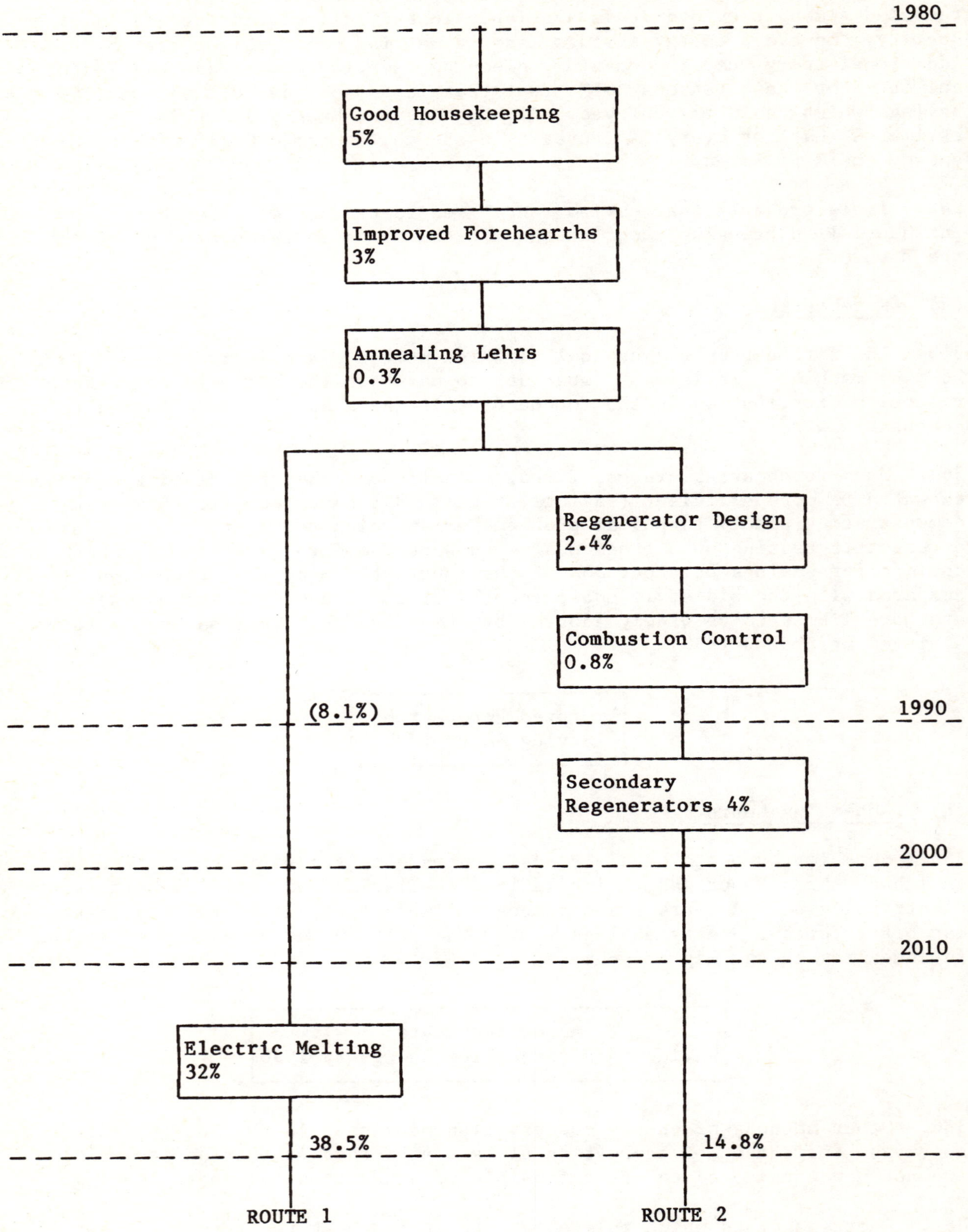

ROUTE 1

ROUTE 2

towards electric melting. The greater efficiency of electric melting is primarily achieved because energy is not lost in combustion gases, and combustion control and recuperative burners would no longer be relevant.

368. There could, therefore, be two alternative routes to be taken, which are illustrated in Figure 18. One involves a switch to all electric melting, with its associated large energy savings to the consumer (energy measured on a heat supplied basis), and the reduction of energy consumption in forehearths, annealing lehrs and general good housekeeping. The other route brings lower energy savings through improvements in fossil fuel combustion and exhaust heat recovery, again with improvements to forehearths, lehrs and good housekeeping.

369. Account must be taken of the fact that, once savings have been made through good housekeeping, the total amount of energy used is reduced by 5%. Savings made using electric forehearths are thus reduced and the resulting energy consumption is (0.97 x 0.95)% of the initial, 1980 consumption. Similarly, savings through improvements to annealing lehrs are reduced but are not affected by savings made by using electric forehearths, as the processes are independent.

370. However, the effects of improved regenerator design, combustion control and secondary regenerators are cumulative, all reducing the amount of energy used in melting furnaces. Therefore, the total saving achieved via route two is calculated as follows:

Savings in energy used for melting
= [1 - (0.976 x 0.992 x 0.96)] ÷ 100 = 7.05%

Savings in melting, annealing and forehearths
= (7.05 + 0.3 + 3.0) ÷ 100 = 10.35%

Total savings via route two
= 1 - [0.95 (1 - 0.1035)] ÷ 100

= 14.83%.

371. The savings achieved via route one are calculated in a similar way, taking account of the fact that energy use in the melting furnace, forehearth and annealing lehr are independent, but that savings through good housekeeping may influence all three. The calculation is, therefore, as follows:

Savings in melting, annealing and forehearths
= (32 + 0.3 + 3.0)%
= 35.3%

Total savings via route one
= [1 - (0.95 x 0.647)] ÷ 100

= 38.5%.

372. However, as route one is strongly dependent on future price relativities between electricity and gas, it may not affect energy consumption until the end of the period covered by this study. Route two will therefore be taken as the most probable out-turn for the future of the glass industry, with savings of around 14.8% of total energy consumption by 2000.

PROJECTED LEVELS OF ENERGY CONSUMPTION

373. As described above, it has been assumed that the introduction of electric melting, on a widespread basis, will not take place until the end of the period being studied, and this depends upon the relative prices of fuels in the future.

374. The technologies which will affect energy consumption between 1980 and 2010 are electric forehearths, improved annealing lehrs, combustion control, improved and secondary regenerators and general good housekeeping measures. These will all be taken up by the industry over a period of time, and a simple diffusion model has been used to determine the energy savings which may be achieved. Figure 19 shows the diffusion curve used to model the uptake of technologies over a period of 20 years, and Table 64 shows the contributions to energy savings which may be expected from each of the technologies considered.

375. The future mean energy consumption per tonne of product is given for both scenarios in Table 65. This falls from 14.66 GJ/tonne in 1980 to 12.49 GJ/tonne in 2000. This represents a decrease of 14.8% overall, or a compound decrease of 0.8% per year over the period. The projected rate of energy conservation in the glass industry would appear to be much lower than the rate achieved over the past decade, which amounted to 3.1% per year from 1970 to 1980. However, the paucity of information on the potential for energy conservation in several areas of the glass industry leads to this apparently conservative estimate of future savings.

376. The future energy requirements of the glass industry will be affected by any change in the balance of output from each sector of the industry. For example, a move away from container and flat glass production currently consuming $\sim$ 8-14 GJ/tonne, and making up 86% of output, towards the production of crystal, domestic glassware and fibres, consuming $\sim$ 22-86 GJ/tonne, could result in an overall increase in the specific energy consumption of the industry.

377. However, there is no information available on how the balance of output from the glass industry may change in the future. The container manufacturers are suffering pressure from substitutes such as plastic bottles and cans, while the manufacturers producing domestic and ornamental glassware are suffering pressure from imports. The export market for flat glass has been reduced as manufacturers have established production facilities abroad, so most areas of the glass industry are suffering pressure on their market from some quarter.

378. It is possible that any overall increase in the output of the glass industry, in future years, will be predominantly from certain sectors of the industry. This could lead to a change in the mean specific energy consumption (either an increase or decrease) the magnitude of which could be dependent upon the future level of output. However, there is no information to suggest how the balance of output may change under different scenarios of the future, nor how this could influence the SEC. It is therefore considered inappropriate to introduce any arbitrary difference in specific energy consumption for different scenarios of the future.

Figure 19. Diffusion curve for technologies taking 20 years to reach
maximum penetration of the glass industry

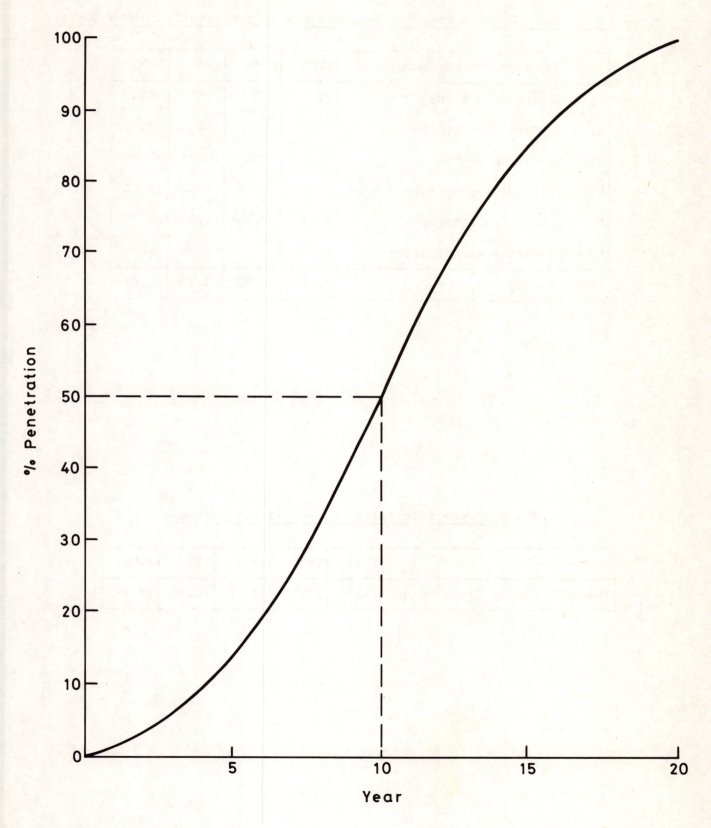

Table 64. Reduction in energy requirements due to conservation technologies

Conservation Measure	1985	1990	1995	2000
1) Good housekeeping	2.5%	5%	5%	5%
2) Improved forehearths	1.5%	3%	3%	3%
3) Annealing lehrs	0.15%	0.3%	0.3%	0.3%
4) Improved regenerator design	1.2%	2.4%	2.4%	2.4%
5) Combustion control	0.4%	0.8%	0.8%	0.8%
6) Secondary regenerators	0.6%	2%	3.4%	4%
Total saving	6.2%	13.0%	14.3%	14.8%

Table 65. Specific energy consumption. GJ/tonne

	1980	1985	1990	1995	2000
Specific Energy Consumption	14.66	13.75	12.76	12.56	12.49

CONCLUSIONS

379. When considered from the point of view of energy use, the future of the UK glass industry is somewhat uncertain. If the price of electricity relative to fossil fuels, particularly gas, falls in the future then electric melting could be introduced on a broad scale, offering large energy savings to the consumer. On the other hand, an anticipated, though as yet unspecified, radical change in glass melting technology could also offer large energy savings.

380. However, on the basis of the information which is currently available, it appears that specific energy consumption could be reduced from 14.66 GJ/tonne in 1980 to 12.49 GJ/tonne by 2000. This amounts to an overall reduction of 14.8% or a compound reduction of 0.8% per year. This is much lower than the average 3.1% per year saving achieved between 1970 and 1980, and may well be increased by the introduction of new technologies such as electric melting, if fuel price relativities are attractive.

381. There is growing interest within the industry in coal firing of furnaces, but as yet there is insufficient information available to anticipate what effect this will have on the future energy consumption of the industry. For this reason, and as the future move to electric melting is difficult to anticipate, no estimate has been made at this stage, of the future fuel requirements of the glass industry.

ENERGY USE AND ENERGY EFFICIENCY IN UK MANUFACTURING INDUSTRY

UP TO THE YEAR 2000

2.6   MISCELLANEOUS BUILDING MATERIALS

2.6 MISCELLANEOUS BUILDING MATERIALS

382. Miscellaneous building materials covers a diverse range of industries which, taken together, represent a substantial proportion of the total energy use in the ceramic materials sector. It includes groups 243-246 of the 1980 Standard Industrial Classification, viz:

- building products of concrete, cement or plaster (2436, 2437),

- asbestos goods (2440),

- working of stone and other non-metallic minerals not elsewhere specified (2450),

- abrasive products (2460).

383. There is very little information on energy use in this subsector. The only available sources are:

- the 1979 BSO Purchases Inquiry [10],

- two reports in the IETS series which cover parts of the subsector [16, 17].

There is some difficulty, however, in relating these two sets of data to each other, because the latter is based on the 1968 Standard Industrial Classification. Furthermore, the energy consumption suggested by the BSO data is difficult to reconcile with an analysis of data in the Digest of Energy Statistics [20]. The analysis of this subsector is therefore incomplete and uncertain.

ENERGY USE

384. Table 66 shows the energy use by the miscellaneous building materials subsector as reported in the 1979 BSO Purchases Inquiry [10]. The total energy consumption is 54.2 PJ. However, it should be noted that there is some uncertainty surrounding this value. An analysis of the data in the Digest for Energy Statistics under the category of 'bricks, fireclay and other building material' would suggest a value of 20 PJ.

385. The subsector is a heavy user of oil, which accounts for 63% of the total energy supplied. Electricity accounts for 12.9% of total energy use for the subsector as a whole. Within the subsector, however, the abrasive products industry appears to be a heavy user of electricity (41%).

Building Products and Abrasives (SIC 2436, 2437, 2460)

386. An IETS report on the miscellaneous building materials and abrasives industries [16], prepared by the British Ceramic Research Association, is based on visits to 51 sites, covering 19% of the sites known to be engaged in these industries. However, the coverage relates to MLH 469 of the 1968 SIC, and includes five sites engaged in the production of coated bitumen products. This activity is now separated under 2450 of the new classification, but the IETS data is not provided in a form which would allow a separation of data relating to it (see paragraph 393).

387. Energy use by the sites visited is shown in Table 67. Since the report does not include the total number of employees or output of the sites visited, it is not possible to gross up the energy use to give an accurate estimate of the total energy use. However, a rough estimate of the total can be obtained by grossing up by the number of sites visited, which gives a value of 17 PJ for the two industries combined. This should be compared with the value of 50.3 PJ for the three activities in Table 66 which correspond to MLH 469. The discrepancy is quite large, and suggests a predominance of small sites in the IETS sample - 27 of the sites visited employed fewer than 100 people.

388. The fuel mix for the sites visited by B Ceram RA corresponds quite well with the BSO Purchases Inquiry, except for two values:

- coal consumption in the building products industry is 21% in the B Ceram RA sample compared with 10.4% in the Purchases Inquiry data;

- electricity consumption in the abrasive products industry is 15% in the B Ceram RA sample compared with the surprisingly high value of 41% in the Purchases Inquiry data.

389. Seventy per cent of the sites visited by B Ceram RA had a drying or curing process, and 20% had a firing or calcining process. In a few of the sites visited, process heat accounted for up to 90% of total energy use. However, in the sector as a whole the main use of energy is for space heating; although the amount is not given in reference 2, a rough estimate of 50% has been obtained from industry sources, with 22% process heat ~ 300°C and 15% process heat ~ 300°C.

Asbestos Goods (SIC 2440)

390. The asbestos goods industry has been covered by the IETS report on Miscellaneous Textiles carried out by the Shirley Institute [17]. The data reported by the Shirley Institute is based on 12 site visits, covering 50% of the total number of sites in the industry, and is shown in Table 67. Comparison with the Fuel Purchases Inquiry data [10] shows close correspondence in both the grossed-up total energy use and the fuel mix.

391. The major processes in the asbestos industry are textile processes - blending, carding, spinning, weaving, etc. Although some process heat is required for drying and curing the main use of oil and gas is for space heating. Electricity is used for motive power for process machinery, and for general lighting.

Working of Stone and Other Non-Metallic Minerals (SIC 2450)

392. There is no source of data on energy use in SIC 2450 other than that contained in the Purchases Inquiry [10]. However, the data in Table 66 suggests that it is a larger user of energy (28 PJ) than the brick industry (19 PJ).

393. The Standard Industrial Classification (1980) [18] describes SIC 2450 as:

- ground, crushed and coated stone; ground and prepared chalk and clay;

- articles of worked stone and marble, cutting of grindstones and millstones from natural stone, carving of ornamental and funerary stonework;

- articles of worked slate and fabricated articles of slate;

- mineral sound and heat insulating materials (not glass fibre), mica and other mineral electrical insulating products; articles of peat and other minerals not elsewhere specified.

It is difficult to infer from this description of the group how such a significant quantity of energy is used.

ENERGY CONSERVATION POTENTIAL

394.   Both of the IETS reports [16, 17] found extensive opportunities for improved energy utilisation on the sites visited. The Shirley Institute found that potential savings using existing technology amount to 18% of energy used in the asbestos industry [16]. Table 68 summarises their findings for a typical site.

395.   The report on the miscellaneous buildings industries [17] lists similar opportunities in most of the sites visited. For all 51 sites visited, a total of 340 specific energy-saving recommendations were made, with lighting and space heating being the most frequently cited areas. Energy intensive processes, such as firing or calcining, could also save significant amounts of energy.

396.   Many of the energy saving opportunities could be undertaken with little or no capital expenditure. B Ceram RA found that a frequent reason for not taking action was unawareness of the opportunity; their report estimates that 'good housekeeping' management measures could save 3% of the total energy use. The report does not quantify the energy savings which could be achieved by additional expenditure on, e.g. insulation or waste heat recovery. However, B Ceram RA suggest that a saving of 15% overall is realistically achievable b 2000 [18].

PROJECTED LEVELS OF ENERGY CONSUMPTION

397.   Table 69 shows the future specific energy consumption (SEC) which is projected for the miscellaneous buildings material sectors. In view of the lack of information, these projections are necessarily speculative.

398.   The specific energy consumption is assumed to fall by 7% by 1990 and 15% by 2000. These rates of improvement are consistent with the available information on the industry and with findings for similar activities in the engineering sector, but are not based on a detailed economic analysis of the energy saving opportunities.

399.   It has been assumed that there will be a progressive shift from oil to coal, since there is no reason to believe that the use of coal in these industries would prove difficult. Gas is assumed to maintain its existing market share, and electricity to increase its share marginally, due to a proportionate increase in the demand for motive power as process and space heating demands are reduced.

Table 66. Energy Use in Miscellaneous Building Materials in
1979 Fuel Purchases Inquiry [12]

	Gross* Output (£M)	Net* Output (£M)	Total Energy (PJ)	SEC (GJ/£k gross)	Fuel Mix (%)				
					Coal	Oil	LPG	Gas	Elec
Building products in concrete, lime, etc.	1,738	815	20.2	11.6	10.4	70.7	0.6	1.6	10.4
Abrasive products	219	113	2.2	10.0	4.5	22.7	–	31.8	40.9
Working of stone, etc.	1,228	701	27.9	22.7	7.5	71.3	2.5	6.1	12.5
Asbestos goods	295	159	3.9	13.3	–	43.6	0.1	43.6	12.8
Total	3,478	1,788	54.2	15.6	7.9	62.7	5.9	10.5	12.9

* 1980 money.

Table 67. Energy Use by Sites Visited in IETS Reports [18, 19]

	No of Sites Visited	% of Total Sites	Energy Use (PJ)		Fuel Mix (%)				
			Sites Visited	Total	Coal	Oil	LPG	Gas	Elec
Building products in concrete, lime, etc.	46	)	2.84	)	21.0	65.4	1.4	2.8	9.4
		) 19		) ca) 17					
Abrasive products	5	)	0.46	)	–	62.9	–	22.0	15.1
Asbestos goods	12	50	1.55	3.37	–	39.0	–	46.0	15.0

Table 68.   Potential Annual Energy Saving at a Typical Asbestos Site
[from reference 17]

Nature of Energy Saving Opportunity	% of Total Energy Use
Good housekeeping	6.0
Improved space heating control	4.8
Boiler replacement, improved controls and steam services	2.6
Improved use of process heat	1.9
Waste heat recovery	1.7
Improved building insulation	0.8
Lighting improvements	0.2
Total	18.0

Table 69.   Projections of Future Specific Energy Consumption

	1980	1990	2000
SEC (GJ/£k)	15.6	14.5	13.3
Fuel Mix (%): Solid Fuel	8.0	20	40
Oil + LPG	68.6	51	25
Natural Gas	10.6	15	20
Electricity	12.8	14	15

CONCLUSIONS

400. This subsector is particularly diverse in the range of industries it covers, and very little information is available on energy use by these industries. However, it has been estimated that the specific energy requirements of the group may fall by $\sim 7\%$ between 1980 and 1990 (0.7% per year) from 15.6 GJ/£k to 14.5 GJ/£k and by a further 8% between 1990 and 2000 (0.9% per year) to 13.3 GJ/£k. However, in view of the lack of information available for this subsector, such projections are somewhat speculative.

REFERENCES

1.	Department of Energy, Proof of Evidence for the Sizewell 'B' Public Inquiry. October 1982.

2.	Energy Audit Series No 11. The Cement Industry. Issued jointly by the Departments of Energy and Industry. July 1980.

3.	Annual Abstract of Statistics. Central Statistics Office, 1981. HMSO 1981.

4.	Digest of UK Energy Statistics 1982. Department of Energy. HMSO 1983.

5.	Fuel Usage in the Manufacture of Building Bricks (3). Technical Note No 337 - C N Walley and H W H West. British Ceramic Research Association Ltd. August 1982.

6.	Energy Audit Series No 2. Building Brick Industry. Issued jointly by the Departments of Energy and Industry.

7.	Industrial Energy Thrift Scheme No 3 - Energy Use in the Bricks, Fireclay and Refractory Goods Industry. Department of Industry. 1978.

8.	Energy Audit Series No 4. Bulk Refractories Industry. Issued jointly by the Departments of Energy and Industry. November 1978.

9.	Energy Efficiency Demonstration Scheme: A Review. Energy Efficiency Series 1. Energy Efficiency Office. HMSO 1984.

10.	1979 Census of Production and Purchases Inquiry, Business Monitor PA 1002.1. Business Statistics Office. HMSO 1983.

11.	British Ceramic Manufacturers' Federation Fuel and Power Census 1980.

12.	Industrial Energy Thrift Scheme. Report No 2. Energy Use in the Pottery Industry. Department of Industry. November 1978.

13.	Energy Audit Series No 7. The Pottery Industry. Issued jointly by the Departments of Energy and Industry. August 1979.

14.	Energy Audit Series No 5. Glass Industry. Issued jointly by the Departments of Energy and Industry. June 1979.

15.	Industrial Energy Thrift Scheme No 23. Energy Use in the Glass Industry, Department of Industry. December 1980.

16.	IETS Report No 5. Energy Use in Abrasives and Miscellaneous Building Materials Industry. Department of Industry, 1979.

17.	IETS Report No 44. Energy Use in the Miscellaneous Textiles Industry. Department of Trade and Industry. 1984.

18.	Standard Industrial Classification, Revised 1980. HMSO, 1979.

ENERGY USE AND ENERGY EFFICIENCY IN UK MANUFACTURING INDUSTRY

UP TO THE YEAR 2000

SECTOR 3. THE CHEMICAL INDUSTRY

J W Coleman

	Paragraph No
Introduction	1
Development of the Chemical Industry	5
Structure of the Chemical Industry	10
Feedstocks in the Chemical Industry	18
Output and Energy Consumption in the Chemical Industry	37
Prospects for the Chemical Industry	59
Use of Energy In Chemical Plant	72
Energy Targetting and Monitoring	86
Energy Conservation Potential	91
Key Energy Saving Measures in the Chemical Industry	118
Steam Usage and Fuel Switching	131
Combined Heat and Power	143
Projected Fuel Demands	160
Commentary and Conclusions	165
References	
Tables	
Figures	
Annex 1: Inorganics	
Annex 2: Organics	

SECTOR 3. THE CHEMICAL INDUSTRY

Introduction

1.   In 1980 the Chemical Industry used 620×10^6 GJ of energy (including feedstocks – gas and petroleum). It is thus the largest energy consuming sector in industry, accounting for some 28% of all industrial energy use, and 10% of the total national energy requirement. If feedstocks are excluded it is still the second largest industrial energy user, with a consumption of 320×10^6 GJ in 1980. The major fuels used are natural gas (250×10^6 GJ including feedstocks or 14% of the total for the UK) and petroleum (300×10^6 GJ including feedstocks or 11% of the total for the UK).

2.   Net output in 1980 was £5,000 M, which gives an energy intensity of 124 MJ/£ output (including feedstocks) or 64 MJ/£ output (excluding feedstocks). It thus ranks as one of the most energy intensive industries overall, although there are large variations within it. Gross output in 1980 was £16,500 M. Total expenditure on energy was around £2,000 M, split almost equally between fuel and feedstocks. The cost of fuel and feedstocks thus represent some 40% of the industry's net output, or 12% of gross output.

3.   In terms of sales volume the UK Chemical Industry is the fifth largest in the world, outside the Eastern bloc, after the USA, Japan, West Germany and France. There is a strong international element in its operations. Exports account for some 45% of total sales, while imports provide about 35% of home demand. The proportion of both exports and imports has risen steadily over the last 10 years but the balance has remained roughly constant. However the balance varies widely within the industry from one subsector to another. There is strong international competition both for the markets and for feedstocks and raw materials.

4.   It is a highly complex industry – really a grouping of several different industries, producing an extremely diverse range of products – and it has some unique characteristics which are not typical of the rest of UK industry. Its present pattern of energy use and the scope for further change have been heavily influenced by the recent development of the industry. It is therefore helpful to consider briefly the structure and growth of the industry before examining the use of energy in more detail.

Development of the Chemical Industry

5.   The chemical industry is very much a phenomenon of the 20th century. Until the early years of the century its output was largely confined to a few basic inorganic chemicals, such as sulphuric acid and caustic soda, used in other industries. There was virtually no production of organic chemicals, except for a few specific materials such as dyestuffs, based on coal. During the first 30 years of this century the industry grew slowly, and then more rapidly up to the 1950s, still based largely on coal, via the tars and other by-products from the manufacture of coal gas and coke. Whole new business sectors have subsequently been developed in plastics, synthetic fibres, solvents, detergents, pharmaceuticals and many more.

6.   From the mid 1950s onwards petroleum products became available in large quantity at low cost, the production of coal gas declined, and with it the availability of by-products for chemical manufacture. The chemical industry

turned to oil as its main feedstock (and increasingly as a fuel) which provided a cheaper route to many of the products derived from coal, and opened up whole new ranges of products. This stimulated the dramatic expansion of the chemical industry in the 1960s and 1970s when its growth rate was over 6% p.a. - the highest of all manufacturing industry. From about 1970 onwards natural gas started to be used in the industry, further displacing coal as a fuel and replacing oil as feedstock in two specific processes. The effect on overall development of the industry, however, was slight, compared to that of the transition from coal to oil. Purchased electricity has remained substantially constant, somewhat below 20% of the total energy supplied, but the industry generates a large amount of its own electricity, and has become the largest private generator of electricity in the UK.

7.   The growth of the industry's output and the consumption of energy and feedstocks are shown in Figures 1 and 2. Feedstocks are considered in more detail in paragraphs 18-36 and the industry's overall use of energy in paragraphs 37-58. In this report the term 'gas' refers to natural gas (i.e. predominantly methane in the UK) unless otherwise qualified. The term 'petroleum' embraces all products from refining of crude oil, such as naphtha, gas oil, fuel oil, etc., and the products known as natural gas liquids (NGLs) or liquefied petroleum gases (LPGs) which include ethane, propane and butane.

8.   During the growth period in the 1960s and 1970s the major chemical companies were building more and bigger plants in the expectation of continued growth and the desire to achieve economies of scale. Most of the present plant in the industry dates from this period. A similar pattern of growth occurred throughout the developed world. Even the oil price rise of 1973 caused only a temporary fall in production (in 1975), and by 1976 output was increasing again, albeit at a slower rate, reaching its highest level in 1979. The second oil crisis in 1979 resulted in further rises in oil prices and, more importantly, the general recession in the economies of the western world. Thus the industry was hit simultaneously by a further increase in its fuel and feedstock costs and a collapse of many of its markets at home and abroad. Output dropped sharply in 1980 and 1981, and it is only now showing signs of recovery.

9.   Undoubtedly the market will grow again, and there will be further developments in processes, products and plant, but the industry is now considered to be relatively mature in terms of market development, and is unlikely ever to show the same growth rate as in the 1960s and early 1970s. The UK chemical industry is now faced with an excess of plant capacity for many major chemicals, in common with the whole of the European chemical industry, and there have been many plant closures and agreements between companies to rationalise manufacture. The prospects for the main sectors of the industry are discussed in paragraphs 59-71.

Structure of the Chemical Industry

10.   The Standard Industrial Classification (SIC) divides the chemical industry into a number of subsectors according to type of product, process, etc. This study has been based on the revised SIC of 1980 (ref. 4), with some of the minor subsectors grouped together for simplicity. The full list of subsectors and their grouping for this study are shown in Table 1 with

data on energy consumption and output, to give an indication of their relative importance. The derivation of this data is explained in paragraph 53. In practice there is less distinction between some of these subsectors than might appear from the official designation of the industry's activities, as discussed in paragraph 13.

11.   Two of these subsectors - Basic Organic and Inorganic Chemicals - are largely concerned with the conversion of feedstocks into base chemicals for the other subsectors. Some of these products are supplied to other industries, but the bulk of the inorganic and organic chemicals are used elsewhere within the chemical industry itself. The main raw material for organic chemicals is petroleum-derived naphtha, yielding base chemicals such as ethylene, propylene and benzene, which are used for the production of plastics, pharmaceuticals, synthetic detergents, etc. Typical raw materials for inorganic chemicals are sulphur, salt, minerals and metals from which are derived acids, alkalis and many other compounds used in the production of fertilisers, metal treatment solutions, pigments, explosives, photographic materials, etc.

12.   Many of these base chemicals are produced on a high tonnage scale (e.g. sulphuric acid 3 m tonne/year, ethylene over 1 m tonne/year). The plants are correspondingly large, they are highly complex, and represent a major capital investment. Production in the bulk organic and inorganic sectors is therefore dominated by a few major companies such as ICI, BP and Shell. Most of these companies are also active in some of the downstream sectors such as plastics, paints or pesticides, and there are many smaller companies as well, producing specialised products. The uses of energy and the opportunities for improving efficiency of its use therefore differ widely throughout the industry.

13.   There are a number of multi-plant sites in the industry, making products which are classified into different industry subsectors, often sharing common energy supplies. This makes it difficult to produce reliable statistics for energy use for specific products, and the distinction between subsectors is not clear cut. Indeed, in some areas the division between the SIC subsectors is quite artificial. For example, some parts of the inorganics and fertiliser sectors operate as an integrated whole.

14.   This is particularly the case in the petrochemical field, which embraces a large part of the organic chemicals subsector, synthetic resins, plastics and rubber, paints and other products. Several major companies in the petrochemical field are subsidiaries of oil companies or closely associated with them. The two operations (oil refining and chemical manufacture) are, to a large extent, separate, but closely linked, with considerable transfer of feedstocks, fuels and by-products between a refinery and a chemical works. In basic petrochemicals manufacture there is considerable use of by-products as fuel, which do not appear in official statistics for energy consumption.

15.   The industry has depended for its growth - indeed its existence, on technological development. It is therefore by nature a technically sophisticated industry, and most companies have their own R&D effort or technical department. Their prime concern is normally technical development of the products and process, but this usually involves energy considerations as well, so the industry has probably paid more attention to energy use than

most others.   The trends in energy consumption relative to output are
considered in paragraphs 37-58.

16.   Many of the larger companies are part of major multi-nationals with
extensive R&D and other back-up services and hence are capable of developing
techniques and transferring the technology across a number of plants, or
indeed from one country to another.   They are also capable of shifting
production from one country to another if conditions, energy prices or other
factors are more favourable.   This contributes an additional element of
uncertainty to the forecasting of production levels and hence UK energy
requirements.   In the past, when most plants were operated at or near full
capacity, a substantial shift in production required the planning and
building of new plant, with lead times of several years.   Now that there is
much spare capacity, as discussed in paragraph 61, shifts in production will
be easier and quicker to effect.

17.   Because of this strong emphasis on research and technical development
within individual companies, and the confidentiality about developments,
there is less scope for an industry research association than in other
industries, such as food or paper.   There are research associations or
research laboratories for particular sectors of the industry, mainly those in
which a number of smaller companies operate, such as rubber and plastics, or
paints.   The main source of information for the industry as a whole, apart
from official (Government) statistics, is the Chemical Industries Association
(CIA) which represents around 230 companies, who between them are responsible
for some three quarters of the industry's turnover, investment and energy.

Feedstocks in the Chemical Industry

18.   The feedstocks used in the chemical industry - natural gas and
petroleum products - are materials which would otherwise be available as
fuels, and they are used in very large quantities (around 300 PJ in 1980, or
about half the total energy consumption of the industry).   Their influence on
the total energy use in the industry is even greater because large amounts of
energy are used in their conversion.   As indicated in paragraph 6 above, the
pattern of feedstock use in the chemical industry has changed over the past
30 years even more dramatically than for general energy consumption.   It is
likely to change again in the future, over a similar timescale, though the
extent of the change is unclear at present.

19.   It is therefore important to understand how feedstocks are used in the
chemical industry, in relation to fuel usage, and how they are accounted for
in published statistics.   This is highly relevant to an understanding of past
trends in energy consumption.   Two main feedstocks are currently used by the
chemical industry: natural gas and naphtha.   Natural gas is used almost
exclusively for the production of ammonia and methanol - basic chemicals in
the inorganic and organic sectors respectively.

20.   The main petroleum product used as feedstock is naphtha or 'light
distillate feedstock' (LDF) which is a similar product to that used for the
production of motor spirit.   Because of pressure on the light end of the
barrel, heavier products are now being used as chemical feedstock (mainly gas
oil) but naphtha is still dominant (almost 90% of petroleum feedstock in
1980).   Lighter products (LPG and NGLs - liquefied petroleum gases and
natural gas liquids) are also used where they are economically available,

and are likely to become more important in future. Petroleum feedstocks are used for the production of basic organic chemicals such as ethylene, propylene, benzene and butadiene.

21.   Gas and petroleum are treated differently in most official statistics on energy consumption, and some of the information is regarded as commercially confidential, which makes it difficult to obtain reliable figures for fuel and feedstock separately in some cases.

22.   The Digest of UK Energy Statistics shows the petroleum used by the chemical industry for energy uses, and lists separately that used as feedstock. However, the corresponding figures for energy uses of gas <u>include</u> that supplied as feedstock, and no separate figures are given for the feedstock gas. This gas is supplied by BGC under long term contracts with two major chemical companies (ICI and UKF). Neither the companies nor BGC disclose the quantities supplied for these purposes, so independent estimates have to be made of this part of the total gas consumption.

23.   There is, in fact, a fundamental difference between the use of gas and petroleum as feedstocks, which makes it more difficult to distinguish the feedstock and fuel uses of gas. The petroleum products used as feedstocks are chemically and physically distinct from those normally used as fuels, and are bought, transported and stored as separate products, but gas used as feedstock is identical to that used as fuel. Furthermore, an ammonia or methanol plant is so highly integrated that any allocation of energy requirement between fuel and feedstock is somewhat arbitrary, and depends on the operating conditions. It is therefore customary in the industry to consider the total gas consumption per tonne of product. The gas can be apportioned between fuel and feedstock on a theoretical basis, for comparative purposes, but this does not represent a realistic separation on the plant. Throughout this report, where estimated figures are given for 'feedstock' gas, this refers to the total gas supplied to ammonia or methanol plants, unless otherwise stated.

24.   It is arguable whether feedstock uses of gas and petroleum should be included with or excluded from total consumption figures when considering the industry's energy in relation to output. Most studies have excluded feedstocks from this kind of analysis, on the grounds that feedstock can be regarded simply as one of the raw material inputs to the industry, along with minerals, metals, salt, etc. It is thus conceptually distinct from the fuels used for providing heat and power to the chemical plant, and if we are examining the industry's performance with regard to energy efficiency it is primarily this use of energy which is being considered. The quantity used as feedstock is now so large that it could distort the picture of energy use in the industry when looking at trends over a long period during which the feedstock component has grown substantially.

25.   However, this quantity of gas and petroleum certainly cannot be ignored. It is part of the total energy requirement of the industry, and there are opportunities for improving the efficiency of its conversion, just as for the efficiency of fuel usage. There is also, to some extent, the possibility of substitution of one feedstock for another, which could chang the pattern of energy use substantially in the long term. This is discussed further in paragraph 29. Of course, if a major plant closes (e.g. in the

face of competition from abroad) the effect on both fuel and feedstock would be even more significant.

26.   Therefore, wherever possible in this report the feedstock uses are identified and the energy consumption is shown with and without feedstocks. However, it should be appreciated that the total energy consumption figures are the more reliable, since the figures net of feedstocks depend on estimates of the gas feedstock usage which, as noted above (para. 22), are still somewhat uncertain. Furthermore, a significant part of the feedstock is eventually burnt as fuel, when it is rejected from processes as a by-product (see paragraph 42).

27.   We have already seen how the industry's pattern of feedstock use has changed from coal to light petroleum products, now supplemented with natural gas. This has been in response to the changing availability and cost (i.e. feedstock price and cost of processing) of the various alternatives. Other European countries have followed a similar pattern of development, though not necessarily with the same shift to gas, and are now all largely dependent on naphtha for chemical feedstocks.

28.   Several countries outside Europe have a quite different pattern of feedstock usage. For example, in the USA, ethane (a major constituent of their natural gas) is much more widely used as a chemical feedstock. A few plants have been built recently using coal as a feedstock, but it is doubtful whether coal will be economic except in special circumstances. In South Africa, because of their unique supply situation, coal provides the basis of their chemical industry, but in plants which are very different from those used in the UK in the 1950s.

29.   In Europe, and particularly in the UK, there is likely to be a broadening of the range of feedstocks in response to further changes in their availability and price. The main stimulus for this change will be any increase in oil price and a further shift in the demand towards the lighter end of the barrel. This may be exacerbated by a move towards lead-free petrol in Europe if this is met by an increase in reforming to produce an acceptable motor spirit (since reforming requires more feedstock per tonne of product). Some of the by-products from other processes which are currently used for aromatics production are also likely to be in demand as blending components to increase octane number. Alternative feedstocks would then be required for aromatics.

30.   The first significant change in feedstock use in the UK is already taking place. This is the use of natural gas liquids (NGLs - ethane, propane and butane) as a feedstock for olefin plants. Shell/Esso are building an ethylene plant at Mossmorran, Scotland, to run on ethane feedstock (due to come on stream in 1985). The ICI/BP ethylene plant at Wilton (Olefins 6) was built, at extra cost, to accept NGLs as well as naphtha, and BP has converted its cracker at Grangemouth for the same reason. These changes will not significantly affect the energy requirements for processing, but will clearly reduce the demand for naphtha feedstock if NGLs are used on a large scale.

31.   The industry could also turn to heavier petroleum products - gas oil, residual fuel oil or raw crude oil - but the capital cost of plant, the process energy requirements and hence the overall costs would be considerably higher than for naphtha or NGLs. Gas oil is already used to a small extent

(see paragraph 20) but the heavy petroleum products seem unlikely to be used as feedstock on a significant scale in the UK within the timescale of this study, though they may begin to be adopted elsewhere in Europe.

32.   There has been much discussion about the prospect of a return to coal as the basis for the chemical industry. This is clearly a technical possibility now, as demonstrated in South Africa for many years, and more recently in the USA, but it is not yet economic in Europe. Coal based chemical plants may be introduced first in other European countries, such as Germany or Italy, which do not have indigenous oil and gas, but have large coal reserves (particularly of low grade coal), or are prepared to import coal if necessary. The timing is very uncertain, depending largely on the movement of oil prices relative to coal. Indeed, it is questionable whether coal will be a viable feedstock for chemicals production in Europe while there is competition from Middle Eastern plants with the advantage of cheap and abundant gas and petroleum feedstocks. If it does, it may become economic on a shorter timescale than SNG production from coal, but it is most unlikely to make a significant contribution in the UK before 2000. The effect would be to displace oil and possibly gas as feedstock and fuel, but the overall energy requirement for a given output would be greater.

33.   More generally, there are other possibilities for using any source of carbon and hydrogen with a source of heat, which could be nuclear, to produce a whole range of chemicals, but these are far from technically or economically viable at present.

34.   A more immediate possibility is the production of methanol in areas with abundant and cheap natural gas, such as the Middle East, and its import into Europe as a component of motor spirit or as a versatile chemical feedstock. This is already occurring on a limited scale, and seems likely to have a significant impact before 2000, but the extent is very unclear at present.

35.   The commercial implications for the chemical industry are discussed in more detail in paragraphs 59-71 below, and the potential effect on energy requirements is considered in the Annex.

36.   There have been several recent reviews of the past trends and future prospects for feedstocks in the European chemical industry which discuss the various options and the energy implications [refs. 14, 15, 16].

Output and Energy Consumption in the Chemical Industry

37.   The chemical industry produces some 30,000 chemical compounds, ranging from high value speciality products with an annual production of a few tons to low value basic chemicals produced on the million ton scale. The key products in terms of energy consumption are identified in the Annex.

38.   Output data for the chemical industry are very variable in quality and accessibility. Such information is often treated as commercially confidential, particularly for products which are produced by only one or two companies. Annual output figures are listed in Business Monitor and similar publications for some of the major products and a number of relatively minor ones. Chemfacts UK [ref. 17] is a useful source of data on the major producers, output, plant sizes and location, but information on some of the

most energy intensive materials such as ammonia and methanol is very limited.

39.   In this report output and Specific Energy Consumption (SEC) are therefore normally expressed in money terms for the industry as a whole and for the subsectors (gross value of output and MJ/£ for SEC). Physical measures of output are used where available for single products or classes of products, and have been taken into consideration in assessing changes in the subsectors. Where trends in output over a number of years are considered (i.e. historic data or projections) these are expressed in money terms at constant prices.

40.   The chemical industry more than doubled its output from 1960 to 1981. This is much higher than manufacturing industry as a whole, which over the same period only increased its output by 20%, having peaked at a 50% increase in 1973.

41.   Energy use in the chemical industry over the period 1960-81 is plotted in Figure 1, which shows clearly the displacement of coal by petroleum and gas as fuels and the massive growth of petroleum and gas as feedstocks. Feedstocks now account for around half the total energy consumption of the industry as compared to a quarter in 1960. Energy use excluding feedstocks (i.e. the industry's requirement for heat and power) has remained substantially constant over the period, despite the doubling of output.

42.   It should be noted that the published energy statistics understate the true energy input to the chemical industry by neglecting the energy content of two important sources. One is the energy of raw materials such as sulphur and phosphorus which are burnt as part of the production process (e.g. for sulphuric acid and phosphoric acid respectively). The other is the energy content of by-products used as fuels (from material which enters the site mainly as petroleum feedstock). These amounts are not formally reported and are very uncertain, but they can be substantial. Estimates show that the total non-conventional fuels used for steam raising is about one fifth of the total recorded energy consumption for the chemicals industry [ref. 8] and that in the organic chemicals sector non-conventional fuels may supply as much as half the total energy requirement [IETS Report No 35 - ref. 6]. In the absence of any reliable annual data on these energy sources they have been omitted from the time series data, on the assumption that they represent a roughly constant proportion of the total, and will not significantly affect the index of consumption.

43.   The changes in energy consumption in relation to output and the resulting change in SEC are shown in Table 3 and in Figure 2 in the form of indices based on 1960 = 100. It can be seen that the SEC (excluding feedstocks) has fallen to less than half its 1960 level. This also compares favourably with manufacturing industry as a whole, in which the SEC has fallen by only 25% in the same period.

44.   Other European chemical industries have shown a similar pattern of gradually falling SEC, and since 1970 the UK's performance has been around the average for EEC countries. Comparison with the USA is reported to show that the SEC of the UK chemical industry is about 60% of that in the US [ref. 16].

45.   There are several reasons for this improvement in energy efficiency in the UK. The change from coal to oil and gas has undoubtedly contributed, but a similar changeover has taken place in other industries without such a marked improvement in efficiency.

46.   Changes in product mix may have had an effect on the SEC of the industry overall, but it is difficult to discern in which direction. The 1960s and 70s saw a rapid expansion in both petrochemicals and pharmaceuticals, which are among the most energy intensive and energy non-intensive sectors of the industry, respectively.

47.   The main cause of the reduction in SEC has been the introduction of new and more energy efficient plant and processes. During the expansion of the industry in the 1960s and early 70s much new plant was built, normally incorporating the latest available technology to make the best use of all inputs - energy, raw materials and labour. Energy efficiency was thus an incidental gain from the developments in the industry at the time, and the SEC fell rapidly up to 1974.

48.   Since the recession in the chemical industry there has been much less new plant building, and hence fewer opportunities for further major improvements in energy efficiency. Energy saving measures have been applied to existing plant and the SEC for the industry overall may have been reduced by closure of older and generally less energy efficient plant, but this is offset to some extent if plants are run at less than full capacity, which normally increases the SEC. The net result has been a continued fall in SEC for the industry, but at a slower rate than during the industry's vigorous growth period.

49.   Despite the apparently impressive record of the industry in improving its energy efficiency, there is still considerable scope for further reduction in SEC, as noted in paragraphs 91-117. However, the opportunities for making such improvements will now be more limited due to the uncertain prospects for many products, the slow down in plant building and the general shortage of finance for 'optional' projects. The past trends in SEC and projections for the future are discussed in more detail in paragraphs 104-117.

50.   Most of the chemical industry has now been covered by reports under the Industrial Energy Thrift Scheme [ref. 53] and by one Energy Audit [ref. 7]. Unfortunately some of the information is rather limited, partly for reasons of confidentiality, especially in the bulk organic, inorganic, fertiliser and plastics sectors, where there are few major producers. In these sectors also the analysis of energy use and the assessment of conservation potential have generally been in terms of complete chemical sites, rather than individual processes or products, reflecting the difficulty of analysing energy flows in large, multi-product, fully integrated chemical sites. The data is also now somewhat dated, since most of the visits were carried out during the period 1976-77. Nevertheless, the Thrift Scheme represents the major published source of information on energy use and conservation potential in the industry, and it has been used extensively in producing this report.

51.   The base year for the present study is 1980. The only recent breakdown of energy use at subsector level is contained in the BSO Purchases Enquiry for 1979 [ref. 2] which was based on the 1980 Standard Industrial

Classification (SIC). It has already been noted (paras. 13, 14) that the energy consumption figures suffer from the lack of real distinction between some of the subsectors. The figures are further complicated by the usual problem of accounting for feedstocks, particularly gas, even though there is provision in the Enquiry for reporting gas used other than as fuel. When compared with the figures reported in the Digest of UK Energy Statistics [ref. 1] it is apparent that the Purchases Enquiry has accounted for less than half of the gas used by the chemical industry for fuel and non-fuel purposes. It has been assumed that the gas reported in the Purchases Enquiry has failed to capture any significant part of the feedstock uses of gas.

52. The petroleum products reported as fuels in the Purchases Enquiry amount to some 25% more than that recorded in the Digest, which may be due to inclusion of some element of feedstock usage. The Purchases Enquiry does include a reported figure for purchases of petroleum feedstocks (by value only). However, the calculated tonnage which this represents is more than 50% greater than the Digest figure for feedstocks. Part of this discrepancy may be due to differences in accounting for fuels, feedstocks and by-products passing between refineries and chemical works on neighbouring sites. Solid fuels and electricity are both within 10% of the Digest figures.

53. In order to derive a consistent set of figures for the base year, 1980, the total energy consumption figures from the Digest for that year have been used. These have been adjusted for the estimated feedstock uses of gas and then pro-rated across the 20 activities of the 1980 SIC in proportion to the consumption recorded in the 1979 Purchases Enquiry, adjusted for the difference in output levels for the two years. Gross output for 1980 has been taken from BSO statistics [ref. 3] sub-divided, where necessary, at the level of individual activities in proportion to the 1979 outputs. The resultant breakdown of output and energy use for 1980 is shown in Tables 1 and 2.

54. The relative proportions of gas and oil in the industry as a whole thus derived for 1980 then differ substantially from those shown by the Purchases Enquiry (44% gas, 37% oil in Table 2, compared with 30% gas, 50% oil in the Purchases Enquiry). To some extent this reflects a genuine change in the fuel mix between the two years - the Digest shows a continuing fall in oil consumption from 1977 and a growth in gas consumption - but the magnitude of the difference between the figures is probably partly due to the adjustments which have been made for feedstocks. The totals for the industry shown in Table 2 are in line with those published by the CIA [ref. 13].

55. It should also be noted that the pattern of output derived from the BSO data differs significantly in some areas from figures reported on the previous Standard Industrial Classification (1968) based on Minimum List Headings. The change in classification and the adjustments which have to be made for feedstocks etc. also invalidate any comparison with previous Purchase Enquiries (in 1974 and 1969).

56. The general problem within the industry of obtaining reliable and consistent data on energy use and output makes it extremely difficult to carry out studies at a disaggregated level over an extended period of time. Data on long term trends in energy consumption and output have therefore only been produced for the industry as a whole, as shown in Table 3. The uncertainties in the data on current use of energy for fuel and feedstocks are, of course, carried through to any estimates for future energy use and

this should be borne in mind when considering the projections later in this report.

57.   It can be seen from Table 1 that almost 60% of the energy use is concentrated in the two main subsectors covering basic organic chemicals and inorganic chemicals. The energy consumption is also dominated by a few large companies - for example about 20 of them account for at least three quarters of total energy use. Changes in this relatively restricted area therefore have a large effect on the industry as a whole.

58.   The main purpose of the present study is to examine the effects of technical changes on energy consumption, so major structural changes in the industry have not been considered in any detail. Nonetheless, it is apparent from the trends noted above (paragraphs 47, 48) that change in SEC is strongly influenced by the commercial activity in the industry (and probably by the expectations of the industry, as well as by its immediate performance), so some consideration of the prospects of the industry is necessary as a background to the assessment of technical change.

Prospects for the Chemical Industry

59.   For the purposes of the present study, two scenarios have been adopted for future output, designated high and low. These are based on the central projections of the Department of Energy's 1982 Energy Projections [ref. 5]. Both scenarios relate to a 1½% growth rate in GDP, but differ in the relative contributions which different industries make to this growth rate. In the low case the energy intensive manufacturing industries make a lower contribution to the economy than in the high case, with the balance taken up by the less energy intensive industrial and service sectors.

60.   The 1982 Energy Projections make explicit assumptions about the overall output of the chemical industry, which have been adopted for this study. In the high case output (on a constant price basis) doubles between 1980 and 2000 (average growth rate 3.6% pa) and in the low case output increases by 50% (average growth rate 2% pa). For the purposes of this study these overall industry figures have been interpreted for the main subsectors, consistent with the general industrial background of the two cases, from an assessment of the commercial prospects for the industry, as outlined below.

61.   Opinions within the industry differ about the detailed prospects, and perceptions vary from one sector to another, but the general view is that the chemical industry is at a crossroads in its commercial development. After 15-20 years of continual growth it has been hit by a sharp fall in demand, rising costs for fuel and feedstocks, and declining profitability. Many markets which have been growth areas because of substitution of other materials by chemical products such as plastics, synthetic fibres etc., have reached maturity and there is little scope for further substitution. Despite numerous plant closures there is still acute over-capacity throughout the western world. In these respects the UK chemical industry is in the same position as that in the rest of Europe. In addition the industry has complained that it has been suffering from exchange rate pressures and high energy costs relative to other European countries.

62.   A threat which has hardly yet begun to take effect is the development of chemical industries in countries which have abundant and cheap sources of

energy and feedstocks - particularly natural gas and associated liquids. There are many such projects in progress in the Middle East, Africa and Asia, often with government backing and the benefits of cheap capital or favourable investment criteria, and some with European, American or Japanese involvement. Few have yet come on stream. Some are in the planning phase and a proportion will no doubt fail to come to fruition, but many are under construction at present and will boost world capacity substantially over the next 5-10 years.

63.   These projects invariably concentrate on the production of bulk commodity products (basic inorganic chemicals, fertilisers and petrochemicals) which have a high energy input and low added value. Since many of these countries are favourably placed with regard to potential markets in the developing countries these plants will have strong advantages over those in Europe. Part of their output, probably over half, is destined for their own home markets, but eventually there will be large surpluses on the world markets. Initially these are likely to erode the potential export markets for European producers, but increasingly they will penetrate European home markets, particularly those from the Middle East. Major developments of this kind in Eastern Europe, also based on cheap natural gas, pose an additional threat, though it is not clear how much will be available for export.

64.   The prospects for these major sectors of the industry therefore look bleak, though the timescale and extent of the competition from these new plants are not at all clear. The major uncertainties are: when and how many of these projects will be completed; how well they are kept on stream; and how much of the production is available for export. There has been much discussion within the industry and in the technical press about future prospects, but there is great uncertainty and very little in the way of projected output figures - certainly not on the 20 year timescale required for this study. Most companies are having enough difficulty with projection over the next five years.

65.   NEDO has published a Long Term Assessment of the chemical industry, prepared by the Chemicals EDC [ref. 9]. This follows on from a previous review of the problems and opportunities facing the industry - 'Chemicals - Contraction or Growth?' published in 1981 [ref. 10]. These documents between them probably represent the most up to date and authoritative published views on the future development of the chemical industry.

66.   The Long Term Assessment (LTA) covers a period of about 10 years - i.e. up to the early 90s. It does not attempt to quantify future demand or output, but examines the prospects for the key sectors of the industry in qualitative terms against an assumed growth rate for the UK economy of somewhat below 1.5% p.a.

67.   For the purpose of the present study, the trends identified in the LTA have been used as the basis for different growth rates and output figures for each sub-sector, for the high and low cases such that the total output for the industry conforms to that shown in the DEn's Energy Projections. In general the trends have been continued for the 90s, in line with the assumptions used for the Energy Projections of a somewhat lower GDP growth rate in the 80s and a higher rate in the 90s. The resultant output assumptions are shown in Table 4. For most subsectors the implied growth

rates are rather higher than those now expected within the industry and reflected in the LTA.

68.   The LTA identifies four sectors (Paints, Soap and Detergents, Dyestuffs and Pigments, Agrochemicals) which are essentially static - i.e. expected to maintain roughly the same relative contribution to the economy as at present. These are the sectors which are closest to the 'consumer' (they are used either directly in the form of consumer products or via other industries making use of finished products such as paints or dyestuffs in manufacturing consumer products). Their fortunes on the home market can therefore be expected to be closely linked to the economy as a whole - i.e. to indicators such as personal disposable income and industrial output. Any additional growth would depend on exports. All these sectors are net exporters, but the present trend is to a declining, or at best stable, balance of trade. The markets are technically mature, with the possible exception of Agrochemicals.

69.   Two sectors are considered to have above average prospects - Pharmaceuticals and Specialised Organics. Pharmaceuticals is another strongly consumer-oriented market. The Specialised Organics sector is linked to others by providing intermediates (e.g. for pharmaceuticals) or special constituents such as additives, inhibitors, plasticisers, textile conditioners etc. These are technically sophisticated markets with a vigorous international trade dominated by the developed countries, both as consumers and producers. The UK has a favourable balance of trade in these products, based on a strong R&D effort. The LTA highlights a number of factors which may affect this position and several conditions which must be fulfilled if this country is to maintain its strength in these areas. If all goes well there is scope for continued growth in these sectors. The energy intensity of these products is generally low in relation to value (though for some low volume products it may be very high on a tonnage basis). Changes in output in these sectors will therefore have relatively little effect on total energy consumption. However, many of these products depend on energy intensive organic or inorganic chemicals as precursors or intermediates in their manufacture. If there is a serious decline in the output of these materials the specialised products would depend on imports for their manufacture, making these sectors more vulnerable to pricing and supply variations outside the UK.

70.   The LTA identifies three sectors which are suffering serious problems in the present economic climate, and face the prospect of decline. These are Inorganics, Fertilisers and Petrochemicals (which covers bulk organic chemicals, plastics and synthetic rubber). Petrochemicals was the great growth area of the 60s and early 70s, and contributed largely to the popular image of 'chemicals' as a high technology, high value-added industry which would continue to grow at a faster rate than the economy as a whole. This is no longer the case, and all three sectors face the prospect of a sharp decline, unless growth rates in the economy as a whole increase substantially. The reasons are basically those outlined at the start of this section - a mature market, rising costs, and increasing competition from abroad. It has been suggested within the industry that parts of the bulk petrochemicals sector could be liable to collapse in certain circumstances - e.g. large scale import of low cost bulk products from the Middle East, and protection of chemical industries on the continent without similar action in the UK.

71.   The growth (or contraction) rates are particularly difficult to predict at present because there are so many imponderables, but they are highly significant because these sectors are so energy intensive. A 10% change in the Organic chemicals sector changes the output in the chemicals industry as a whole by only 2%, but changes the total energy requirement by almost 6%. The present uncertainty about the prospects for these sectors is also affecting investment in energy conservation measures. These aspects are discussed in more detail in the appropriate sections later in this report.

Use of Energy In Chemical Plant

72.   The way in which energy is used in chemical plant differs widely from one process to another and is often highly complex, as indicated below. Data on the detailed breakdown of energy use are therefore extremely sparse, and previous investigators have commented on the difficulty of obtaining an analysis of energy use comparable to that available for other major industries.

73.   One such study was carried out by Grant [ref. 11] which provided a breakdown between heat, electrolysis and motive power. ETSU has also published a more detailed breakdown [ref. 12] based on information provided by NIFES (National Industrial Fuel Efficiency Service). Both these analyses depended largely on estimates of energy use deduced indirectly from a variety of sources.

74.   The Thrift Scheme visits, although primarily concerned with energy saving measures, gathered much detailed data on energy consumption at plant level across a large proportion of the chemical industry. The sites visited account for about half the energy used in the industry, though they only represent some 12% of the total number of sites in the industry and 33% of the employees, so they are biased towards the larger sites and bigger energy users. Within these limitations the Thrift Scheme represents a unique source of information on actual usage of energy on chemical plant. Little of this data has yet appeared in the published reports, which concentrated mainly on the overall pattern of energy consumption in each sector and the opportunities for basic energy conservation measures.

75.   In a follow-up study the data from the Thrift Scheme visits is being analysed to provide a breakdown of energy use in chemical plant (under headings such as distillation, drying, compression, etc.) which will help in identifying those areas with the greatest scope for technical improvements, demonstration projects etc. It is intended to publish this analysis in due course, but only preliminary results are available so far, based on the larger users (sites with annual energy consumption > 90 TJ). This is shown in Table 5. The smaller energy users will have a somewhat different pattern of energy use, and this is reflected in the published reports on the Thrift Scheme [ref. 6] which show some minor differences in the limited data on breakdown of energy use which they provide. Electrochemical energy is not specifically identified in Table 5, because the sites covered did not include any major electrochemical producers. Otherwise Table 5 has been taken as representative of the industry as a whole, modified by data from the published reports where appropriate. In the course of further studies the data may be refined and a modified pattern of use be produced.

76.   Table 5 shows that the major use of energy in the chemical industry is for process heating and associated operations such as distillation,

evaporation and drying. Energy requirements for space heating are low (less than 10%) compared with other sectors of industry, because manpower levels in the chemical industry are relatively low. Many large plants are operated in the open air, especially in the bulk organic and inorganic sectors, and space heating areas are confined to buildings such as control rooms, administration blocks, workshops, canteens etc. In some of the less energy intensive sectors such as pharmaceuticals and paints, space heating can be a significant part of their total energy requirement, but for the industry as a whole it is less important.

77.   The bulk of the process energy is supplied by steam, the remainder by direct firing of kilns or furnaces, or by electrical energy. It has been estimated that about 80% of the fossil fuel burnt in the chemical industry is used for raising steam [ref. 8]. Chemicals is the biggest single steam-using sector, accounting for about a quarter of all fuel used for steam-raising in industry. It is also the only sector in which steam usage has increased over the past 30 years. Combustion of fuels and the raising, distribution, application and recovery of steam are therefore important areas for energy efficiency and for fuel switching, as discussed in paragraphs 131-142.

78.   The estimate of the energy used for steam raising has been used in conjunction with the Thrift Scheme data and other data on specific products, as shown in the Annexes, to provide a breakdown of energy in the categories used in Volume 1 of this report, as shown in Table 6.

79.   Much of the steam is produced in conjunction with electricity generation in CHP plant, or used in direct drives for compressors, pumps, etc. The chemical industry is also by far the largest private producer of electricity, with 37% of industry's private generation in 1981, and 46% of the CHP. This, then, represents a major contribution to energy efficiencye though the proportion of electricity generated in CHP plant has remained static or declined in recent years.

80.   For chemical processes to be carried out efficiently they require temperatures and pressures to be maintained within fairly narrow limits, and process streams often need to be heated, cooled, compressed, blended or separated several times at different stages. Some processes are exothermic - i.e. potentially net suppliers of heat (e.g. sulphuric acid production) though there may be no suitable use for the energy in the plant itself.

81.   Chemical plants are therefore normally designed with a high degree of waste heat recovery in mind, and energy is cascaded from high temperature processes to lower temperatures. Typical chemical plants are highly integrated, with close inter-linking of the energy for heat, power and feedstock within the plant, and sometimes between plants. For example, at the ICI Wilton petrochemical complex, it is estimated that the heat input to the site is re-used three times before it is finally rejected to the cooling system [ref. 18].

82.   This close integration of chemical plant has a number of consequences for energy efficiency. Plants normally operate most efficiently at their full design output on a continuous basis. Operation at reduced output or with frequent start-up and shut downs usually gives rise to a higher reject rate of inferior product and to a lower energy efficiency, which is part of the reason why the earlier marked reduction in SEC has not been maintained in

recent years since the chemical industry has been in recession. The close integration of plants also means that it may be difficult to shut down one plant, even temporarily, without affecting others, which imposes a further constraint on the operation of chemical plant.

83.   The need for continuous operation also tends to inhibit retrofitting. When plants are operated at full capacity the operators are reluctant to shut them down to make improvements unless required for production reasons, and when they are shut down due to a fall in demand, there is often a shortage of capital and lack of business confidence to invest in improvement. Retrofits for energy improvements are, of course, made on large continuous plant, but they are more likely to be associated with modifications carried out for other reasons, e.g. debottlenecking.

84.   Large, continuously operated chemical plants, although representing the major energy consumption in the chemical industry are by no means universal. Sectors such as pharmaceuticals, dyestuffs, paints and other speciality products produce much of their output in batch processes. Many of these products could, in principle, be produced in continuous processes, but the output does not warrant a major dedicated plant. Batch process plant generally has a lower capital cost and tends to be more versatile, such that the same equipment can be used for more than one product.

85.   The types of plant and processes and the scale of operation thus differ greatly throughout the industry, and the opportunities for improving energy efficiency are correspondingly varied. However, it appears that on many sites these opportunities are not being identified due to lack of information on energy usage.

Energy Targetting and Monitoring

86.   The integration of process streams, heat and power make it very difficult to analyse energy flows in chemical plants, and there is much scope for improving information on the breakdown of energy use in the plant. Monitoring of energy use is therefore the first step in improving energy efficiency and there are many reports of the effectiveness of even the simplest energy monitoring system in conjunction with targetting [refs. 6, 11, 20, 22, 23]. Systems of this kind not only provide the basis for determining where improvements should be made and for measuring their effectiveness, but they help to get the commitment of management and staff to sustain the effort on energy conservation.

87.   The appointment of energy managers, energy wardens or other representatives for the workforce and promotion of consultation within the company are all important measures in making a sustained and coordinated effort on energy conservation. The companies which have achieved most in improving energy efficiency have set up organisations along these lines, but many other companies have yet to take this approach seriously. It was remarked in the Thrift Scheme reports that the attitudes towards energy management and the implementation of energy saving measures are generally better in the larger companies than in the small ones.

88.   Even some of the companies who are active in energy conservation matters, undertaking energy audits etc., only measure the energy inputs to particular buildings or areas of the site, so they do not necessarily have a

clear picture of how energy is used in the process. This requires more
detailed metering and logging on individual plants.

89.   On a large integrated site the volume of data to be collected and
processed is enormous, and computer based monitoring systems may be required.
ICI have commented that data loggers for energy monitoring are amongst the
most effective energy saving investment [ref. 23] and they are installing a
computer based energy monitoring system at the Billingham site at a cost of
some £4M [ref. 22]. Indeed, ICI's experience with such systems is to be
assessed and promoted within the industry by means of a collaborative project
between the company and the Rubber and Plastics Research Association (RAPRA),
supported by a contract from the Department of Energy.

90.   Until recently such sophisticated energy monitoring systems have been
confined to the larger companies with the scope to use them fully and the
resources to apply them. However, developments in microelectronics are
starting to make an impact in this area, in producing cheaper and more
versatile monitoring and control devices which can be more widely used on
chemical plant of all kinds and sizes. It is not possible to assess the
'energy saving potential' of such measures in isolation, but they contribute
to the effectiveness of the other measures described in the following
sections, and indicate where they can best be applied. They will also
gradually improve the understanding of energy use in the industry overall.

Energy Conservation Potential

91.   Quantitative assessment of the scope for further energy conservation in
the chemical industry is particularly difficult because of the complexity
of the industry and the lack of detailed information on energy consumption at
plant level - partly through commercial secrecy, and partly because the
companies simply do not have the data themselves, as indicated above.

92.   For a single chemical process or type of plant, it should be possible,
in principle, to establish what is the best current practice (i.e. Specific
Energy Consumption) and the average performance of existing plant, and make
some assumptions about the rate at which the average will approach the best
practice by replacement of plant or retrofitting.

93.   A good example is ammonia - a single product, produced in largely
self-contained plant by a clearly defined process. Data are available on the
average SEC of existing plant and the best new process (see Annex 1,
paragraph 14). This shows a potential reduction of some 35% in going from
average to best practice. However, plans for building the new type of plant
have been shelved, because of the current unfavourable commercial prospects,
and there is no indication when, if ever, this plant will be built in the UK.
Any improvement to existing plant would require retrofitting, for which the
technical potential is much lower, and the economic case uncertain.

94.   For most other types of plant the situation is even less well defined.
Many plants of different type, age and performance are integrated with others
on a multi-product site, sharing common services. Plants have often been
modified several times since initial construction, so the original design
performance, even if known, is usually invalid. The SEC of individual
processes cannot be identified, and sites differ too much in processes,
product mix and inputs for meaningful comparisons to be made between them,

even if the data were available. Estimates of SEC improvements have therefore had to be more broadly based on data for the industry as a whole, and on the achievements or targets of individual companies, backed up with specific examples where possible.

95.   The Industrial Energy Thrift Scheme reports [ref. 6] appear to provide a quantitative basis for estimating the potential for energy conservation in the industry, and they have been used for this purpose. However, there are several reasons why these can be no more than a guide.

96.   As noted in paragraph 74 the data from the Thrift Scheme are biased towards the larger sites and bigger energy users. It was noted in at least one of the reports (No 35) that these are the more energy conscious companies, so they can be assumed to have already adopted more of the simple energy saving measures than the rest of the industry. They will also have a higher proportion of large sites with continuous process plant in the open air, so the opportunities for energy saving may be different from the smaller sites with simpler plant housed in buildings.

97.   The Thrift Scheme also concentrated mainly on measures of a housekeeping nature and relatively simple improvements to existing plant (e.g. insulation, improved controls and steam systems, waste heat recovery) which could be identified during a one-day visit. These studies have not attempted to identify the improvements which may result from major re-design of plants or new proceses, which have made the biggest contribution to reduction in SEC in the past.

98.   Within these limitations the Thrift Scheme gives an indication of the potential for improvements of this kind, although this is likely to understate the potential for the industry as a whole, for the reasons given above.

99.   The savings identified in the Thrift Scheme amount to about 12% of the total energy use in the sites visited (ranging from 6% in Pharmaceuticals to 15% in Soap and Detergents). The original visits were made mainly during the period 1976-78 (i.e. before the second oil price shock) so much has changed since then. Brief follow-up contacts have been made by telephone (in 1982 for most of the sites) to find out what measures have been taken and savings made. The estimates of savings cannot be regarded as very precise, and it has been difficult to derive much reliable quantitative information. Some of the companies have gone out of business, some have changed in ways which make comparison with the original study invalid, and others were unable to provide any useful information. Those who did reply in a meaningful way reported savings which ranged between 2% and 60% of their total energy usage. The average saving was around 12%, achieved over a period of 4-5 years since the original visit (around 2.5% p.a.).

100.   As with the original Thrift Scheme visits, it is not clear how representative these figures are of the industry as a whole, but it is reasonable to assume, again, that it is the more energy conscious firms who are in a position to give a quantitative reply. Thus this achievement is likely to be better than the average for the chemical industry. This is borne out by comparison with the change in SEC for the chemical industry as a whole, which has fallen by 1.8% p.a. since 1974 (see paragraph 106).

101.   The average saving of 12% found in the follow ups happens to be the same as the average saving identified for all the sites covered by the original Thrift Scheme vists, so it might appear that this has exhausted the potential for energy savings of this kind in these sites.   However, the measures adopted to achieve this saving of 12% represent only a proportion of those originally identified in the Thrift Scheme visits.   Part of the savings have come from measures which were not covered by the original studies (e.g. process changes).   Therefore some part of the original savings remains to be achieved.

102.   In some sectors the savings being achieved are considerably higher than those identified in the original study (e.g. Pharmaceuticals - 6% in the original Thrift Scheme study; 30% average among the sites followed up. Plastics - 10% in the original study; 15% in the follow up).   None are showing lower savings than the Thrift Scheme study.   Thus the original studies may have underestimated the savings which could be made by simple measures, or substantial savings are being made by other means.   There is insufficient information to show the savings that can be attributed to individual measures in the majority of these cases.

103.   An alternative approach to assessing the savings in the industry as a whole is from an analysis of the trends in SEC for the industry over a long period.   Projections have been made on this basis, as follows.

104.   The figures for SEC in Table 3 which are plotted in Figure 2 show that the trend in SEC since 1960 falls into two distinct periods.

105.   The first period is from 1960 to 1974 - i.e. up to the first oil price shock (or the time when it seriously began to affect the industry).   Output was rising at 6% p.a., the industry was expanding rapidly, and much new plant was being built.   During this 14 year period the SEC fell by 50% - i.e. an average rate of fall of 4.8% p.a.

106.   The second period is from 1974 to 1981, when the effects of two oil price shocks and the industrial recession hit the industry, output fluctuated widely, but net growth was nil.   During this seven year period the SEC fell by 12% - an average rate of fall of 1.8% p.a.

107.   The changes in SEC result from a combination of effects, including the introduction of new plant, retrofitting of old plant, general housekeeping measures, change in the fuel mix, structural change in the industry, and variations in plant operating levels.   It is impossible to distinguish the contribution of all these factors, but it is apparent that the overall effect has been for the SEC to fall faster when the industry was growing strongly than when it was retrenching in the face of higher energy prices and reduced demand.

108.   There are still many opportunities for improving energy efficiency, by new plant building or retrofitting, as discussed below.   It is therefore assumed that the SEC will continue to fall throughout the industry, and that this will be at a higher rate in high growth situations than in low growth, as in the past.   The projections for SEC in the future have been based on the relationship between SEC and industry growth rate in the two periods discussed above - i.e. 1.8% p.a. fall in SEC at nil growth rate, and 4.8% p.a. fall in SEC at 6% p.a. growth rate.

109.   However, we cannot expect the reduction in SEC to continue at the same rate indefinitely, even if growth rate remains constant. As the 'easy options' or the most cost effective measures are taken up, the scope for further reductions in SEC becomes more limited.

110.   To represent this 'diminishing returns' effect it is assumed that, from 1980 to 2000 the SEC falls at only half the rates noted above - i.e. at 0.9% p.a. at nil growth rate, and 2.4% p.a. at a growth rate of 6% p.a. Intermediate points have been interpolated. The growth rates for each subsector, as shown in Table 4, have then been used to derive an appropriate figure for the rate of fall in SEC. The resultant SEC projections are shown in Table 7, and the percentage reduction in SEC is shown in Table 8. It should be appreciated, from what has already been said about the reliability of the basic data at subsector level (paragraphs 13 and 56), that these projections can be regarded as no more than broad indications of likely trends.

111.   When viewed against the reduction in SEC achieved in the past these assumptions appear relatively modest. If the industry were to return to growth rates like those seen in the 1960s and seriously applied the available energy saving technologies, the SEC could be expected to fall more rapidly. However, the implementation of further energy saving measures will now be more dependent on capital investment.

112.   The view within the industry is that the companies which are most active in energy conservation, which tend to be the larger energy users, have now taken most of the low cost housekeeping measures. There are still many companies, particularly the smaller ones, which have the potential to make large percentage savings individually, but they account for a relatively small proportion of the energy use between them. Among the larger users there are further opportunities to make substantial savings, as discussed in the next section, but they will require more capital investment.

113.   Investment criteria for energy saving projects differ widely between companies, and even from one division to another within the same company, depending on performance. It is difficult to generalise, but it is reported that companies are typically looking for paybacks of 1-2 years, and less than 12 months in some cases. In general the more secure the company is in a particular sector, the longer the payback that will be accepted. In some cases the availability of effort at the right level is as much of a limitation as capital. The time of management and professional staff is required to identify the opportunities and see the project through effectively. When a company is hard pressed and suffering staff reductions this effort may be severely restricted.

114.   The CIA carried out a survey among their member companies which indicates that 8-9% of total capital investment is currently identifiable with energy saving projects [ref. 13]. With total capital expenditure in the industry of £1,100 M in 1982, and a similar level in 1983, this indicates expenditure on energy saving projects of around £90 M per year at present. However, the reduction in SEC of 1.8% p.a. represents an energy saving of some 5.5×10^6 GJ per year, equivalent to 130,000 te of fuel oil, worth around £17 M. Clearly this does not accord with the short payback periods reportedly sought on energy-saving projects. It is now thought that the

investment had been overstated, and probably represents the total cost of projects which include significant energy savings.

115. A number of companies have reported savings or set targets which indicate that substantial savings are being achieved at the individual company level, as follows:

- ICI have achieved a reduction of 4% p.a. over the past 10 years, and are aiming to reduce energy consumption by 5% p.a. at present [ref. 22];

- BP have a group target of 10% reduction over five years - i.e. 2% p.a. [ref. 23];

- Pfizer set an initial target of 20% reduction in five years, which they did not quite achieve, but they still set a target of a further 10% reduction in the next five years [ref. 19].

These reports, too, are puzzling because, if the major energy users are achieving savings of this order while the industry as a whole is reducing SEC by only 1.8% p.a., one wonders what the rest of the industry is doing by way of energy saving.

116. These apparent inconsistencies in the implications of the energy data currently available highlight the difficulty of obtaining a consistent picture of what is happening in the industry and the need for more reliable statistics.

117. The major technical changes and other developments which will contribute to the projected reduction in SEC shown in Table 7 are summarised in the next section. Some of the more important are discussed in more detail in subsequent sections.

Key Energy Saving Measures in the Chemical Industry

118. Steam raising is a technology which cuts across the whole of the chemical industry and accounts for the bulk of the fossil fuel used in the industry. Potential improvements in the generation and distribution of steam could give rise to energy savings of at least 8% by 2000. Opportunities are discussed in more detail in paragraphs 131-141. Steam usage also gives the greatest scope for fuel switching. Although this may not significantly affect the SEC it will be important in assessing the future pattern of fuel use in the industry. This aspect is discussed in paragraph 136.

119. Waste heat recovery is recognised as one of the major energy conservation measures in the chemical industry. This has always been an important aspect of the design of chemical plant, and the basic technology is well established, but technical developments and the changing economics in the light of higher energy costs have given rise to greater opportunities, particularly in the design of new plant, for which energy savings of 20% relative to similar plant designed 20 years ago can be readily achieved. Retrofits on existing plant seldom give such large savings, but there are examples of savings of 5-10% from additional heat recovery equipment. Grant [ref. 11] gives a useful review of heat recovery technology, with further references to more detailed sources.

120. Waste heat recovery is, of course, only viable when the heat recovered
can be used effectively. Conventional approaches to waste heat recovery have
tended to consider the heat content of process streams individually, and to
look for ways of recovering that heat and using it elsewhere in the process
(e.g. for pre-heating feed streams) or sometimes on another plant. A new
approach, developed over the last few years, aims to avoid generating the
excess heat in the first place.

121. This method, termed 'Process Integration', examines the overall heat
balances in the process from fundamental thermodynamic principles,
establishes the minimum practicable heat requirement to carry out the
process, and then enables the designer to optimise the heat input and capital
cost of the plant as a whole. These techniques are most effective in the
design of new plant, for which energy savings of up to 50 or 60% against
conventional designs have been achieved, often with reduced capital costs.
Retrofit projects on existing plants have achieved savings of 20-30% or
sometimes more, with paybacks of 6-18 months. The techniques were developed
by Bodo Linnhoff at ICI, and so far have hardly been applied in other firms
in this country, although they are now being more widely publicised and are
being promoted by a programme of demonstration projects. Over the next few
years these techniques can be expected to become the norm for designing both
new plants and retrofits, and to give rise to substantial improvements in
efficiency. Further explanation and examples are given in reference 24.

122. Ultimately most of the energy input to the process is rejected as low
grade heat, which is at too low a temperature to be recovered by conventional
means, and is normally dumped in the cooling water or air. In principle it
is possible to recover much of this heat by means of heat pumps or to use it
to generate electric power by Rankine cycle power generators. Such systems
have seldom been used up to now because this is a relatively unfamiliar area
of technology, requiring specialised hardware. Companies are unwilling to
complicate plants with equipment which may prove unreliable or reduce
operational flexibility. Technical advances are improving the technology,
but equipment of this kind will not be generally incorporated into plant
until appropriate packaged units are available. Equipment manufacturers are
only just starting to meet this requirement for heat pumps, and hardly at all
yet for Rankine cycle generators. This is an area with great technical
potential - as noted above, a large part of the process energy input to the
industry - but it is impossible at present to know how much will be
economically recoverable.

123. CHP generation represents a major energy conservation measure which is
already extensively applied by the chemical industry, as detailed in
paragraph 79. There is the potential for a substantial increase (probably at
least 20%) in the amount of electricity generated by the industry in CHP
systems and there are no serious technical problems to be overcome. However,
it is not at all clear how much will prove economically viable with current
investment criteria, since CHP plant generally has longer payback times than
the industry is prepared to accept in the present uncertain economic climate.
A study is currently being mounted to assess the present and potential use of
CHP in the industry. Opportunities are discussed in more detail in
paragraphs 143-159.

124. The three 'separation processes' - distillation, evaporation and drying
- between them consume about one third of the total energy in the chemical

industry (see Table 5). The technology of distillation and evaporation are reasonably well understood, and techniques for energy saving in these processes are well established. These include mechanical vapour recompression, thermal recompression and multiple effect evaporation. Energy savings of 50% or more against basic plant have been identified, but these may only be economically attractive for specific applications - mainly on new plant. The opportunities for economic retrofits on existing plant are more limited. This is an area where new plant building offers scope for substantial savings.

125. Drying, which consumes around 5% of the energy in the industry, is a more empirical and less well understood process which offers more potential for retrofit projects, e.g. heat recovery from exhaust air. Heat pumps offer potential advantages here, but the technology needs further development, particularly in increasing the 'output' temperature that can be achieved, and in scaling up the size of commercially available units.

126. Overall it would be reasonable to assume that savings of 10% can be made in the energy requirements of the separation technologies by 2000 - i.e. 3% of the total energy requirement of the industry.

127. Automatic process control can give substantial improvements in energy efficiency as well as labour efficiency and product quality. This has been a relatively static area of technology in the past, but it has received more attention in recent years, with the stimulus of cheaper and more versatile microcomputer systems. It is not possible to estimate the savings from this technology, but it is expected to contribute to the effectiveness of other measures, as noted for monitoring and targeting (paragraph 90).

128. Space heating accounts for less than 10% of the total energy use in the industry. This is a very dispersed load, and in some areas the energy is provided by low grade waste heat which has no other economic use, so there is little incentive to improve its efficiency. There will be little effect on total energy use in the industry from changes in this area, although in some subsectors with high space heating requirements, such as pharmaceuticals, considerable improvements are possible on individual sites.

129. New process routes, catalyst developments, etc. are potential sources of further reduction in SEC. Experience suggests that such advances will continue, because this is part of the business of the industry, but their effect cannot be predicted. Biochemical processes, which are already in use in some sectors, notably pharmaceuticals, may become more widespread, but they seem unlikely to have an impact on the more energy intensive processes. The possibility of a major breakthrough - e.g. in the biological fixation of nitrogen on an industrial scale - cannot be ruled out, but no account of such unpredictable developments has been taken in this study.

130. The contribution of the various energy saving measures have been classified into several broad categories, defined in Volume 1, as shown in Table 8.

Steam Usage and Fuel Switching

131. As noted above (paragraph 77) steam raising accounts for the bulk of the coal, oil and gas used in the chemical industry. Steam technology is

therefore of great importance to the industry in energy terms, and boilers offer by far the greatest opportunity for fuel switching.

132.   The uses of steam include process heating and (to a much lesser extent) space heating; driving turbines to generate electricity or as direct mechanical drives (e.g. for compressors); producing vacuum conditions by means of steam ejectors; participating directly in chemical reactions, as a raw material (e.g. in producing ammonia or methanol) or as a diluent (in producing ethylene).   The steam may be used on the plant for more than one purpose - e.g. passed through a turbine and then used for process heating, and finally space heating.

133.   The SPRU survey of the steam boiler stock in industry [ref. 8] estimates that around 80% of the fossil fuel burnt in the chemical industry is used for steam raising (73% of natural gas, 82% of coal, 90% of oil, and perhaps 80% of waste materials - off-gases, liquid by-products, etc.).   The average boiler size is large and load factors high compared to industry as a whole.   The boiler stock is also relatively modern, with 68% of the fuel used in boilers installed since 1960 (and over half of these since 1970) reflecting both the surge of growth in the chemical industry in the 1960s and 70s and the incentive to scrap older plant (mainly coal fired) to get the benefits of cheap natural gas during this period.   This is a major factor in considering the need for replacement of boiler plant and the possibility of fuel switching.

134.   Most of this recent boiler plant is oil or gas fired, and the large boilers, which account for most of the fuel use, can be expected to have a life of 30-40 years, so few of them would normally be replaced before 2000. Their design efficiency is very little lower than that available now, so there would be little benefit in replacing or converting prematurely unless there are substantial incentives (e.g. specially favourable coal prices, or the recent Government grant scheme to encourage use of coal).

135.   Some of the older plant which will need replacement over the next 20 years is already coal fired, but there is a substantial amount of plant which was originally coal fired, then converted to oil or gas, and which could be converted back.   This re-conversion is normally more difficult and costly than the original conversion, because plants have seldom retained the coal handling and storage equipment.   The economic case for conversion to coal is difficult to assess and will depend critically on the expectation of relative prices and incentives such as grants towards the capital costs.   At a time of economic uncertainty, companies are unwilling to commit large capital sums and additional management effort to projects with an uncertain return which are peripheral to the main business activity.   The choice of coal for new plant, involving higher capital costs than other fuels, is subject to similar considerations, although the economics are more favourable to coal.   Use of coal will therefore be strongly dependent on growth in the industry.

136.   The Government's coal conversion scheme, which provides grants of up to 25% of the capital cost to assist conversion to coal has encouraged many firms to examine this option.   Schemes already approved (as at October 1983) will take some 450,000 tonne/year of coal.   Others under consideration represent a further 200-300,000 tonne/year.   In addition ICI are examining the conversion of the Wilton power station back to coal (it was originally converted from coal to oil and gas firing).   Other conversions outside the

grant scheme are less well documented, but there are reports of some increased use of coal, even in sectors where the use has been minimal up to now, such as pharmaceuticals. It is therefore difficult to estimate the future pattern of fuel use at present. On the basis of the above it is assumed that, in the high case, an additional 1.5 million tonne/year of coal (say 40 PJ) will be used by 1990, and perhaps a further 1 million tonne/year by 2000, the changeover coming mainly from oil. The low case assumes an additional 1 million tonne/year by 2000, most of this by 1990.

137. Replacement of old plant will continue to give a small improvement in energy efficiency (perhaps 1-2% of total boiler fuel over 20 years), but otherwise none of the potential changes in fuel use would have much effect on the total energy requirement, since the design efficiency of modern boiler plant differs little between the fuels (around 80-90%, depending on size and type of boiler). Of more significance is the potential improvement in operating efficiency of the boiler plant, in practice, through better maintenance and control, and in the application and use of the steam.

138. A recurring theme in the IETS reports and other papers on this subject is that most boilers operate at well below their design efficiency. A NIFES study [ref. 21] showed that typically 3-5% of efficiency is lost and sometimes much more through a variety of minor causes. The greatest of these is incorrect combustion conditions (too much or too little air). The SPRU boiler survey [ref. 8] reviews the main methods which can be used to improve boiler efficiency. These range from basic cleaning and adjustment at negligible cost to installation of automatic monitoring and control equipment. In the past such equipment has been costly, inconvenient to use and unreliable, but advances in microelectronics have brought down the costs and increased the capabilities such that these systems are now being applied more widely. Cost does not increase significantly with boiler size, so returns are better for larger boilers with high load factors. Paybacks of 6-12 months can be achieved for boilers over 15,000 lb/hr capacity, which is well below the average size for the chemical industry. This type of equipment can therefore be expected to be introduced quite rapidly, particularly if capital costs continue to fall relative to fuel costs. Savings of at least 2% in boiler fuel could be achievable by this means - say 1% by 1990 and a further 1% by 2000.

139. A major cause of loss of efficiency of boilers (or other chemical plant) is operation at part load, or with intermittent start-up and shut-downs. The SPRU boiler survey [ref. 8] summarises the reasons and reports that efficiency penalties of 10% or more can be attributed to this cause. However, in many cases it is difficult to do anything to improve the situation, at least in the short term. Steam demand depends on the highly variable needs of the production processes, on shift work patterns, etc., and these are unlikely to be radically altered to save a few per cent on boiler fuel. The present over-capacity in the industry also contributes to the part load problem, but this should be gradually improved, with further rationalisation, so that capacity comes more into line with demand. When boilers are replaced on an existing plant the replacements are generally of a lower capacity than the originals. (This has been noted particularly on the coal conversion scheme.) Therefore a small improvement in efficiency can be expected over the next 20 years as the part load problem is reduced. In the high case this is assumed to be 1% by 1990 and 2% by 2000.

140. So far we have considered only the generation of steam at the boiler.
The transfer and distribution of steam offer a further significant potential
area for savings. The IETS reports highlight a number of measures including
insulation of steam pipes, recovery of condensate and better maintenance of
steam systems. (The latter covers such items as sealing leaky joints and
glands, maintenance of steam traps etc.) In some cases the measures are not
practicable - e.g. condensate may be unusable because of contamination. In
many companies the more readily applied measure will have already been taken,
but recent visits to chemicals sites indicate that there is still the
potential for low cost measures of this kind, with a possible saving of up to
2% by 2000.

141. The net effect of the improvements discussed above (boiler replacement,
improved load factor, automatic monitoring and control of boilers, insulation
of steam lines, condensate return etc.) is a reduction of demand for boiler
fuel, relative to 1980 by 4% in 1990 and 8% by 2000. However, the greatest
scope for improving the efficiency of steam usage is in its application on
the process plant, by measures such as those discussed in paragraphs
119-127.

142. An alternative approach to the use of steam is direct heating by means
of electricity or gas. Steam is the traditional means of supplying heat to
process plant, and has the advantages of flexibility in use and of being a
well established technology. However, the overall efficiency of energy
supply via steam systems is commonly less than 50%, due to the accumulation
of losses in generation, distribution and application, as indicated above.
Direct heating by electricity may be economic where heating is required
intermittently for short periods, for which steam is particularly
inefficient. Direct fired gas heaters can be used in immersion tubes for
heating liquids in vats, tanks, etc., with an overall efficiency of 80%.
Changes of this kind are being considered at the time of replacement or major
overhaul of steam systems. The approach is relatively recent, and the extent
of its effect is difficult to assess at present. It is likely to result in
continued or increased use of gas and electricity, despite more competitive
coal pricing, and a further small improvement in the energy efficiency of the
industry.

Combined Heat and Power

143. Because of a high requirement for both steam and electricity, much of
it on big continuously operated plant, the chemical industry has always
generated a large amount of electricity in CHP plant. The chemicals sector
is in fact the largest private producer of electricity in the UK, with some
37% of industry's private generation in 1981 and 46% of the CHP. Around 60%
of fossil fuel is used for steam raising in conjunction with electricity
generation. In addition a substantial amount of power is produced as direct
mechanical drives (e.g. for compressors, pumps etc.) via steam which is used
for process heating, but this cannot be quantified with any certainty. Data
have been drawn from the Department of Energy's 1977 Inquiry into Private
Generation [ref. 25] and a limited survey to update the information to 1981.

144. In 1981 the chemical industry generated some 4.6 TWh of electricity, or
27% of its total electricity consumption, of which CHP plants accounted for
3.6 TWh, or 21% of total consumption. The proportion of CHP has been around
this level for the past 25 years, but has declined in recent years. In 1977

private generation amounted to 33% of total electricity consumption, and CHP 24%.

145. A similar decline in the proportion of CHP has occurred in all sectors of industry except oil refining. In fact the fall in chemicals has been less than in most other industries. It may appear surprising that there should be a reduction in CHP at all, since it apparently offers higher energy efficiency, and industry has been vocal in its complaint about the cost of electricity from the public supply. The reasons for the decline in CHP and the prospects for its future use in the industry are complex.

146. First, the ratio of cost of purchased electricity to cost of fossil fuel has fallen over the past ten years, which reduces the economic viability of both existing and new CHP systems. This also contributes to the uncertainty about trends in fuel prices in the future, making it more difficult to assess the long term viability of CHP schemes.

147. Second, some CHP plant has been shut down with other plant in the course of the closures and cutbacks which have taken place in the industry. The bulk chemical sectors, which have the largest potential for CHP, have been hardest hit by the recession.

148. The major factor affecting CHP plant has been the change in steam/power ratio which has occurred on many chemical sites. On one major petrochemical site, for example, the steam/power ratio has changed over 20 years from 3:1 to 2:1. This reduction in steam demand has arisen partly through successful energy conservation measures (it has generally proved easier to save steam than to save electricity) and partly through the shut-down of some complete processes with a high steam demand - either through a change in product mix or technical obsolescence.

149. In the chemical industry most of the CHP plant is back-pressure steam turbine type, for which the steam/power ratio is fixed (normally in the range 5:1 to 10:1) when the plant is built. The ratio cannot then be altered significantly without major plant changes. Thus a CHP plant built to meet a site's requirements 20 years ago will probably be producing more steam than the site can use now.

150. On some sites, where CHP plant is kept running to meet power requirements or to burn off waste products, large quantities of steam are blown to waste or, at best, plants of marginal utility are kept in operation as a 'sink' for surplus steam. The alternatives, without further capital investment, are either to shut down the CHP plant altogether and purchase all electricity from the public supply, or to run the plant at reduced output to meet the steam requirements, producing some electricity, and purchase the balance. However, running at reduced output generally involves some penalty in terms of lower overall thermal efficiency. The economics of the various alternatives depend critically on the tariff structure, including any peak load charges, so there may be a strong incentive to keep the CHP plant running, even though steam is wasted. Thus established CHP systems, although in principle making a major contribution to energy efficiency, may no longer provide the expected savings in practice, or may reduce the flexibility of operating the rest of the plant.

151. Several methods are available to improve the heat/power ratio and energy efficiency of CHP systems, either by building new plant or by modifications to existing plant. The technology for new plant is well established, but there have, as yet, been very few retrofits on existing plant.

152. If new plant is being built, gas turbines or diesel engines offer a variable heat/power ratio in certain circumstances and a higher proportion of the output as electrical or mechanical power than steam systems. The characteristics of these main types of plant are compared in simplified form in Table 11. The variations in heat/power ratio are achieved by different degrees of after-burning and waste heat recovery. Capital costs for gas turbine and diesel systems are generally lower than for steam plant, and paybacks can be much shorter - possibly as low as two years in favourable circumstances, but usually four years and upwards.

153. There are currently very few gas turbine or diesel installations in the chemical industry. Gas turbines have been used mainly where waste gases or light distillates are available as fuel on a large scale with a sufficiently consistent quality. The former pricing policy for natural gas (in which the interruptible tariff was related to the cost of the distillate standby fuel) was reported to have inhibited the installation of turbines running on natural gas. Gas supplies for turbines are now available on the normal firm or interruptible tariff, but it is too soon to say whether this will have any effect on numbers of installations.

154. Several possibilities are available for retrofit projects to improve the heat/power ratio and efficiency of existing steam plant. One method is to install additional heat recovery equipment to 'recycle' surplus heat, e.g. for preheating the boiler feed. Another possibility is to install a gas turbine or diesel engine in conjunction with existing steam plant to form a combined cycle system, in which the exhaust from the additional engine is used to boost the power available from the steam plant. It is claimed that such retrofit projects can have paybacks of 2-3 years or less, but so far very few cases have been put into practice.

155. Thus there are several obstacles to the further use of CHP in practice. Although the technology is available and most is relatively straightforward, it is still unfamiliar to many companies, and there is some reluctance to add the complication of CHP schemes, particularly if they appear to reduce operational flexibility or to require specialised skills to operate them.

156. Payback periods on new plant (usually over four years) are generally longer than companies are prepared to accept for a project which is not part of the main stream activity and may be seen as a relatively high risk venture, particularly when the economics are susceptible to changes in fuel prices. Paybacks on retrofit projects are claimed to be shorter, but there is little practical experience to serve as an example.

157. There may also be institutional barriers which affect the economics of the project or add to the uncertainty. The pricing policy for gas supplies has been mentioned above. The electricity tariffs for standby supplies or for buy-back of surplus electricity, and the technical requirements for grid connection have all been cited as factors inhibiting further use of CHP. Some of these may be more perceived than real, but even if they do not affect

the economics they help to feed the doubts which a company has about a CHP project. The Energy Act 1983 should help to clarify this area and remove some of the uncertainties. It also opens the possibility of using the public supply grid to transmit electricity generated at one site to another, which may widen the scope for economic CHP schemes in large companies with several widely scattered plants.

158. A study is therefore being mounted by the Department of Energy and ETSU to update the knowledge of CHP in all sectors of industry and assess the scope for further use of CHP. It is difficult at present to assess the potential in the chemical industry, but some indication could be gained from previous studies on the amount of plant and use of fuels for CHP. The 1977 Inquiry into Private Generation [ref. 25] showed that the chemical industry used 62% of its fossil fuel in CHP plant. With the decline in use of CHP mentioned above this would be around 60% by 1980. Clearly the industry cannot burn all its fossil fuel in CHP schemes and it seems unlikely that more than, say, 75% of fuel could be used in this way. If this was achieved it would represent an increase of 25% on present usage in CHP plant. The actual achievement by 2000 is likely to be considerably less than this.

159. Any increase in CHP in the industry will not necessarily result in a lower requirement for delivered energy unless electricity can be generated from heat which is wasted at present. It is unlikely that there will be many opportunities for this since most waste heat is relatively low grade and unsuitable for economic electricity generation. Generally CHP schemes wil result in an increased requirement for fossil fuel roughly equivalent to the electric power generated, depending on the type and efficiency of plant involved. The benefit to the company will be in a cost saving on purchased electricity. The energy saving will appear as a reduced fuel requirement for the public electricity supply, which could mean an indirect displacement of coal by oil or gas.

Projected Fuel Demands

160. The projected changes in SEC (Table 7) have been applied to the output assumptions (Table 4) to produce the projected energy demand shown in Tables 9 and 10. The split between fuels has been based on the assumptions about fuel switching outlined in paragraphs 131-142.

161. In both the high and low cases there is a move from oil to coal, mainly in the period up to 1990, under the stimulus of the coal conversion scheme (see paragraph 136). In the high case much of the growth in energy demand has been taken by coal, with a smaller growth in gas and electricity. In this case there will be more opportunities for new plant building and a higher level of business confidence, so the longer payback times needed for the higher capital cost of coal fired plant will be more acceptable.

162. Much of the boiler stock in the chemical industry is relatively recent, and is unlikely to be replaced prematurely unless a wide differential opens up between coal and oil or gas prices, as discussed in paragraphs 133-134. Therefore, in the low case, after the initial conversions under the stimulus of the coal conversion scheme it is assumed there will be only slow further growth in coal usage.

163. Coal has been assigned to subsectors on the basis of what is known of the coal conversions at present and on the existence of large steam raising plant, which offers the greatest opportunity for coal usage - predominantly in the bulk organic and inorganic sectors. The pharmaceutical sector has used virtually no coal in the past, but two major companies are installing coal fired plant, and this may stimulate others to consider coal in the future. Therefore some coal has been assigned to the growth in energy demand projected for this sector.

164. Most of the shift to coal is at the expense of oil. There may be some transfer from gas to coal fired plant, but this is expected to be compensated by the opportunities for direct firing in place of steam systems, as discussed in paragraph 142. Gas is therefore expected to at least maintain its contribution in terms of energy supplies, increasing in the high case, though its share falls slightly. Electricity is expected to increase its energy supplied and its share in both cases, slightly more in the high case than the low case.

Commentary and Conclusions

165. The chemical industry has a good record in improving energy efficiency. In 20 years its Specific Energy Consumption (SEC), excluding feedstocks, has been reduced to less than half its 1960 level - considerably better than in manufacturing industry as a whole. The trend in SEC has fallen into two distinct periods, underlining the importance of growth and new investment in reducing SEC. Up to 1974 - the first oil price shock - when the industry was expanding at 6% p.a., SEC fell at 4.8% p.a. Since then, while the industry has shown no net growth, SEC has fallen at 1.8% p.a.

166. For the chemical industry as a whole the SEC is projected to continue falling from its 1980 level of 19.6 MJ/£ of gross output to a level in 2000 of 12.6 in a high growth case or 13.4 in a low growth case - a reduction of 36% in the high case or 32% in the low, corresponding to rates of fall of 2.2% p.a. or 1.9% p.a. respectively.

167. This fall in SEC for the whole industry is greater than for any of the individual subsectors except for pharmaceuticals. The overall fall in SEC also shows only a small difference between the high and low cases, despite the fact that, for most of the subsectors, the SEC shows a considerably larger difference between the two cases (see Tables 7 and 8).

168. The change in SEC for the chemical industry as a whole can be attributed to two main factors:

(a) Technical change - improvements in energy efficiency at the individual process/plant level, as summarised in paragraphs 118-130.

(b) Structural change - different relative growth rates in areas of high and low energy intensity.

169. Both these factors are working to reduce SEC in the industry, but they are working at different rates in the two growth cases:

(a)   Technical change is greater in the high growth case, due to a
        higher level of activity in new plant building, plant replacement
        or retrofitting, all of which tend to improve energy efficiency,
        whether this is a primary objective or not.

(b)   Structural change is greater in the low growth case, since the
        more energy intensive bulk sectors tend to be hardest hit in
        times of economic stagnation.

170.   The net result, as noted above, is to drive down the SEC for the whole
industry more rapidly than would be expected on the basis of technical change
alone, and to draw together the high and low growth cases in terms of the
rate of fall of SEC. However, the technical component is dominant, and the
SEC is still projected to fall more rapidly in the high growth case, due to
the higher level of technical change.

171.   When looking at these projections it is important to recognise the
limitations of the basic data and the assumptions on which they are based.

172.   It has been noted in this report that the quality of the data on
current energy use in the chemical industry leaves much to be desired. There
are substantial discrepancies between the two official sources of statistics
– the BSO Purchases Enquiry and the Digest of UK Energy Statistics. This may
be partly due to the problem of accounting for feedstocks, both gas and oil,
and partly to difficulties in classifying parts of the industry, especially
at the borderline between the petrochemical and oil refining industries. In
addition to the discrepancies between the Digest and BSO data, there is a
further major source of energy, in the form of by-products and waste
materials, which does not appear in official statistics at all, and for
which estimates are highly speculative. The existence of this source has
been noted and an estimate given (paragraph 42 and Table 2) but, because of
the uncertainty, it has not been counted in deriving the figures for SEC.
The total energy consumption used in this report has been based on the Digest
figures, adjusted only for gas used as feedstock, as described in paragraph
53.

173.   The breakdown of energy use by individual subsector, for which the only
source of data is the BSO Purchases Enquiry, is subject to further
uncertainty. One of the main difficulties here is that many chemical sites
carry out operations which span several different categories of the Standard
Industrial Classification (SIC), sharing common energy supplies, etc. It is
therefore often not possible to relate energy use reliably to individual
products or classes of product. Unfortunately it is in the most energy
intensive sectors that the greatest difficulty lies in making this
distinction and in accounting for feedstocks and use of by-products. Changes
in the SIC have also made it extremely difficult to derive reliable
time-series data for the subsectors. It is clear that the subsector
breakdowns and projections are considerably less reliable than the totals for
the industry, but it is still helpful to work at this level of disaggregation
to build up a better picture of what is happening in the industry as a
whole.

174.   To improve the reliability of the basic data would require further
investigation of these discrepancies, but at present there appear to be no
other accessible sources of information on energy use in the industry. The

chemical industry does not independently produce such data of its own, as do many other industries, which is a serious handicap to studies of this kind. Considering that the industry consumes some 10% of the total national energy requirement this leaves a major gap in our detailed knowledge of energy consumption.

175. These uncertainties affect the absolute quantity of the energy used in the industry, its breakdown between subsectors and the SEC figures for 1980, but they do not seriously affect the percentage change in SEC projected for the future. This is dependent on the growth rates, the structural change and technical change referred to above.

176. The overall industry growth rates are not forecasts, or projections, but merely assumptions, adopted to keep this study consistent with others, as discussed in paragraphs 59 and 60. The differential growth rates for the subsectors (Table 4) are likewise assumptions, consistent with the overall growth of the industry and with a distillation of published views on the prospects for the industry. Comments from the industry indicate that the high case would now be considered highly optimistic, but the low case is a reasonable representation of the chemical industry's likely pattern of future development. Given particularly favourable circumstances it could do better, but there is at least an equal chance that it could do worse.

177. Fortunately the overall change in SEC is not too sensitive to the growth rate assumptions, as indicated in paragraph 166 above, but this is because of the structural change built into the assumptions. A different pattern of structural change could produce a different result, but the pattern adopted looks qualitatively consistent with industry views.

178. The most important determinant of the reduction in SEC is technical change. The major technical opportunities have been identified and an indication given of the scale of their potential impact on energy consumption (paragraphs 118-130). The most important are waste heat recovery, improvements in steam technology, and a novel approach to process design to make the optimum use of heat flows, known as process integration. This last approach looks particularly promising. Further use of CHP could improve energy efficiency on a national scale, and may give cost savings to the chemical industry, but would not significantly affect the industry's total energy consumption, as discussed in paragraphs 143-159.

179. Energy targetting and monitoring schemes and automatic process control will make an increasingly important contribution to improvements in energy efficiency by identifying and quantifying the scope for technical improvements, and by ensuring the effectiveness of the measures taken. Information arising from such schemes, if it could be made generally accessible, would also be valuable in supplementing the limited official statistics on energy use, and in resolving discrepancies between those already available.

180. Most of the technical measures can be applied either as retrofits to existing plant or incorporated in the design of new plant. Their impact on energy consumption and their cost effectiveness are, of course, normally greater when applied to new plant, but there are many examples of good, cost-effective retrofits.

181. It is very difficult to quantify the technical potential of these measures for the industry as a whole, because of the diversity of the industry and the lack of information on the performance of existing plant - i.e. we do not know the starting point from which the improvements will be made. It is even more difficult to assess the extent to which the technical potential will be achieved in practice. This depends not only on the growth rates and level of investment in the industry, as noted above, but also on the response to changes in energy prices, availability of alternative fuels etc.

182. For these reasons it has not been feasible to estimate the rate of uptake of individual measures. The projection of SEC changes has therefore been made on a more global level, based on the past record of the industry in reducing SEC during the periods of growth and stagnation indicated above. The projections thus represent an assumed average effect of a large number of companies applying a wide diversity of measures to different degrees.

183. There is undoubtedly the potential for a greater reduction in SEC than that projected, if the technical measures available were applied extensively throughout the industry. However, evidence from various contacts with the industry indicates that there has been a wide variation in the uptake of such measures in the past, and there is likely to be so in the future. There are also some inconsistencies in the performance of the industry relative to that of individual companies as discussed in paragraph 115. Several of the major companies, which together account for well over half the total energy use in the industry, are apparently making reductions of the order 3-5% p.a. in their SEC, yet the industry as a whole is reducing SEC by less than 2% p.a. This is another area which would benefit from further investigation and better statistics.

184. This report has endeavoured to make the best use of currently available information, but it is apparent that the detailed breakdown of energy use and the projections for the future are subject to considerable uncertainty - perhaps more than for other industries. This reflects the diversity of the chemical industry, the wide range of technical options and the limitations of current statistics on energy use. The value of the report would be enhanced if it stimulates the provision of better data on the chemical industry. Comments on the analysis, or information which would help to refine any part of it would be welcome.

REFERENCES

1. 'Digest of United Kingdom Energy Statistics, 1981', Department of Energy, HMSO, London, 1981.

2. '1979 Census of Production and Purchases Inquiry Analysis of Production Industries by Standard Industrial Classification, Revised 1980', Department of Industry Business Statistics Office, Business Monitor PA 1002.1, HMSO, London, 1983.

3. 'Business Monitor. Report on the Census of Production 1980', PA 251, 255, 256, 257, 258, 259. Department of Industry Business Statistics Office, HMSO, 1982.

4. 'Standard Industrial Classification, Revised 1980', HMSO, London 1979.

5. 'Energy Projections 1982', Proof of Evidence to the Sizewell 'B' Public Inquiry, Department of Energy, 1982.

6. 'Industrial Energy Thrift Scheme Reports', Department of Trade and Industry, 1979-1983.

   No 10, Energy use in the soap and detergents industry.
   No 17, Energy use in the plastics industry.
   No 18, Energy use in the miscellaneous chemical manufacturing
         industries.
   No 19, Energy use in the pharmaceutical chemicals and toilet
         preparations sectors.
   No 24, Energy use in the dyestuffs and pigments sectors.
   No 33, Energy use in the fertilizer industry.
   No 35, Energy use in the general chemicals industry.
   No 43, Energy use in the paint manufacturing industry.

7. Energy Audit Series No 13. The Fertilizer Industry. Issued jointly by the Departments of Energy and Industry, 1981.

8. 'UK Industrial Energy Demand: Economic and Technical Change in the Steam Boiler Stock', J Chesshire and M Robson, Science Policy Research Unit Occasional Paper No 19, University of Sussex, 1983.

9. 'Chemicals Industry: Report to the National Economic Development Council', National Economic Development Office, 1983.

10. 'Chemicals - Contraction or Growth?' National Economic Development Office, 1981.

11. 'Energy Conservation in the Chemical and Process Industries', C D Grant, The Institution of Chemical Engineers, 1979.

12. 'The Pattern of Energy Use in the UK - 1976', R P Bush and B J Matthews, ETSU R7, Energy Technology Division, Harwell, 1979.

13. 'Energy Bulletin No 8, March 1983', Chemical Industries Association Ltd.

14. 'Energy and Feedstocks in the Chemical Industry', Ed A Stratton, Society of Chemical Industry/Ellis Harwood Ltd, 1983.

15. 'Trends in feedstock for the chemical industry', A Stratton, Endeavour, New Series, 7, No 2, 1983.

16. 'Chemicals and Energy - The next 25 years', P G Caudle, Futures, October 1978, 361.

17. 'Chemfacts United Kingdom', Chemical Data Services, Industrial Press, 1983.

18. 'Exploiting waste heat recovery in the chemical industry', T A Kantyka, Chemistry and Industry, 1 September 1979, 571.

19. 'Organising energy saving activities in the small/medium size firms', P C Baker, Chemistry and Industry, 1 September 1979, 580.

20. 'Continuous monitoring of energy in a large chemical plant', P Moon and G Tasker, Chemistry and Industry, 1 September 1979, 585.

21. 'Energy savings in plant utilities and services', H B Weston, Chemistry and Industry, 1 September 1979, 589.

22. 'Energy Monitoring - A Company View', G V Ellis and D Boland. Paper presented at CBI/DEn conference 'Gain Control of Your Energy', 30 June 1982.

23. 'An ICC Survey of the Most Effective Energy Conservation Practices in Selected Industries. Section 1: The Chemical Industry.' International Chamber of Commerce, Commission on Energy, Document No 222/34, 1982.

24. 'User Guide on Process Integration for the Efficient Use of Energy', The Institution of Chemical Engineers, 1983.

25. 'Inquiry into Private Generation of Electricity in Great Britain, 1977', Department of Energy, Economics and Statistics Division.

26. Financial Times, 3 May 1983.

27. Financial Times, 5 January 1984.

28. 'Heat Recovery from the Absorption Stage of a Sulphuric Acid Plant'. Energy Conservation Demonstration Projects Scheme, Project Profile No 1, ETSU, Harwell.

29. European Chemical News, 12 March 1984, 14.

30. Financial Times, 26 April 1983.

Table 1. UK Chemical Industry 1980, Output and Energy Consumption

STANDARD INDUSTRIAL CLASSIFICATION (1980)		SUBSECTOR IN THIS REPORT	GROSS OUTPUT		ENERGY CONSUMPTION (Excl. Feedstocks)		SEC
No.	Activity		£M	%	PJ	%	MJ/£
2511	Inorganic chemicals	INORGANIC	1298	7.9	55.5	17.1	42.8
2512	Basic organic chemicals	ORGANIC	3410	20.6	133.9	41.4	39.3
2513	Fertilisers	FERTILISERS	957	5.8	23.6	7.3	24.7
2514	Synthetic resins & plastics		1504	9.1	15.6	4.8	10.4
2515	Synthetic rubber		145	0.9	4.0	1.2	27.6
		POLYMERS	(1649)	(10.0)	(19.6)	(6.0)	(11.9)
2516	Dyestuffs & pigments	DYESTUFFS	674	4.1	20.1	6.2	29.8
2551	Paints, varnish, etc.		1048	6.3	4.2	1.3	4.0
2552	Printing ink		136	0.8	0.5	0.2	3.7
		PAINTS	(1185)	(7.1)	(4.7)	(1.5)	(4.0)
2570	Pharmaceutical products	PHARMACEUT'S.	2442	14.8	15.4	4.8	6.3
2581	Soap & synthetic detergents		850	5.1	3.3	1.0	3.9
2582	Perfumes, cosmetics, etc.		679	4.1	1.6	0.5	2.4
		DETERGENTS	(1529)	(9.2)	(4.9)	(1.5)	(3.2)
2562	Form. adhesives etc.		275	1.7	3.9	1.2	14.2
2563	Chemical treatment of oils & fats		108	0.6	1.6	0.5	14.8
2564	Essential oils & flavourings		185	1.1	1.0	0.3	5.4
2565	Explosives		169	1.0	3.6	1.1	21.3
2567	Misc. industrial chemicals		1483	9.0	30.9	9.5	20.8
2568	Form. pesticides		433	2.6	1.4	0.4	3.2
2569	Adhesive film, cloth, etc.		94	0.6	0.8	0.2	8.5
2591	Photographic materials		414	2.5	2.4	0.7	5.8
2599	Chemical products NES		224	1.4	0.6	0.2	2.7
		MISCELLANEOUS	(3385)	(20.5)	(46.2)	(14.2)	(13.6)
		TOTAL	16529	100.0	323.7	100.0	19.6
					INCL. FEEDSTOCKS		
2511	Inorganic chemicals	INORGANIC	1298	7.9	150.5	24.1	115.9
2512	Basic organic chemicals	ORGANIC	3410	20.6	337.6	54.2	99.0
		TOTAL	16529	100.0	622.4	100.0	37.7

Sources: As for Table 2.

Table 2. UK Chemical Industry 1980, Energy Supplied - Fuels and Feedstocks

ENERGY SUPPLIED PJ

SUB-SECTOR	CONVENTIONAL ENERGY SOURCES					OTHER SOURCES	FEEDSTOCKS	
	Total	Coal & Coke	Oil	Gas	Elec	Waste & Raw Materials	Oil	Gas
Inorganic	55.5	3.1	23.3	18.2	10.9	12	7.0	88.0
Organic	133.9	0.7	58.7	66.8	7.7	80	178.5	25.2
Fertilisers	23.6	0.4	4.2	14.3	4.6	1		
Polymers	19.6	1.4	8.3	4.3	5.7	7		
Dyestuffs	20.1	4.9	6.8	6.4	2.0			
Paints	4.7	0.1	1.9	1.9	0.8			
Pharmaceuticals	15.4	-	6.2	6.1	3.0			
Detergents	4.9	0.4	1.7	1.9	0.8			
Miscellaneous	46.2	2.5	9.4	22.5	11.8			
TOTAL	323.7	13.5	120.5	142.4	47.3	100	185.5	113.2

Sources:

Conventional Energy — Business Monitor 1979 Purchases Enquiry (PA 1002.1) re-calculated in line with 1980 output data and Digest of UK Energy Statistics energy data (see paragraph 53).

Waste and Raw Materials — IETS Reports (broad estimates only).

Feedstocks — Oil - Digest of UK Energy Statistics
Gas - Estimated from output data.

Table 3. UK Chemical Industry - Output and Energy Consumption 1960-1981

| YEAR | Output | INDEX 1960 = 100 | | | |
| | | Energy Consumption | | Specific Energy Consumption (Energy per unit output) | |
		Excluding Feedstocks	Including Feedstocks	Excluding Feedstocks	Including Feedstocks
1960	100	100.0	100.0	100.0	100.0
1961	101				
1962	105	99.0	104.4	94.3	99.4
1963	115	96.6	112.9	84.0	98.2
1964	125	97.6	119.5	78.1	95.6
1965	134	99.9	124.1	74.6	92.6
1966	141	103.3	130.8	73.3	92.8
1967	150	107.6	163.9	71.7	109.3
1968	161	108.9	158.7	67.6	98.6
1969	170	105.9	167.6	62.3	98.6
1970	179	110.4	174.8	61.7	97.7
1971	183	115.9	184.6	63.3	100.9
1972	194	108.3	194.1	55.8	100.0
1973	217	118.7	223.1	54.7	102.8
1974	225	111.8	220.8	49.7	98.1
1975	206	103.3	183.5	50.1	89.1
1976	230	109.2	204.0	47.5	88.7
1977	238	114.7	209.2	48.2	87.9
1978	241	112.7	208.8	46.8	86.6
1979	245	117.2	212.8	47.8	86.9
1980	226	103.3	173.3	45.7	76.7
1981	223	97.7	178.7	43.8	80.1

Sources:

Output - Annual Abstract of Statistics.

Energy - Digest of UK Energy Statistics
(previously Ministry of Power Statistical Digest).

Table 4. UK Chemical Industry, Growth Rate and Output Assumptions

SUB-SECTOR	BASE 1980 Gross Output £M	HIGH CASE (a) 1990 Growth Rate % p.a.	HIGH CASE (a) 1990 Gross Output £M	HIGH CASE (a) 2000 Growth Rate % p.a.	HIGH CASE (a) 2000 Gross Output £M	LOW CASE (b) 1990 Growth Rate % p.a.	LOW CASE (b) 1990 Gross Output £M	LOW CASE (b) 2000 Growth Rate % p.a.	LOW CASE (b) 2000 Gross Output £M
Inorganic	1298	2.0	1582	2.5	2121	0.5	1364	0.5	1434
Organic	3410	2.5	4365	3.0	6141	1.0	3767	1.0	4161
Fertilisers	957	2.0	1167	2.5	1563	0.5	1006	0.5	1057
Polymers	1649	2.5	2111	3.0	2969	1.0	1822	1.0	2012
Dyestuffs	674	2.0	822	2.5	1101	1.0	745	1.5	908
Paints	1185	2.0	1445	2.5	1936	1.0	1309	1.5	1596
Pharmaceuticals	2442	5.0	3973	6.0	7810	4.0	3615	4.5	5889
Detergents	1529	2.5	1957	3.5	3033	1.0	1689	1.5	2059
Miscellaneous	3385	3.0	4549	3.5	6714	2.0	4126	2.5	5547
TOTAL	16529	2.9	21971	3.6	33388	1.6	19422	2.0	24663

(a) HIGH CASE: GDP growth rate - average 1.5% p.a.
               High industrial growth

(b) LOW CASE: GDP growth rate - average 1.5% p.a.
               Low industrial growth

For basis of growth rate and output assumptions - see paragraph 60.

Gross Output is on a constant price basis.

Table 5. Use of Energy by Chemical Process Plant

Process	% of Total Energy Use
Space heating	6
Process heating	37
Evaporation/distillation	30
Drying	5
Refrigeration	5
Fans/blowers	2
Drive motors	2
Compressed air	5
Mixing	2
Pumping	5
Grinding	1

Source:

Industrial Energy Thrift Scheme Studies
(major energy users - i.e. > 90 TJ per site).
Provisional data only - see paragraph 75.

Table 6. UK Chemical Industry – Energy Demand by Final Use, 1980

Subsector	Total	Electrochemical and Process Heat > 300°C		Process Heat < 300°C		Space and Water Heating		Motive Power and Lighting	
	PJ	%	PJ	%	PJ	%	PJ	%	PJ
Inorganic	55.5	42.0	23.3	35.0	19.4	3.0	1.7	20.0	11.1
Organic	133.9	18.0	24.1	64.0	85.7	4.0	5.3	14.0	18.8
Fertilisers	23.6	9.0	2.1	61.0	14.4	10.0	2.4	20.0	4.7
Polymers	19.6	4.0	0.8	60.0	11.8	20.0	3.9	16.0	3.1
Dyes	20.1	5.0	1.0	52.0	10.5	20.0	4.0	23.0	4.6
Paints	4.7	0	0	33.0	1.6	45.0	2.1	22.0	1.0
Pharmaceuticals	15.4	0	0	52.0	8.0	25.0	3.9	23.0	3.5
Detergents	4.9	0	0	77.0	3.8	15.0	0.7	8.0	0.4
Miscellaneous	46.2	15.0	6.9	50.0	23.1	15.0	6.9	20.0	9.3
TOTAL	323.7	18.0	58.2	55.1	178.3	9.5	30.9	17.4	56.5

Source: Primarily from IETS data – see paragraph 78.

Table 7. UK Chemical Industry, Changes in SEC Projected

| Subsector | SEC MJ/£ | | | | |
| | Base | High Case | | Low Case | |
	1980	1990	2000	1990	2000
Inorganic	42.8	37.2	31.2	38.5	34.7
Organic	39.3	33.8	27.5	35.0	31.0
Fertilisers	24.7	21.5	18.0	22.2	20.0
Polymers	11.9	10.2	8.4	10.6	9.4
Dyestuffs	29.8	25.9	21.2	26.5	22.9
Paints	4.0	3.5	2.9	3.6	3.1
Pharmaceuticals	6.3	5.0	4.0	5.2	4.2
Detergents	3.2	2.8	2.2	2.8	2.5
Miscellaneous	13.6	11.6	9.5	11.8	9.9

Basis of Assumptions:

As indicated in paragraph 109, the changes in SEC are based on what has been achieved already, company targets for energy saving, known process improvements and the potential for general approaches to energy saving, such as process integration.

Reductions in the high case are generally greater than in the low case because of three main effects:

. more capital available for retrofits, etc;

. more new plant building (as additional capacity or replacement) with greater energy efficiency;

. some response to the higher energy prices in this case.

Reductions are greater in sectors with higher growth rates because of the higher rate of new plant building.

Table 8. UK Chemical Industry – Reduction in SEC Categorised by Energy Efficiency Measure (%)

SUB-SECTOR	TOTAL 2000		MANAGEMENT MEASURES (a) 2000		ADDITIONAL EQUIPMENT MEASURES (b) 2000		REPLACEMENT EQUIPMENT MEASURES (c) 2000		NEW PROCESS TECHNOLOGIES (d) 2000	
	High	Low	High	Low	High	Low	High	Low	High	Low
Inorganic	27	19	4	4	10	8	10	6	3	1
Organic	29	21	4	4	10	8	12	7	3	2
Fertilisers	27	19	5	5	10	7	10	6	2	1
Polymers	29	21	4	4	10	8	12	7	3	2
Dyes	27	23	5	5	10	9	10	8	2	1
Paints	27	23	5	5	10	9	10	8	2	1
Pharmaceuticals	38	34	3	3	9	9	20	17	6	5
Detergents	30	23	5	5	10	9	12	8	3	1
Miscellaneous	30	27	4	4	11	10	12	11	3	2

(a) Any measures which can be taken by management action with little or no capital expenditure, including switching off plant when not in use, better scheduling of processes, adjustment and maintenance of plant.

(b) Capital investment on measures specifically for energy saving, such as insulation, waste heat recovery etc. May be retrofitted to existing plant or 'optional extras' on new plant.

(c) Replacement of worn out or obsolete plant with new plant of modern design (not necessarily for energy saving reasons). Includes all energy saving from closure of old plant and building new plant, whether as a direct replacement or not.

(d) As for replacement equipment, but involving radical change in process concerned.

Table 9. UK Chemical Industry - Projected Fuel Demand (Energy Supplied Basis) PJ - High Case

SUB-SECTOR	COAL			OIL			GAS			ELECTRICITY			TOTAL		
	1980	1990	2000	1980	1990	2000	1980	1990	2000	1980	1990	2000	1980	1990	2000
Inorganic	3.1	11.5	17.1	23.3	17.5	17.5	18.2	18.4	19.2	10.9	11.5	12.5	55.5	58.9	66.3
Organic	0.7	15.5	25.0	58.7	48.7	48.9	66.8	73.3	81.0	7.7	10.0	14.0	133.9	147.5	168.9
Fertilisers	0.4	1.5	2.5	4.2	3.9	3.5	14.3	14.4	15.9	4.6	5.3	6.3	23.6	25.1	28.2
Polymers	1.4	5.5	7.8	8.3	5.2	4.2	4.3	4.9	5.6	5.7	6.2	7.5	19.6	21.6	25.1
Dyestuffs	4.9	6.0	7.2	6.8	5.8	5.3	6.4	7.2	7.3	2.0	2.3	3.5	20.1	21.3	23.3
Paints	0.1	0.3	0.5	1.9	1.5	1.3	1.9	2.2	2.6	0.8	1.0	1.3	4.7	5.0	5.7
Pharmaceuticals	-	4.5	8.0	6.2	5.1	5.5	6.1	6.4	10.8	3.0	4.0	6.7	15.4	20.0	31.0
Detergents	0.4	0.8	1.3	1.7	1.3	1.3	1.9	2.3	2.9	0.8	1.0	1.3	4.9	5.4	6.8
Miscellaneous	2.5	7.5	12.0	9.4	7.4	7.5	22.5	24.1	28.2	11.8	13.6	16.2	46.2	52.6	63.9
TOTAL	13.5	53.1	81.4	120.5	96.4	95.0	142.4	153.2	173.5	47.3	54.9	69.5	323.7	357.4	419.2
%	4.2	14.9	19.4	37.2	27.0	22.7	44.0	42.9	41.4	14.6	15.4	16.6			
SEC MJ/£ (based on output in Table 4)													19.6	16.3	12.6

Table 10. UK Chemical Industry – Projected Fuel Demand (Energy Supplied Basis) PJ – Low Case

SUB-SECTOR	TOTAL			COAL			OIL			GAS			ELECTRICITY		
	1980	1990	2000	1980	1990	2000	1980	1990	2000	1980	1990	2000	1980	1990	2000
Inorganic	55.5	52.5	49.7	3.1	8.5	8.5	23.3	17.2	15.2	18.2	16.8	17.0	10.9	10.0	9.0
Organic	133.9	131.8	129.2	0.7	5.5	6.3	58.7	51.7	48.3	66.8	66.9	67.1	7.7	7.7	7.5
Fertilisers	23.6	22.4	21.1	0.4	0.7	0.7	4.2	3.2	2.8	14.3	14.1	13.4	4.6	4.4	4.2
Polymers	19.6	19.3	18.9	1.4	4.0	3.9	8.3	5.2	4.8	4.3	4.1	4.3	5.7	6.0	5.9
Dyestuffs	20.1	19.8	20.8	4.9	5.1	5.4	6.8	5.9	6.0	6.4	6.8	7.2	2.0	2.0	2.2
Paints	4.7	4.7	4.9	0.1	0.2	0.2	1.9	1.6	1.6	1.9	2.1	2.2	0.8	0.8	0.9
Pharmaceuticals	15.4	18.9	24.5	-	3.5	6.3	6.2	5.4	5.6	6.1	6.0	6.6	3.0	4.0	6.0
Detergents	4.9	4.8	5.1	0.4	0.5	0.5	1.7	1.4	1.4	1.9	2.1	2.4	0.8	0.8	0.9
Miscellaneous	46.2	48.8	55.1	2.5	5.5	8.0	9.4	8.0	8.1	22.5	22.5	24.1	11.8	12.8	14.9
TOTAL	323.7	323.0	329.3	13.5	33.5	39.8	120.5	99.6	93.8	142.4	141.4	144.3	47.3	48.5	51.5
%				4.2	10.4	12.1	37.2	30.8	28.5	44.0	43.8	43.8	14.6	15.0	15.6
SEC MJ/£ (based on output in Table 4)	19.6	16.6	13.4												

- 45 -

Table 11. Comparison of CHP Prime Movers

	Steam Turbine	Gas Turbine	Diesel Engine
Heat/Power Ratio	> 5:1	1.5:1 with WHR 1.5-15:1 with after-burning	1:1 with WHR 1-5:1 with after-burning
Thermal Efficiency	80%	60-85%	75-85%
Typical Installed Costs (New System)	> £1,000 kWe	£200-400 kWe	£300-600 kWe
Simple Payback (New System)	5-10 years	2-10 years*	2-4 years
Size Range	10-100 MWe	10-30 MWe	0.5-10 MWe
Flexibility	Heat:Power ratio 'fixed' Turndown poor	Heat:Power ratio variable with after-burning Turndown poor	Heat:Power ratio variable with after-burning Turndown poor
Fuel Required	Any including coal and waste	Gas or light distillate fuel oil - waste gas	Heavy fuel oil
Number of Sets in Chemical Industry (1977)	119	7	4

* Depends on fuel pricing - see text.

Data is indicative only. Performance and cost of actual plant may lie outside these ranges, depending on circumstances.

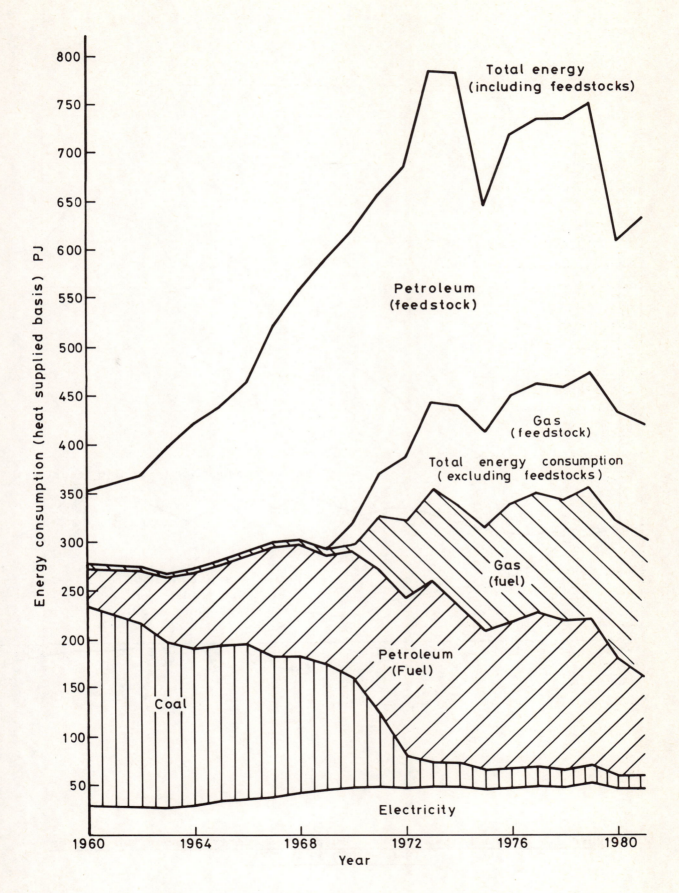

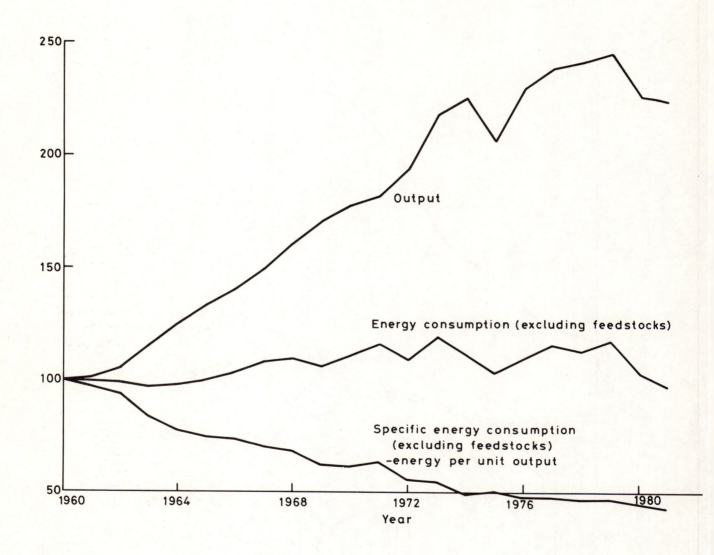

Figure 2. UK Chemical Industry. Output and Energy Consumption Indices
(1960 = 100) 1960-1981

INORGANICS

1.   Key data for 1980 (part estimated):

    Gross output   £1298 M
    Net output   £ 367 M

    No. of employees   25,000
    No. of establishments 179

Total energy consumption (heat supplied basis) PJ:

	TOTAL	COAL & COKE	OIL & LPG	NATURAL GAS	ELECTRICITY
Fuel	55.5	3.1	23.3	18.2	10.9
Feed	95.0		7.0	88.0	

Principal Products:

PRODUCT	OUTPUT 000 tonnes	ENERGY REQUIREMENT (APPROX)		RAW MATERIALS OR FEEDSTOCKS
		SEC GJ/tonne	TOTAL PJ	
Ammonia	2000	44	88	Natural gas and steam (all counted as feed- stock - see para 12).
Nitric acid	2830	- 1.5	- 4	Ammonia.
Sulphuric acid	3380	- 3	- 10	Sulphur.
Phosphoric acid	500	3.5	2	Phosphate rock Sulphuric acid.
Sodium carbonate	1300	10	13	Brine, ammonia, limestone.
Carbon black	170	80	14	Fuel oil feedstock (half of total energy requirement - see para 57).
Chlorine (& caustic soda)	910 (1050)	14	13	Brine.
Titanium dioxide	180	40	7	Minerals - rutile, ilmenite.

2.   This sector is primarily engaged in processing raw materials to produce basic chemicals on a large scale as intermediates for other parts of the chemical industry. The fertiliser sector is one of the major 'customers' for inorganic chemicals. In fact there is no real distinction between the sectors in practice. Most of the larger manufacturers operate integrated sites on which ammonia and the bulk acids are produced and feed directly into plants producing fertilisers. This makes it extremely difficult to attribute energy use reliably to the two classes of products in official statistics, so the data derived from these statistics is subject to some uncertainty. Indeed, it is doubtful whether even the production of chemicals is always accurately accounted in the appropriate subsector, particularly on sites whose major product is fertiliser, but which also produce inorganic chemicals for captive use in the process.

3.   Another major part of the inorganic chemical output goes to the petrochemical sector - e.g. ammonia for nylon production, chlorine for PVC. The bulk acids are also widely used in the petrochemical sector and elsewhere in the industry. Carbon black is used in making tyres.

4.   The following sections give a brief account of the major processes in the inorganic sector, with details on energy requirements and the scope for improving SEC where these are available. Much of the data on plant and production has been obtained from Chemfacts UK [ref. 17]. Data on process energy requirements have been obtained from the Energy Audit on the fertiliser industry [ref. 7], supplemented by estimates derived from a variety of published sources and private communications. In many cases different sources yield different data, so some judgement has been applied in deriving the figures quoted.

Ammonia

5.   Ammonia production is by far the largest single energy user in the inorganic chemical sector, and one of the largest in the whole chemical industry. It uses more energy than, for example, the non-ferrous metals industry.

6.   Three quarters of the ammonia output is used in fertiliser manufacture, mainly via the conversion of part of the output to nitric acid - see paragraph 23 below. Much of the remainder is used for production of plastics and fibres such as nylon, melamine and urea-formaldehyde resins. Ammonia can be used as a fertiliser by applying it directly to the land, but there is very little used in this way in the UK.

7.   There are only two manufacturers in the UK - ICI (the largest) and UKF (formerly Shellstar). Total plant capacity is somewhat above 2 million tonnes per annum, the plants dating from the early 60s to mid 70s. Plant size ranges from about 300 to 1,000 tonne per day. All the plants run on natural gas, though at least one of them was built to run on naphtha feed and later converted to natural gas. In the past ammonia has been made from coal, but there are no such plants operating in the UK now.

8.   The process is a simple one in principle - the reaction of hydrogen with atmospheric nitrogen over a catalyst at high pressure, known as the Haber process. The hydrogen is obtained from natural gas by reaction with steam at high temperature (steam reforming) to produce a mixture of hydrogen

and carbon monoxide (synthesis gas) which is further treated to yield hydrogen and carbon dioxide. The latter may be recovered as a by-product. The synthesis gas may be produced from any carbon or hydrocarbon feed such as coal or naphtha, but the process is more complex and less efficient from these heavier feeds and the capital cost of the plant is higher. Thus natural gas is the preferred feedstock. (Synthesis gas also provides the basis for the production of a wide range of organic chemicals, including methanol and other alcohols.)

9.   In practice the plants are extremely complex. The process streams must be purified, heated, compressed, recycled, cooled and separated in a multi-stage process under optimum temperature and pressure conditions at each stage. Most stages require a large energy input for heating or compression, but some are exothermic.

10.   The natural gas has three main functions in the process - as a feedstock for producing the synthesis gas; as a fuel for heating the primary reformer to drive the steam reforming reaction; as a fuel to raise the steam and generate the power required. Similarly the steam serves both as a feedstock for producing the synthesis gas and as a medium for heat transfer and power generation. In order to purge impurities from the reactors part of the product stream is also bled off and burnt as additional fuel.

11.   A modern ammonia plant is built as a highly integrated unit with extensive waste heat recovery. It is usually self-contained from an energy standpoint, with all mechanical and electrical energy requirements generated within the plant. In fact there is normally some provision for exchange of steam or power with neighbouring plants to provide operational flexibility, particularly during start-up or shut-down. A typical 1,000 tonne/day plant has a heat output of 150 MW, of which around 35 MW is used to drive the compressors. Indeed, a modern ammonia plant has been described as a large power station producing ammonia as a by-product.

12.   The allocation of the energy requirement is somewhat arbitrary, and depends on the operating conditions of the plant. The energy can be apportioned between feedstock and fuel on a theoretical basis (approximately 65% feed, 35% fuel), but there is not such a clear distinction between these functions on the plant, and it is customary in the industry to express plant or process performance in terms of total energy requirement per tonne of ammonia produced. This total energy requirement has all been identified as 'feedstock gas' in the tables, as indicated in the main report, para. 26.

13.   The energy requirement has been reduced from around 100 GJ/tonne for early coal based plants to around 35-40 GJ/tonne for modern gas fed plants. The labour requirement has fallen even more dramatically. Most of this improvement has derived from the simpler process route using gas as a feedstock, but the efficiency of gas plants has also been improved by greater attention to the heat integration of plants and economies of scale. It is now felt that plants have reached the largest practicable size in relation to the markets they serve (1,000-1,500 tonne/day) so there is no more to be gained from this approach.

14.   Most plants operating in the UK at present have an energy requirement of 42-45 GJ/tonne, which is not as good as the best current designs (down to 35 GJ/tonne) because even the most recent plants are now almost 10 years old.

Several companies have been trying to develop even lower energy processes. ICI now have such a process, called AMV, based on a new catalyst and other improvements, which is claimed to reduce the energy requirement to about 28 GJ/tonne. This is close to the theoretical minimum of 22 GJ/tonne, which is set by the thermodynamics of the process. A plant of this type was planned for the ICI Billingham site (initially in conjunction with a methanol plant, since the process technology is very similar). The plans have now been shelved because of the recession in the chemical industry, though a plant using the same technology is being build in Canada under licence. It is now doubtful whether even one of these plants will be built in this country because their viability is questionable, even at this level of efficiency, in view of the present overcapacity among Western countries and the potential competition from Middle Eastern countries with cheap natural gas and an easier financial environment, as discussed in paragraph 62 in the main report.

15.   This illustrates the dilemma facing the chemical industry at present. There is the potential for a reduction of perhaps one third of the energy requirement for one of the biggest energy using processes in the industry (i.e. a saving of up to 30 PJ at present production levels) but it is most unlikely to be achieved because of the uncertainty in the market and lack of finance to build the plants.

16.   There is the possibility of some improvement in existing plants, although because of the highly integrated design it has generally been considered that energy-saving retrofits would not be cost effective. However examples are now starting to appear which indicate that there may be more potential here than was originally thought. The technology of the AMV process may become available as a retrofit package, but it will not provide the same energy reduction as on new plant designed specifically for this process.

17.   The Audit on the fertiliser industry [ref. 7] identified two potential improvements which do not appear to have been applied yet. Hydrogen recovery from the purge gas was considered to give a potential energy reduction of about 1 GJ/tonne (i.e. a total saving, if applied to all plants of around 2 PJ). Recovery of low grade waste heat was estimated to give a maximum potential saving of some 18 PJ across all plants. However the problems in doing this are the same as those in recovering low grade waste heat from any large power plant, and there seems no prospect of achieving this on any significant scale.

18.   There is scope for improvement of the heat balance in ammonia plants using process integration techniques, and there seems to be some prospect that savings in energy requirement of around 10% could be achieved with paybacks of less than two years. Because of the size and complexity of the plants, such modifications would represent a major investment which could be of the order of £5M. In the present uncertain state of the market, companies are reluctant to invest on this scale, even with an apparently good return.

19.   How much of this will actually be achieved is very difficult to assess. With only two companies operating in this market they are unwilling to reveal much of their intentions, and they are probably genuinely uncertain about their plans themselves, for the reasons indicated above. Continued ammonia production also depends on continued availability of natural gas at a price

which enables the industry to compete with imports. Although in principle the plants could be converted to naphtha or NGL feedstocks, this is most unlikely to be economic for permanent operation, and restriction of gas supplies would almost certainly close the UK ammonia industry.

20.   The future for ammonia production in Europe as a whole is uncertain. Hoechst have announced that they intend to buy supplies from the Middle East rather than build any more plant in Germany [ref. 26] and this is probably typical of many companies. ICI recently indicated that they intend to build a methanol plant in the Middle East rather than the UK, because of favourable gas prices [ref. 27], and similar considerations can be expected to apply to ammonia plant. It is therefore unlikely that any further ammonia plant will be built in the UK unless particularly favourable feedstock supply conditions arise. If new plant is built, then older plant would probably become uneconomic, and would be shut down. A gradual closure of older plants as they become uneconomic in the face of competition from imported ammonia seems likely in any case.

21.   It is therefore assumed in the high case that output is maintained at the present level to 2000, and in the low case that output is reduced by 10% by 1990 and a further 10% by 2000. It is also assumed that existing plant will be retrofitted to improve energy consumption. This, coupled with closure of the least efficient plant will reduce SEC by 10% by 1990 and a further 10% by 2000 (to 40 and 36 GJ/tonne respectively).

22.   It should be recognised that much of this energy is sequestered in the ammonia product, and is carried through to subsequent processes (e.g. production of nitric acid, urea, etc). Improvements in utilisation of ammonia in these processes therefore contribute to the overall energy efficiency of the chemical industry. Because of the highly integrated nature of the industry, and especially the inorganics and fertilisers sectors, it may be difficult to distinguish from published statistics where any savings in energy use arise.

Nitric Acid

23.   Nitric acid is one of the major bulk products of the inorganic chemical sector - production has been rising in recent years, and in 1982 had reached 3 million tonnes per year. The process is a net energy producer, but only a small proportion of the energy can be usefully recovered.

24.   Some 70% of the output is used in fertiliser manufacture, either for ammonium nitrate (involving reaction with ammonia) or for compound fertilisers. The remainder is used in the manufacture of explosives, dyestuffs, pharmaceuticals and a variety of organic chemicals.

25.   There are six major manufacturers in the UK, of which ICI is by far the largest. The plants are spread over about ten different locations with site capacities which range from 50,000 to one million tonnes/year.

26.   The process is basically a simple one - ammonia is burnt in air over a catalyst in a two stage process and the gases are absorbed in water. The product is a dilute acid which is suitable for fertiliser manufacture. For most other purposes a concentrated acid is required, which is normally produced by distillation, although direct routes are available. The tail

gases from the plant contain oxides of nitrogen which are subject to emission standards – more stringent for new plants than for existing ones.

27.   Different temperatures and pressures are required in the two oxidation stages and the absorption stage to optimise the reaction rate and yield, and to meet the emissions standards. Ideally, the absorption stage should be carried out at higher pressure than the oxidation stage, but this requires an intermediate compressor which must withstand highly corrosive conditions and increases the cost of the plant. In practice different types of plant are built, making different compromises between operating conditions and cost.

28.   All the chemical reactions in the process are exothermic. Much of the heat is generated at the ammonia burner at high temperature, and part can be recovered by generating high pressure steam which can be used to generate electricity or for mechanical drives. The remainder of the heat is released at low temperature (to optimise reaction rates and minimise corrosion) and is largely dissipated in cooling water. The total economically recoverable heat from most plants is around 1–2 GJ/tonne of product. In principle more heat could be recovered, but this would be at lower temperatures, and is unlikely to be economic except in special circumstances. Utilisation of the energy i entirely dependent on funding suitable opportunities elsewhere on site, and it may be that even the high grade heat cannot be fully utilised.

29.   The largest single element in the energy balance of the process is the energy content of the ammonia feedstock. It is therefore important to ensure that this is utilised fully. However, utilisation efficiency is already around 96–98%, so the scope for further improvement is limited.

30.   Overall, the nitric acid process itself offers little scope for anything other than marginal improvements, but on some sites there may be opportunities for better utilisation of the heat released. The process is therefore unlikely to contribute significantly to any change in SEC for the inorganic chemicals sector. Production is expected to continue, at least at the present level, and probably to rise. ICI is building a new plant at Billingham, and new plants have recently been built by Norsk Hydro at Immingham and by SAI at Leith.

Sulphuric Acid

31.   Sulphuric acid is one of the most important industrial chemicals, and demand for it has long been viewed as an indicator of the state of the chemical industry as a whole. It is used in the manufacture of fertilisers (about 30%), detergents, paints, pigments, dyestuffs, plastics, synthetic fibres, explosives and in metallurgical processing. It has traditionally been the highest tonnage product in the whole of the chemical industry, although it has been overtaken by nitric acid in the last two years. Production was 3.9 m tonnes in 1973, but down to 2.9 m tonnes by 1981, reflecting the decline in the chemical industry. It is now expected to remain stable, or to grow slightly, if there is a recovery in the chemical industry as a whole.

32.   Production is more widely dispersed than for other bulk products with 15–20 manufacturers and over 20 plant locations in the UK. There is no single dominant producer. Some of the largest, all with over 10% of total

capacity; are Albright and Wilson, ICI, Leather's, Norsk Hydro and Tioxide. Plant sizes range from about 50 to 700 tonne/day.

33.   Virtually all the sulphuric acid produced in the UK is made from imported elemental sulphur by the contact process. This involves three distinct steps, all of which generate some heat.

34.   In the first step sulphur is burnt in air to produce sulphur dioxide. This is where most of the heat is produced - some 3 GJ/tonne of sulphuric acid. There is also a small heat release (about 0.3 GJ/tonne) from drying of the air stream by passage through concentrated acid.

35.   In the second step sulphur dioxide is oxidised to sulphur trioxide over a catalyst in a four-stage reactor, generating about 1 GJ/tonne of product. Progressively lower temperatures are used in the four stages to optimise reaction rates and yield.

36.   In the final step the sulphur trioxide is absorbed in medium strength sulphuric acid to produce concentrated acid. Part of this acid (around 1%) is drawn off and the remainder is diluted with water and recycled. This step generates 1.3 GJ/tonne of product. In some more recent plants the absorption step is carried out in two stages, the first of which is integrated with the catalytic conversion step. This gives a higher utilisation of the sulphur, which is necessary to meet current emission standards, but does not significantly affect the energy balance.

37.   The total heat release in the process is 5.7 GJ/tonne of product, of which about 3 GJ can be recovered from the process stream after the sulphur burner and catalytic converter by generating high pressure steam in a waste heat boiler. Much of the remainder is dissipated from the absorption section at low temperatures in cooling water or to the air, and has generally not been recovered. However, it is now being found that plants can be retrofitted to recover some of this heat with good paybacks.

38.   One example, carried out under the Energy Conservation Demonstration Projects Scheme is on the Berk Spencer Acids plant at Stratford, East London [ref. 28]. A heat recovery system has enabled 15% of the heat from the absorber (0.2 GJ/tonne of product or 4% of the total heat generated in the process) to be used to preheat the boiler feedwater. The original project had a payback of 1.4 years and is being extended (by the addition of further heat exchange surface) to recover more of the available heat with a payback of 1.2 years. An important feature is the use of an intermediate closed loop heat transfer circuit to separate the acid stream from the feedwater stream and reduce the risk of contamination. Heat recovery in the same part of the process has also been applied on the Richardson's plant in Belfast.

39.   The total amount of energy saved in these cases is relatively small, and it is not clear whether the same approach can be applied on all plants, but they illustrate several important points. The risk of contamination, which often discourages heat recovery, can be substantially reduced by careful design. A good return can be achieved from utilisation of low grade heat in situations where this has traditionally been regarded as uneconomic. However, the scheme is entirely dependent on there being a use for the energy recovered on site (or a potential sale outside the site).

40.   If the high grade heat is not being fully utilised it will not be economic to recover the low grade heat. For example, on one major site with large scale acids production, large quantities of high pressure steam are regularly blown to waste because it cannot all be absorbed in other plants or the site's turbo-alternator. It is therefore impossible to make reliable estimates of potential savings for the whole industry of individual measures in isolation. For example, there may be interactions with phosphoric acid production, which is often carried out on the same site, as discussed in paragraph 47.

41.   A novel development of the process has been demonstrated by Bayer, using a fluidised bed reactor to replace one or more stages in the catalytic converter. This offers the prospect of higher heat recovery, lower emissions and lower capital cost in new plant. However, there are no plans to build such plant in the UK, and with the present poor outlook and overcapacity no new building is expected in the foreseeable future.

Phosphoric Acid

42.   Phosphoric acid is the third of the bulk inorganic acids - produced on a smaller scale than nitric and sulphuric acids, but with important uses in the chemical industry. Production has been around 500,000 te/year for about 10 years, but with a marked fall since 1980. At least 75% of production is used in fertiliser manufacture. The bulk of the remainder is used in making detergents, with a small amount in foodstuffs. There are five manufacturers in the UK, of which Albright and Wilson is the largest.

43.   Statistics on production quantities, plant capacities, etc., are complicated by the use of two different conventions in the industry. It is customary to express quantities of any phosphatic fertiliser in terms of equivalent phosphorus pentoxide (P_2O_5) content. This convention is often applied to phosphoric acid as well, but it is not always clear which method is being used. One tonne of phosphoric acid (100%) is equivalent to 0.72 tonne P_2O_5.

44.   Over 90% of production is made by the so-called 'wet' process described below. A small amount is made by a dry process in which elemental phosphorus is burnt in air and hydrated to the acid. This is an exothermic process, and thus has a useful heat output, but there is a high energy requirement in producing the phosphorus. It is therefore a relatively high cost route, and is used mainly for acid of high purity required by the foodstuffs industry.

45.   In the basic wet process, known as the dihydrate (DH) process, the imported phosphate rock is ground and digested with sulphuric acid. This yields a slurry of phosphoric acid and calcium sulphate, from which the latter is filtered off. The filtrate - dilute phosphoric acid - must then be concentrated by evaporation to give the required concentration for normal industrial use (50% P_2O_5). The energy and acid requirements vary somewhat with the phosphate rock used. Typical SEC of the basic DH process is 5 GJ/tonne, of which 0.5 GJ is electrical energy for grinding the rock and 4.5 GJ is steam for evaporation.

46.   There are two variations of this process - the hemihydrate (HH) and hemidihydrate (HDH) processes - which reduce the energy requirements dramatically. The variation is in the digestion stage, enabling higher

strength sulphuric acid to be used, and producing phosphoric acid at the
required concentration directly. The evaporation stage is therefore
eliminated, and the total SEC is about 1 GJ/tonne. The main difference
between the two variants is in the phosphorus utilisation efficiency and the
quality of the calcium sulphate by-product (gypsum). The HDH process
produces it in a form suitable for use in the cement or plasterboard
industries without further treatment.

47.   Despite the obvious attraction of these later variants in terms of
energy efficiency, the HDH process only accounts for about one third of all
capacity in the UK, the remainder being still DH plant. The economic case
for replacement of plant to improve energy efficiency is invalidated if the
energy is available 'free' or at low cost. Phosphoric acid is often produced
on the same site as sulphuric acid, and can utilise low grade heat which
would otherwise have no alternative use. This may also discourage other
measures which could be taken on existing DH plants to reduce steam
consumption, such as installing double effect evaporators.

48.   The SEC for phosphoric acid production in the industry as a whole has
been taken to be around 3.5 GJ/tonne, as the weighted average of the two
processes. This could be reduced to 1 GJ/tonne if all production were
switched to the HDH process, but this is most unlikely to be achieved, for
the reasons given above. It is also doubtful whether there will be much new
plant building, since there is a trend towards phosphoric acid production
in areas which presently export phosphate rock, such as Jordan and North
Africa. It is therefore assumed that production in the UK will remain at
about the present level, but that SEC will be reduced to 2.5 GJ/tonne by 2000
as a result of improvements or replacement of existing plant.

49.   The Energy Audit on the Fertiliser Industry [ref. 7] remarked on the
uranium content of phosphate rock and commented that it is possible to
recover uranium from the phosphoric acid in the course of production. The
potential recovery in the UK was estimated to be around 200 tonne/year of
U_3O_8 which could provide, in a thermal reactor, electrical energy equivalent
to that produced by 3 million tonne of coal in a conventional power station.
Uranium recovery is practised at a number of plants in the US and Canada, bu
there appears to be no such operation in this country - possibly for a
combination of economic and political reasons.

Sodium Carbonate (Soda Ash)

50.   Sodium carbonate is another of the bulk inorganics - produced in
smaller quantities than the major bulk acids, but with a relatively high
energy requirement. No firm production statistics are available, but an
output of around 1.5 million tonnes per year has been indicated for the mid
1970s, and probably less by 1980 - say around 1.3 million tonnes per year.
It is used in the production of glass, paper, soap, detergents and a variety
of chemicals. It has also been used in the production of caustic soda, but
little or none is used for this purpose now.

51.   There is only one manufacturer in the UK - ICI, with plants at three
locations near Manchester (Lostock, Winnington and Wallerscote). No detailed
information is available, but the main stages of the process (Solvay process)
are well known. Brine is saturated with ammonia in an exothermic reaction.
Carbon dioxide is dissolved in this liquid in a two stage process which is

also exothermic. Sodium bicarbonate is formed as a precipitate which is
dried and calcined to produce the sodium carbonate. At the same time carbon
dioxide is given off, which is used in the earlier stage. Additional carbon
dioxide is also generated by kilning limestone. The lime so produced is used
to treat the by-product ammonium chloride solution to re-generate ammonia.

52.   This is therefore a tightly integrated process in which by-products are
treated and recycled to an earlier stage in the process. Although the early
stages are exothermic the heat produced is at relatively low temperature.
There are large requirements in the later stages for high temperature heat
for kiln-firing and drying. Energy efficiency will therefore depend largely
on waste heat recovery in several different ways (from kiln gases, reaction
mixtures, driers, etc.) provided the heat can be utilised. The SEC has been
estimated as around 10 GJ/tonne, but there is no indication of the
reliability of this figure, or the scope for improvement.

Carbon Black

53.   Although produced on a much smaller scale than the major inorganic
products (150-200,000 te/year) carbon black represents one of the biggest
single energy consuming processes in the inorganic sector, after ammonia.
The bulk of the output (around 90%) is used as a filler in tyres and other
rubber and plastics products. The remainder is used largely as a pigment in
printing inks, paints, etc., and in dry batteries.

54.   Production has been gradually declining over the past ten years, from
220,000 te in 1973 to 150,000 te in 1981. Exports, which once considerably
exceeded imports, have similarly declined. This decline in the industry
reflects the great increase in tyre life in recent years, despite an increase
in numbers of motor vehicles. There are only two manufacturers in the UK now
- Cabot Carbon at Stanlow and Sevalco at Avonmouth, each with roughly half
the total UK capacity. A third manufacturer, with a small plant at Swansea,
ceased production in 1982.

55.   The process is basically one of incomplete combustion or thermal
cracking of a hydrocarbon feedstock - either natural gas or oil. Several
different processes have been used, yielding different types of carbon black.
The characteristics of the product, such as particle size, density, surface
properties, etc., are crucial to its use in various applications, and these
depend critically on the operating conditions of the process. Indeed,
production probably involves more art than science.

56.   Most of the production in the UK is by the furnace process, using oil
as feedstock (usually a special grade of heavy fuel oil or similar). The oil
is sprayed into a furnace where it is partially oxidised and cracked, using
its own heat of combustion, supplemented with that of auxiliary natural gas.
The product stream is quenched with a water spray, and the carbon black
collected in a system of cyclone separators, filters, etc. The other
process, now little used, is the thermal process, in which the feedstock (oil
or natural gas) is subjected to straight thermal decomposition in a
refractory furnace, operated in a cyclic manner. No quench stage is used,
but separation is similar to the furnace process.

57.   Both processes have intrinsically high energy requirements, and the
approaches to energy efficiency depend on waste heat recovery and utilisation

of the combustible waste gases, possibly involving electric power generation.
However, if there is no suitable use for this energy elsewhere on the site
the viability of energy recovery schemes relies on sales off-site. This
means that the design of plant, its mode of operation and the scope for
improvements in energy efficiency are highly site specific. A modern plant
with a high degree of heat recovery has a SEC of around 75-80 GJ/tonne for
the net energy requirement of the process, of which about half is accounted
for by the feedstock, oil. Actual energy requirements may differ from this,
for the reasons outlined above.

58.   Existing plants are subject to commercial confidentiality, and no data
are available, but it is unlikely that their performance will differ
substantially from the above, or from each other. There may be some further
improvement in energy efficiency, but there seems no prospect of a
substantial change in this sector of the industry, either in level of
production or energy usage.

Chlorine and Sodium Hydroxide (Caustic Soda)

59.   Chlorine and sodium hydroxide are normally produced simultaneously as
co-products of the same process. Their joint production, together with
some peripheral activities, is collectively referred to as the chlor-alkali
industry.

60.   Uses of sodium hydroxide include pulp and paper manufacture, textile
processing, refining of petroleum and vegetable oils, and in the production
of many other chemicals. It is also used in the initial stages of aluminium
production, but not in the UK. Chlorine is used in the production of
chlorinated plastics (e.g. PVC, Neoprene), solvents, degreasing agents,
disinfectants, pesticides, aerosol propellants and in numerous other chemical
processes. It is also used as a bleaching agent, particularly in the paper
and textile industries, and as a disinfectant.

61.   For many years the industry was geared to the production of sodium
hydroxide, and chlorine was available as a by-product. Since about 1970 the
position has been reversed, and chlorine demand determines the production
rates, with sodium hydroxide as the by-product. However the demand for both
products is falling, and the industry is now in a weak state.

62.   This is the result of a number of effects. Some of the user industries
such as textiles and paper have, themselves, declined. Increasing energy
costs have hit the chlor-alkali industry particularly hard because the
process is intrinsically energy intensive, and some of the downstream
products have suffered in competition with less energy intensive materials.
This has happened particularly to PVC, for which the cost effects have been
reinforced by competition from technically superior plastics. Environmental
and safety pressures have also had an effect. Vinyl chloride monomer, the
intermediate in production of PVC, has come under attack on health and safety
grounds, as have chlorinated solvents, pesticides and aerosol propellants.
The main type of plant used in the chlor-alkali industry - the mercury cell -
has come under scrutiny for environmental reasons. Tighter standards have
increased costs and encouraged the development of other types of cell whic
are not without their own problems.

63.   Chlorine production was growing strongly in the 1970s, reaching a peak of over 1 m tonne/year, and is now falling. No figures are available for sodium hydroxide, but production could be some 15% higher than for chlorine, since the two products are normally produced in a constant ratio of about 1.15 to 1.

64.   There are four producers in the UK - ICI, BP, Associated Octel and Stavely, of which ICI is by far the largest. There is a considerable excess of plant capacity (some 40-50% higher than current production) despite the fact that BP recently closed a large plant in Wales, in conjunction with the rationalisation of plastics production between BP and ICI.

65.   The process is a simple one in principle. A concentrated brine solution is electrolysed, producing chlorine gas at the anode and sodium or sodium hydroxide (depending on cell type) at the cathode. For many years virtually all production was in mercury cells, in which the cathode is a pool of mercury. The sodium deposited by electrolysis forms an amalgam with the mercury which reacts with water in a separate part of the cell to produce a concentrated solution of sodium hydroxide, at the correct strength and purity for most industrial purposes. The mercury is returned to the electrolysis section of the cell. The depleted brine solution is also continuously removed, stripped of dissolved chlorine and purified before being replenished with more salt and returned to the cell. Concern about mercury in wastes from the plant have led to tighter standards on emissions, resulting in higher capital cost of plant.

66.   In recent years the industry has therefore turned to other types of cell, though mercury cells still account for the bulk of production in the UK. The alternative cell types are the diaphragm cell and a more recent variant, the membrane cell. In these, a diaphragm/membrane (originally asbestos, but now various plastic materials) is used to separate the products in the cell. The cells produce sodium hydroxide directly at the cathode, but at a lower concentration than in the mercury cell. The diaphragm cell usually also leaves some chloride contamination of the sodium hydroxide. The product from both cell types therefore normally requires a further stage of purification and/or evaporation to produce a saleable material.

67.   The energy requirement in the mercury cell is mainly for electrolysis (around 14 GJ/tonne of chlorine, as electrical energy). The diaphragm cell has a lower energy requirement for electrolysis, but has a large requirement for low temperature heat evaporation and purification (total energy requirement around 21 GJ/tonne of chlorine, of which about half is specifically electrical energy). The energy requirements of the original membrane cells fall between the other two, but the most recent designs can be below that of the mercury cell. Some additional energy is required for liquefaction of the chlorine gas, but hydrogen is produced in all types of cell, which may have a value for other chemical processes on site, or as a saleable by-product, or can be burnt as fuel.

68.   The choice of cell type and the means of energy supply depend on local conditions. If new plant were being built, membrane cells would probably be favoured. Because of the high and continuous electrical requirement there is an obvious case for on-site CHP, provided there is a use elsewhere for the heat generated, since there is little required on the chlorine plant itself except with the diaphragm cell. However, because of the overcapacity and

poor prospects in the industry there is unlikely to be significant investment
in new plant.

69.   In existing plant the opportunities for energy saving are relatively
limited. Newer types of electrode can improve the electrical efficiency of
the cells. Otherwise the opportunities lie mainly in peripheral operations
such as power generation and utilisation of the by-product hydrogen.
Overall, it is doubtful whether SEC can be improved by more than about 10%
without considerable investment, which the industry is not likely to make in
the foreseeable future.

Titanium Dioxide

70.   Titanium dioxide is widely used in the production of pigments and
fillers for paints, printing inks, paper, rubber and plastics. Production
has declined slightly in recent years, from 205,000 tonnes in 1978 to about
170,000 tonnes in 1982. There are two manufacturers in the UK - Tioxide,
with some 70% of the capacity and Laporte with the remainder (sold in 1984 to
SCM of New York). There are plants at three locations on the North East
coast.

71.   Two processes are used, starting from different minerals, each with
roughly equal shares of the production, and each operated by both companies.
In the sulphate process the mineral ilmenite, which also contains iron, is
treated with concentrated sulphuric acid, and ferrous sulphate is removed.
The titanium sulphate solution is hydrolysed with steam to produce titanium
hydroxide, which is calcined to titanium dioxide. In the chloride process
the mineral rutile - an impure form of titanium dioxide - is treated with
chlorine in a reducing atmosphere to produce titanium chloride. This is
oxidised at high temperature to the pure titanium dioxide.

72.   As with many such commercial processes with few producers, no firm
process details are available. An overall SEC of around 40 GJ/tonne has been
indicated for the production of titanium dioxide, apparently as an average
for the two routes. The chloride process has a lower energy requirement in
the process itself, but depends on chlorine which has a high energy
requirement in its production. In both processes the main energy requirement
is for high temperature heating and for drying at lower temperatures, and
there is a significant energy input into grinding. These operations, then,
are the main areas for seeking improvement in energy efficiency.

Inorganic Chemicals - General

73.   The processes described above represent the biggest identifiable
tranches of energy demand in the inorganic sector and, in general, the
biggest tonnage products. Together they account for some 123 PJ or 80% of
the total energy consumption in the sector including feedstocks. The bulk of
this (88 PJ) is the gas used as feedstock for one product - ammonia. Of the
energy used purely as fuel in the inorganic sector (55 PJ) only some 28 PJ
(net of energy released in nitric and sulphuric acid production) is accounted
for, or about 50%.

74.   Thus half of the fuel used in the sector - around 28 PJ - is used in
production of the hundreds or thousands of minor products, by a wide variety
of processes. These will include many pure metals and non-metals and their

compounds such as oxides, sulphides, salts and halogen compounds, in
quantities from a few tonnes to many thousand tonnes. Most of these products
are not identified individually in official statistics, and many are used as
intermediates in the preparation of other materials, so it may be difficult
to account for the products themselves, let alone the energy used in their
manufacture.

75.   The main processes used in their production include calcining,
evaporation and drying; also some of the reactions are exothermic. Therefore
an important technology for energy conservation is waste heat recovery over a
wide temperature range, from high temperature flue gases to low temperature
vapours from driers. Mechanical operations such as grinding, mixing and
pumping are also significant energy users in these processes.

76.   A significant change in the 1980 Standard Industrial Classification is
that the production of industrial gases (including oxygen, nitrogen,
hydrogen, carbon dioxide and the rare gases) has been reclassified from
'Inorganics' to 'Miscellaneous Chemicals for Industrial Use' (Class 2567).
This is a relatively energy intensive group of products, with a high
requirement for electrical/mechanical energy for compression and
liquefaction. The transfer will have caused some discontinuity in the
figures for the inorganic sector, as compared to previous years, which makes
it more difficult to discern trends.

77.   For all the above reasons it is difficult to assess the potential for
improving efficiency in a large part of the energy used for fuel purposes in
this sector. This will depend partly on the growth in the sector and hence
the extent of new plant building or major revamps of existing plant.

78.   The prospects for the bulk products look poor, as discussed above, and
most of these are likely to remain static or decline in the low case, with
the possibility of some growth in the high case. The minor speciality
products have better prospects, and should achieve growth rates of at least a
similar order to the industry as a whole. This will provide opportunities
for the conservation measures discussed in the main report (paragraphs
118-127), which could reduce SEC by perhaps 15-20% in these processes over 20
years. In addition, structural change in this sector away from the energy
intensive products would further reduce the SEC for the sector as a whole.

79.   In the high case there would be more investment in new plant with
higher energy efficiency, but less structural change. Overall the effects on
energy consumption are consistent with the assumptions discussed in the main
report (para. 110) and incorporated in Table 7, which shows a reduction in
SEC of 27% in the high case and 19% in the low case. In addition the gas
used as feedstock for ammonia plants is subject to the changes discussed in
paragraph 21.

ORGANICS

1.   It has not been possible to produce an analysis of the organics subsector in the same detail as for inorganics, through the limitations of both the available information and time. Instead a brief outline of this part of the industry is given, with an indication of the main products, processes and feedstock aspects.

2.   This sector accounts for some 70% of the feedstocks, 40% of the conventional fuels and 80% of the by-products used as fuels in the whole of the chemical industry. It forms the major part of what is loosely termed the petrochemical industry. The major input, other than conventional energy sources used as fuels, is petroleum feedstock, mainly in the form of light distillate naphtha. This is processed to produce a wide range of basic organic chemicals which are used for the production of plastics, synthetic rubber, fibres, solvents, paints, detergents, pharmaceuticals, pesticides, etc.

3.   The organics sector therefore has close links with many others in the chemical industry, especially plastics, whose major processes are often carried out on the same site as basic organic chemical manufacture. There is close integration of plants, including sharing of utilities and use of by-products as fuel, so it is often difficult to assign energy use reliably to basic organic chemical manufacture and other sectors. It is convenient to consider organic chemicals in several major groups, olefins, aromatics and 'the rest', but methanol stands out as a product which merits individual attention.

Methanol

4.   Methanol is somewhat anomalous, in that it is classified as an organic chemical but it is produced (from natural gas) by a process which has much in common with that used for ammonia – one of the main inorganic chemicals (see Annex 1, paragraph 8). Indeed, the only plant in the UK, operated by ICI, is at Billingham, alongside the ammonia plant, and there have been plans to build an integrated ammonia/methanol plant on this site.

5.   With only one producer, detailed production and plant data are confidential. Published data indicate that production is probably of the order 600–700,000 te/year, of which perhaps half is exported. Energy input is estimated to be around 25 PJ of natural gas which fulfils the function of both fuel and feedstock in the same way as for ammonia. All has been counted as feedstock in the tables.

6.   Methanol is used in the production of a number of different plastics, solvents and other basic organic chemicals, such as acetic acid. It is also used specifically by ICI as the feedstock for production of their synthetic protein 'Pruteen'.

7.   There has been much discussion of the potential use of methanol in motor transport, for which it could be used as a fuel in its own right, in specially designed engines, or as an additive to normal motor spirit as an octane improver. A small amount is being used for this purpose in Europe already. It can also be used to produce methyl tertiary butyl ether (MTBE) which offers some advantage as an octane improver. Neither of these uses

have any significant economic advantage over lead compounds as an octane
improver at present, but as legislation forces lead out there could be more
interest in this application.

8.   Methanol is also a potentially versatile chemical feedstock, which
could provide an alternative route to a wide range of products currently
derived from petroleum naphtha. It offers no particular advantage for this
purpose at present, but if the supply/pricing of feedstocks shifts to favour
gas supplies from the Middle East or elsewhere, then methanol could have a
part to play in tapping these resources and providing a relatively cheap and
convenient energy carrier.

9.   As far as the UK is concerned it seems unlikely that any new methanol
plants will be built unless particularly favourable circumstances arise. As
noted above, ICI were planning a methanol plant at Billingham that would have
roughly doubled UK capacity. The project has since been deferred and recent
indications are that ICI is intending to build a plant in one of the Gulf
states, where it would have the advantage of cheaper natural gas feedstock
[ref. 27]. The Chairman of ICI has been reported as stating "it is
inconceivable that we will continue to invest in the UK in methanol. We
can't compete against the Gulf States" [ref. 29].

10.   It is therefore assumed that methanol production in the UK will
continue at roughly the same level as at present, subject to any
de-bottlenecking of existing plant. Any possible improvement in SEC is
assumed to offset any increase in plant output, so requirements for natural
gas feedstock would be largely unchanged.

Olefins

11.   The principal chemicals in this category are ethylene, propylene,
butylenes and butadiene. Total production is well over 2 m te/year, of which
about half is ethylene. These are among the highest tonnage products in the
organic sector, and can be considered as the core of the petrochemical
industry. Between them they form the starting point for most of the
plastics, solvents and synthetic rubber produced by the industry, and for
many other materials. The plants are some of the biggest and most costly in
the chemical industry. There are four producers in the UK - ICI, BP, Shell
and Esso.

12.   The basic process is one of thermal cracking of a hydrocarbon feedstock
at high temperature in the presence of steam. This yields a mixture of
olefins, by-product gases and liquids which are separated by distillation.
The by-product gas is normally used as fuel, on the plant itself or elsewhere
on site. The by-product liquids (hydrocarbons from C_5 upwards) normally have
a high aromatic content, and may be further processed to extract the
aromatics, used as a motor spirit blending component on a neighbouring
refinery, or burnt as fuel.

13.   The yield pattern from an olefin plant depends on the hydrocarbon
feedstock used and the cracking conditions. Typical yield patterns for
fairly severe cracking are shown below.

14.   The major feedstock in the UK at present is naphtha (around 90% of the
total). Some of the crackers are capable of using gas oil, but little is run

at present, and this is more likely to decrease than to increase. The main change in feedstocks which is taking place at present and will continue in in the future is the move towards ethane, propane and butane (Natural Gas Liquids, NGLs) from the North Sea.

Typical Yield from Cracking Hydrocarbon Feedstocks - % weight

	Ethane	Propane	Butane	Naphtha	Gas oil
Ethylene	80-81	44-46	35-38	31-33	24-26
Propylene	1-3	15-17	16-18	14-16	14-16
Butadiene	1	1-2	1-2	4-5	4-5
Butylenes	-	-	-	4-5	4-6
Fuel gas	14-16	28-30	30-32	15-18	10-11
C_5 +	1-2	7-9	12-14	24-27	36-39

15. The ICI/BP cracker at Wilton (Olefins 6) has the flexibility to run on naphtha, propane or butane, though it is understood to have run largely on naphtha so far. Since this represents about one third of current UK capacity it could have a major impact on feedstock demand. The BP cracker at Grangemouth (about 15% of current UK capacity) has been converted to run on ethane, and there have been indications that their Baglan Bay cracker may be adapted to run on a wider range of feedstocks [ref. 30]. The new Esso/Shell cracker at Mossmorran is being built to run on ethane only (almost 30% of current UK ethylene capacity, due on stream in 1985).

16. These changes in plant capacity and feedstock flexibility, together with possible changes in product demand, will have a number of effects which are not easy to predict. First, it is not clear whether the Mossmorran cracker will add to UK total capacity, or whether other plant will be shut down. Europe as a whole is still considered to have excess olefins capacity, despite numerous plant closures, and there are pressures to make further reductions. The situation will be exacerbated when Middle East plants begin to come on stream from about 1985 and start impinging on the European market. Whatever happens over the next few years, it seems likely that, by 1990, total olefins capacity in the UK will be no higher than in 1980, and it is doubtful whether any further plant will be built by 2000. Production is currently only some 60-70% of capacity.

17. With modern plant and the flexibility to use a range of feedstocks, depending on availability and cost, the UK should be in a position to at least defend its own markets, and it is assumed that production of olefins will at least be maintained at the present level. In a high growth scenario production could increase substantially by 1990, but prospects after that are less promising, and it is doubtful whether there would be any further growth up to 2000. Alternatively it has been suggested that cheap imports of bulk chemicals from the Middle East could encourage the concentration of European chemical manufacture in the industrial heartland of Europe, leaving the UK in an exposed and vulnerable position.

18.   The future feedstock requirement is thus difficult to estimate. It is also influenced to some extent by aromatics production (see paragraph 22) but the requirement for olefins is the main determinant. In total it could range from somewhat below the 1980 level (4 m tonnes or 190 PJ) to perhaps 50% greater. The composition of this requirement will certainly shift from the present domination by naphtha towards a position where naphtha may be less than half the total, with ethane providing the greater part of the remainder, and propane/butane the balance. Such estimates are highly speculative however, since the availability of these alternative feedstocks can vary dramatically with new discoveries in the North Sea.

19.   Changes in the feedstocks will affect the proportion of the products, as indicated in paragraph 13 above. In particular, a move to ethane in place of naphtha will substantially reduce the amount of propylene produced, which could affect the industry in several different ways. The industry may readjust its downstream processing on the basis of a lower availability of propylene - eg cutting out the production of some products or switching to other process routes. Alternatively, propylene could be produced from other sources, such as refinery gases, or imported from countries with a surplus - i.e. those still processing heavier feedstocks. A move to gas feedstocks will also reduce the amount of liquid by-products available for production of aromatics (see paragraph 22 below) or for use as fuel.

20.   The implications of all these changes for overall energy demand in the industry cannot be assessed realistically without a much deeper analysis. Energy requirements, other than as feedstock, are primarily for the cracking furnace itself, for generating the steam used as diluent, and for the separation and compression of the product streams. Modern plants are highly integrated from an energy standpoint, with extensive use of CHP and waste heat recovery. New plant will certainly have a higher energy efficiency than the equivalent old plant, but the changes in feedstock, the availability of by-products and the consequent changes in downstream operations may offset this to some extent. The net effect on energy requirement, excluding feedstocks, is taken to be included in the overall reduction in SEC for the sector shown in Tables 7 and 8.

Aromatics

21.   This group of chemicals includes benzene, toluene, and xylenes. Production has fluctuated widely over the past 10 years, reaching a peak in 1979 of over 1.5 m tonnes, and is currently less than 1 m tonnes. Benzene accounts for about two thirds of the total. The aromatics are important basic chemicals, used in the production of many plastics, fibres, detergents, pharmaceuticals, pesticides, explosives and dyestuffs. Major producers are ICI, Shell and BP, plus a few smaller companies.

22.   Production is mainly based on extraction and purification of these compounds from mixed streams derived from other processes - chiefly the liquid by-products from olefin plants, as mentioned above, or refinery reformate streams. Some catalytic reformer plants are run specifically for aromatics production, using a naphtha feedstock. Energy requirements vary widely, being relatively low for the extraction/purification processes and considerably higher for reforming processes - hence the interaction with the energy requirements for olefins production referred to above.

Organic Chemicals - General

23.   There is a vast range of other organic chemicals, mainly produced from
the basic chemicals described above, sometimes in reactions involving
inorganic chemicals, such as chlorine, nitric acid or sulphuric acid. Some
of these are used as products in their own right, such as solvents,
anti-freeze, etc., but many are produced only as intermediates in the
manufacture of plastics, fibres, etc. The range of processes is enormous,
but the energy-using operations fall into a few categories, as summarised in
Table 5 - chiefly heating and cooling of process streams, distillation,
evaporation and compression. The opportunities for improvement in energy
efficiency have therefore been covered in the main report (paragraphs
118-130) and in the SEC changes summarised in Table 7 and 8.

ENERGY USE AND ENERGY EFFICIENCY IN UK MANUFACTURING INDUSTRY

UP TO THE YEAR 2000

SECTOR 4. PAPER, PRINTING AND PUBLISHING

K F Langley and A V Ward

Subsector	Paragraph
Introduction	1
4.1 The Paper and Board Industry	7
4.2 Conversion of Paper and Board	85
4.3 Printing and Publishing	94
References	

SECTOR 4. PAPER, PRINTING AND PUBLISHING

INTRODUCTION

1.   The paper, printing and publishing sector covers a disparate range of industries which have only one common feature: they are all concerned with paper. In terms of energy consumption and process technology, however, they could scarcely be different.

2.   The Paper and Board Industry is concerned with the manufacture of paper and board from pulp. It involves energy intensive processes, especially for drying.

3.   The conversion of paper and board covers a group of industries which convert the output of the paper and board industry into finished products, such as wallpaper, hygiene goods, and packaging containers.

4.   The printing and publishing industry includes the printing of newspaper, magazines, books and printed stationery. It has one of the lowest energy intensities of all the subsectors of manufacturing industry and comprises a very large number of small firms.

5.   Table 1 compares the three subsectors in terms of total energy consumption, number of establishments and energy intensity (defined as energy purchases/net output)

Table 1. Comparison of the Subsectors of Paper, Printing and Publishing

Subsector	Energy# Consumption (PJ)	Number** of Establishments	Energy Purchases** Net Output (%)
Paper and Board	93.3	262	33.4
Conversion	23.4*	1,731	4.0
Printing and Publishing	19.2*	11,740	1.7
Total	135.9	13,733	5.4

\#   1980 data
*   estimated
** 1979 data (ref 1).

6.   In this report, the energy intensive paper and board industry is examined in greater detail than the other sectors, partly because there is more information available. Because of its high energy intensity, it also provides the greatest scope for energy saving. By contrast, there is little information available on the other subsectors. It is more difficult to identify the scope for energy savings and to assess the likely uptake of energy saving measures for these subsectors.

4.1   THE PAPER AND BOARD INDUSTRY

INTRODUCTION

7.   Paper and board manufacture is an energy intensive industry which in 1980 used 93.7 PJ, ie 5.9% of the energy used by UK manufacturing industry. The industry has undergone considerable contraction and structural rationalisation in the recessions of 1974-76 and 1980-82, but the decline has recently stabilised, with some increase in output evident in the past year or so.

8.   Energy use in the paper and board industry has been studied in a report in the IETS series (ref 2) published in 1979, and a report in the Energy Audit series (ref 3) published in 1982; both reports were prepared by PIRA, the research association for the paper and board, printing and packaging industries.

9.   The British Paper and Board Industry Federation (BPBIF) has published an annual energy report (ref 4) for the industry based on an annual survey since 1965, and on a quarterly survey since 1981. It has also conducted a pilot programme of detailed energy monitoring in four paper mills since 1980. It is planned to extend this programme to cover the whole industry in a comprehensive energy monitoring and targeting programme with assistance from the Department of Energy.

10.   The paper and board industry traditionally generates a substantial proportion of its own electricity in combined heat and power installations (CHP). However, many of the CHP plants are old and are not generally being replaced. The proportion of CHP has declined from about 66% in 1965 to 32% in 1982.

11.   This report contains a description of the energy use in the industry in 1980 and of the potential for improving energy use up to 2000. Where data are available, trends in energy use since 1980 are noted and incorporated into the projections of future specific energy consumption.

STRUCTURE AND TRENDS IN THE INDUSTRY

12.   Figure 1 shows the principal processes and material flows in the UK paper and board industry in 1980. The basic raw material for paper making is pulp comprising, on average, a mixture of 52% recycled wastepaper, 40% imported virgin pulp and 8% virgin pulp from domestic pulpwood.

13.   The UK paper and board industry produces approximately half of the total UK consumption of paper and board products; the remainder is met by imports.

14.   There is a wide range of different products covered by the industry. These are generally classified according to the following product groups:

Figure 1. Principal Processes and Material Flows in UK Paper and Board
Industry in 1980 (million tonnes)
(Source: Annual Abstract of Statistics ref 5)

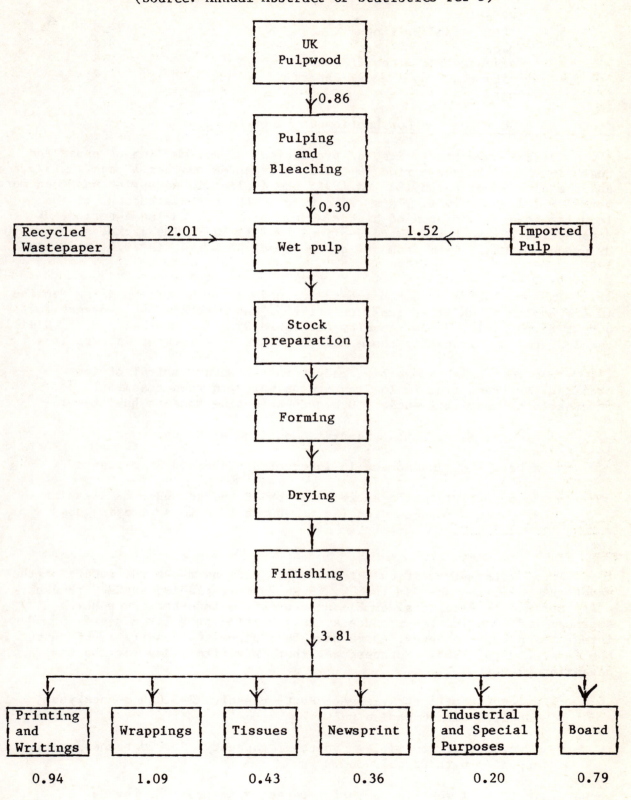

Product Group	Proportion of Output in 1980 (%)
Newsprint	9.5
Printing and Writings	24.5
Wrappings	28.6
Household and other Tissues	11.4
Industrial and Special Papers	5.2
Board	20.8

Trends in Production and Consumption of Paper and Board

15.   Table 2 show statistics for production and consumption of paper and board in the UK in the period 1971-81. Production reached a peak in 1973, and has since declined by 23% to 1987; the decline in newsprint has been more marked - 80% since 1971. This reflects a shift in the structure of the industry towards products of higher added value. In the same period, UK consumption of paper and board declined by only 12%. The market share of the UK producers has fallen from 58% of the total UK market in 1973 to 48% in 1980.

16.   The decline in output has been accompanied by a corresponding decline in the number of UK mills (26% since 1971), and a persistent tendency for output to be 10-15% below capacity (cf Table 3). The average age of UK mills is high and until recently there has been little investment in capacity.

17.   Thus the industry has been caught in a downward spiral of low profitability resulting in low investment which in turn results in poor international competitiveness. Other contributing factors have been:

- high exchange rates vis-à-vis Continental Europe

- substantial investment in integrated plants in Scandinavia

- favourable terms of entry into the UK market under EFTA tariff agreements for Scandinavian producers (due to expire in 1984).

Future Prospects

18.   It is extremely difficult to make projections about the future of the paper and board industry in the UK. Even in an expanding market, producers will continue to face considerable pressure from imports. Both North American and Scandinavian producers are investing on a large scale in modern integrated plant. However, there have been signs of recovery in UK output in the first half of 1983. Furthermore, there are firm plans for two new newsprint plants:

- a rebuilt mill at Ellesmere Port, capacity 245,000 tonnes/year, supplied with Canadian pulp,

- a new integrated mill at Shotton, capacity 180,000 tonnes/year, supplied with UK pulpwood.

These projects will restore newsprint capacity to pre-1973 levels.

19.   Higher value products such as printings and writings, and tissues have increased their share of total output and may continue to do so over the next ten years or so. Wrapping paper and board, however, are under competitive pressure from other packaging materials, and may decline as a proportion of output. It is not clear how the market for industrial and special papers will change.

Production Processes in Paper and Board Making

20.   Paper making is a continuous process. In fully integrated mills, pulpwood (logs, wastewood, chippings, etc) is converted into virgin pulp which is then blended with recycled pulp from waste paper. Most UK mills, however, buy their pulp from overseas sources. There are presently only two fully integrated mills in the UK.

21.   In the following paragraphs, a brief technical description is given of the main steps in paper making. A schematic diagram of the process is given in Figure 2.

22.   <u>Pulping</u> Paper, tissue and board consists of cellulose fibres derived mainly from wood. The wood fibres are separated either by mechanical pulping or chemical pulping. Mechanical pulping consists simply of grinding wet logs until the fibres are torn apart. This provides a high yield of ground wood and is relatively inexpensive. Chemical pulping uses solvents which dissolve the lignin (a resin which binds the cellulose fibres together) causing the fibres to separate without damage. The pulp is then bleached if white paper is required. The chemical process is more expensive and provides a lower yield but permits more refined processing of the pulp to give better quality papers. Where pulp is bought in, as in most mills, it must be re-dispersed in a slusher; the pulp suspended in water is called <u>stock</u>. This is the start of paper making in most UK mills.

23.   <u>Refining</u> The next process is refining, in which the fibres are lacerated in a controlled fashion. To obtain the required paper properties the fibres are bruised to increase flexibility and frayed to increase the surface area. The required strength, fineness and appearance of the finished sheet paper depends upon the degree of flexibility and fraying. Very heavy refining gives dense translucent sheets such as greaseproof or tracing paper.

24.   Refining machinery consists of a barred metal cone rotating inside a barred conical shell which bruises and frays the stock between the two sets of bars as one rotates within the other.

25.   <u>Pulp cleaning</u> The refined stock is then passed through a vortex cleaner in which the stock is diluted with recycled water from the paper machine and screened to remove impurities and unwanted débris. The degree of cleaning depends in part upon the proportions of stock from waste paper and stock derived from pulp mills; in particular de-inking may well be necessary if a large proportion of recycled waste is printed matter. Waste paper stocks need much less refining because the fibres have already been through that process.

26.   <u>Paper Making</u> The stock is turned into paper on the paper machine. This consists of several parts, shown in Figure 2, which are driven by

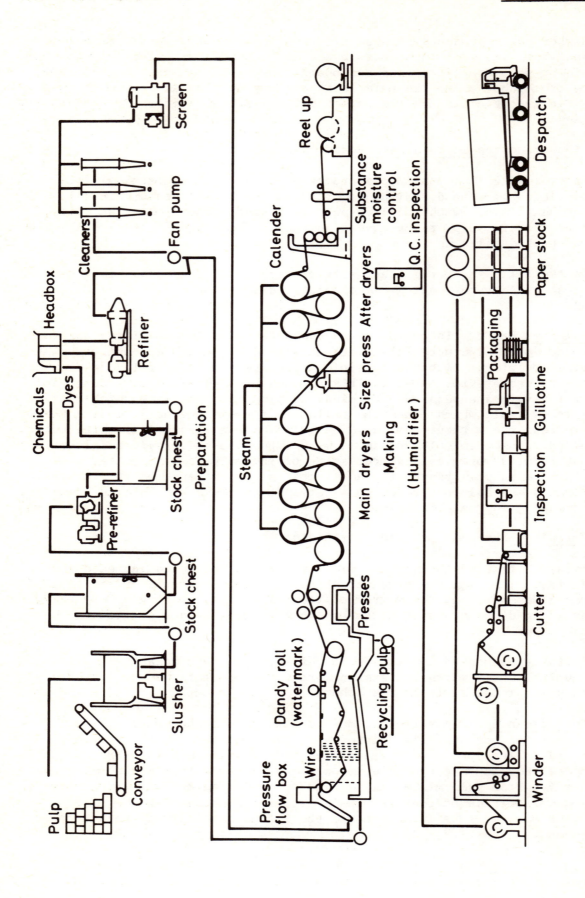

Figure 2. Paper Making Processes

Source: Reference 3

synchronised drives. The main function of the machine is to remove water from the stock and leave the fibres in the form of a sheet. Paper machines are typically 50 metres long and 3-5 metres wide and contain perhaps three presses and 40 drying cylinders. The thin stock (fibres suspended in about 200 times their weight in water) flows out of a narrow slit on to the wet end of the paper machine. Most of the water is drained leaving a layer of fibres felted together which is further squeezed in presses before the sheet enters the dryers where it is held against steam heated cylinders. After passing over about 30 drying cylinders the dry sheet is sometimes passed through a size press which treats both sides with solutions aimed at modifying the surfaces of the sheet. The sheet is finally reeled at the dry end of the paper machine having passed through a calender which smooths the surface of the paper. About two tonnes of water have to be evaporated for each tonne of paper made.

Paper and Board Products

27. This basic process can be varied to produce a very wide range of paper and board products. The variations in paper types have been described by the industry as follows: 'Paper may be impregnated, enamelled, metallised, made to look like parchment, scraped, waterproofed, waxed, glazed, sensitized, bent, turned, folded, twisted, crumpled, cut, torn, dissolved, lacerated, moulded, embossed. It may be coloured, coated, printed, marked and the mark erased. It may be laminated with itself and with fabric, plastic and metal. It may be made opaque, translucent or transparent. It may be made to burn or made fireproof. It may serve as a carrier and barrier or a filter, it may be made tough enough to withstand acid or soft enough for a baby's skin. It may disintegrate or it may be reused' (ref 3).

28. Some papers are coated with a mixture of pigments and binders; the degree of finishing depends on the end use of the paper. Paper is sometimes simply cut and re-wound into smaller reels; other finishing includes embossing or laminating.

29. Making board is a matter of bonding together several plies of paper from stock which is prepared in two or more parallel systems. The wet end is adapted to handle the appropriate number of plies which are merged and bonded in the process. Paper making machines run at speeds of between 100-1,000 m/minute (around 4-40 miles per hour); board machines are slowest and tissue machines fastest.

30. Paper, particularly damp paper, is vulnerable to tears and breaking. Damaged paper is called broke and is recycled within the mill. This reduces product yield and increases the amount of energy used to produce a 'net' tonne of saleable paper or board, which is the usual measure of output taken.

31. The different grades are normally classified into the six product groups referred to previously (paragraph 14). The main characteristics of the product groups are described in the following paragraphs.

32. Newsprint Newsprint has to be suitable for printing at high speed. It is made from mechanical pulp on large, fast machines.

33.   Printing and Writings   These are white papers made from a mixture of
chemical and mechanical pulps with mineral fillers to increase the opacity.
Chemical pulps give high quality paper but at a cost. The paper is used for
books, magazines and stationery in the office, home and school. This area
accounts for 25% of the output of the paper and board industry and meets
about 25% of UK demand; it is also exported in significant quantities.

34.   Wrappings and Packagings   This product group comprises papers for
wrapping, bags and sacks and case materials used for making cases and boxes
(about 80% of the total). Production accounts for about 1/3 of the output of
the paper and board industry and meets about 50% of UK demand for packaging.
Case materials include corrugated board made by a conversion process in which
three layers of paper are laminated together. The middle ply, which is
called fluting, is corrugated during the process; the outer layers, called
liners, are glued to the peaks.

35.   Household Tissues   This market consists mainly of toilet paper but
facial tissues and handkerchiefs also comprise significant quantities of
output. They are made from chemical and mechanical pulps and from selected
wastes which does not require much cleaning. This sector has grown in the
last decade and production plant is relatively modern.

36.   Industrial and Special Papers   Both the range of special purposes and
the methods of manufacture vary so much that it is difficult to generalise.
Most special purpose papers are made from chemcial pulp to give the required
quality and much of the output represents conversion rather than paper
making. It includes most of the industrial fire proofing, water proofing,
oil proofing and similar conversion processes. It accounts for about 7% of
the output tonnage and the contribution to exports is significant.

ENERGY CONSUMPTION IN THE PAPER AND BOARD INDUSTRY

37.   The energy consumption in the paper and board industry for each year
since 1965 is shown in Table 4.

38.   Specific energy consumption has fallen from a peak of 26.7 GJ/tonne in
1968 to 21.8 GJ/tonne in 1982, a decrease of 18.4%. Much of the decrease
occurred in the period 1968-1974; after 1974 the specific energy consumption
increased sharply, to 26 GJ/tonne in 1976. The trend in SEC is shown
graphically in Figure 3, where it can be compared with the trend in load
factor (output/capacity). It can be seen that the rise in SEC in 1974/5 was
accompanied by a sharp fall in load factor; a similar but smaller effect
occurred in 1979/80. Thus there is an inverse correlation between SEC and
load factor. This is because the energy consumption involves standing losses
in keeping plant hot when it is running slow, or stopped between frequent
changes of product run.

39.   Nevertheless, there is an overall downward trend in SEC, which can be
seen by comparing two years when the load factors were almost the same. For
example, between 1971 and 1981, the SEC fell from 25.7 GJ/tonne to 21.9
GJ/tonne. This represents an underlying improvement in energy efficiency of
15%. Part of this improvement is due to the closure of less efficient mills.
Table 3 shows that 44 mills closed between 1971 and 1981.

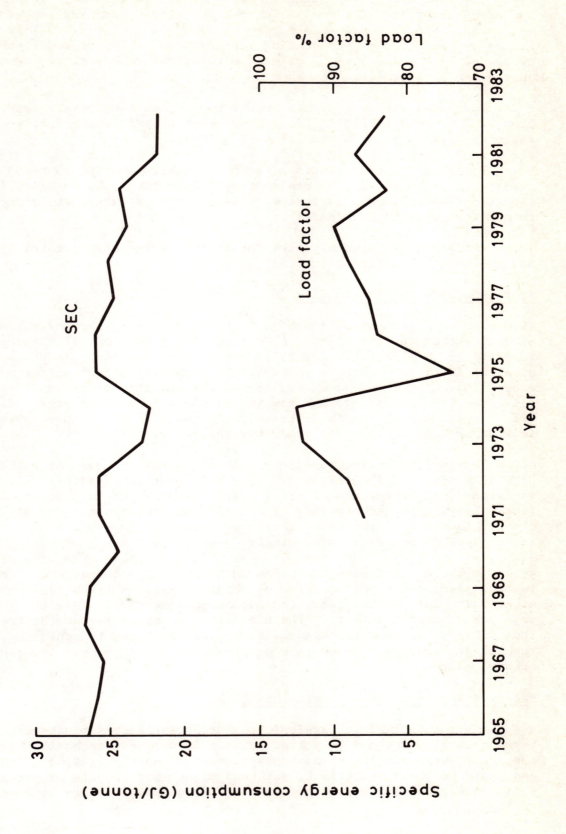

Figure 3. Trend in SEC and Load Factor

40.   The fuel mix in the industry has changed dramatically since 1965, when the predominant fuel was coal. In the late sixties and early seventies there was a rapid shift to oil, and subsequently to natural gas. Oil is still the largest source of heat, accounting for about one third of total energy purchased by the industry. Coal and natural gas each account for about one quarter of the total.

41.   Purchased electricity has increased its share of total energy purchased from 4.2% in 1965 to 12% in 1982. In the same period, the self-generation ratio has fallen from 66% to 32% (cf Table 4).

42.   Including self-generation, the total electricity consumption in the industry is about 17.5% of total energy consumption. Up to half of the electricity requirement is in the refining process, where it is used to rotate the refining machines.

43.   Process heat is mainly used for drying, where steam-heated cylinders consume up to 14 GJ/tonne (ref 3).

Self-generation of Electricity

44.   The self-generation plant in the industry is mainly passout and back-pressure steam turbine plant. According to a survey of self-generation in industry in 1977 (ref 7), the paper industry generated 2072 GWh (ie 7.46 PJ) of electricity from 73.7 PJ of fuel. Thus the electrical efficiency is around 10%. The survey does not record the quantity of process steam delivered, but according to a sample calculation in the energy audit (ref 3), 69% of the input energy is delivered as steam. Thus, the heat:power ratio of self-generation in the paper industry in 1977 was about 7:1.

45.   Passout/condensing turbines which accounted for just over 50% of the total CHP plant in industry in 1977, can be operated flexibly, allowing some variation in the heat:power ratio. Back-pressure turbines do not have this flexibility. The requirements of the industry for steam and power have changed in that the required heat:power ratio is now about 4. Hence, the latter type of plant is increasingly uneconomic.

46.   Replacement of obsolete steam turbine plant with new coal-fired steam turbine plant has a payback time of about 5 years, whereas replacement with a steam-only coal-fired boiler, and purchased electricity from the grid, has a pay-back time of 2-3 years. The economics therefore favour the trend away from CHP. This could be changed if the industry were to adopt new CHP technology based on gas-turbines or diesel sets with boost-firing of the exhaust gases.

Energy Management Demonstration Project

47.   A project aimed at developing an effective energy management system based on monitoring and targeting principals was undertaken in 1980-82 by the BPBIF, with funding from the Department of Energy (ref 6). The project was based on the concept that, although major improvements in energy efficiency require investment in plant and equipment, there is considerable

opportunity for energy saving to be achieved through better management control of energy use within existing plant conditions.

48. Energy management systems have been established at four pilot mills of different characteristics, broadly representative of the whole UK paper and board industry. The systems all involve a three level approach of monitoring and reporting energy use:

- a top level for senior management, on a monthly basis

- a second level for departmental management on a weekly basis

- a third level for continuous detailed control.

Where necessary, additional instrumentation was installed to measure fuel use and other relevant parameters.

49. Preliminary data from the pilot study indicates that savings of 5-13% have been achieved from the application of monitoring and targeting systems. Scope for additional savings has been identified, but requires plant changes involving some investment.

50. An effective monitoring and targeting system can influence energy efficiency because it:

- highlights unfavourable comparisons with other mills, thus providing motivation to make savings

- allows allocation of management responsibility

- identifies and evaluates the opportunities for savings

- provides a feedback on the savings achieved as a result of measures taken.

51. In the next section, the range of specific energy consumption in a number of mills is analysed in order to determine the difference between average and best practice energy consumption. This difference is a measure of the scope for energy saving which could be achieved if the average mill reduced its energy consumption to the current best practice level.

'Best Practice' Energy Consumption

52. There is a wide variation in specific energy consumption by individual mills. Much of this is due to inherent differences in the energy requirement for different products. Table 5 shows a breakdown of SEC by different product groups based on 1982 data collected by the BPBIF from a sample of 43 mills. The average of all mills within a group is weighted by the output of each mill. It varies from 16.5 GJ/tonne for wrapping and packaging (corrugated case materials) to 38.3 GJ/tonne for coated printings and writings.

53. Even within a product group, there are wide variations in SEC for individual mills. For example, for household tissues the lowest energy

consumption was 9.4 GJ/tonne, compared with a weighted average of 19.7 GJ/tonne. In general, the difference between the lowest and the highest SEC within a product group is a factor of 2-4.

54.   The main factors influencing specific energy consumption are:

- the type of paper within the broad category of product groups

- the weight of the paper (as measured in gram/m^2)

- the machine (which will vary in specification by width, speed, hood design and the efficiency of the presses).

55.   Many other factors also influence specific energy consumption. These include:

- degree days which will affect the incoming air temperature to the drying cylinders as well as the space heating requirement

- the composition of the input pulp - different composition of waste paper (proportions of board, news print, fine glossy magazine paper, miscellaneous papers) vary in their refining properties and their capacity to retain or release water during the drying process

- scheduled and unscheduled maintenance

- the renewal frequency and the quality of felts in the presses

- the formation or uniformity of the newly formed paper web - this is important in determining the drying time

- the length of the order or production run. Altering a paper machine to accept a new grade of paper takes some 10-15 minutes and during this time the paper machine is maintained at a fairly high temperature to minimise the start up time, particularly for the steam cylinders. Frequent changes in product will increase SEC.

56.   It is not clear to what extent the wide variation within product groups is due to further differentiation in product type, the nature of the machines, operational procedures or special circumstances. However, it seems reasonable to assume that a value for 'best practice' SEC can be obtained by disregarding exceptionally low SEC values and considering those values which are around the 20th percentile in a ranking by SEC of all mills within a group. The last column in Table 5 shows the values for 'best practice' SEC which are obtained in this way. For some product groups, there were not enough mills in the sample to provide a best practice value based on the 20th percentile criterion. For these groups, the lowest SEC provides a lower bound for the best practice SEC.

57.   Applying these best practice values to the 1980 product mix gives a weighted average best practice for the industry as a whole of about 18 GJ/tonne. Thus for the average SEC to become equal the current best practice, would require an improvement in energy efficiency of 26% compared with the 1980 average (the base year for the present study) or 17% compared with the 1982 average.

58.   Technically, it should be possible for all mills to achieve the best practice SEC. However, for some mills, especially the older ones, it is unrealistic to assume that the best practice can be achieved without major investment.

59.   Some proportion of the best practice potential can be realised by management measures and relatively minor investment in energy saving equipment; some more can be achieved by more intensive investment. Eventually, old mills must be replaced by modern plant, and the average mill will thus achieve the current best practice SEC, while technical improvements will reduce the best practice even further. In the next section, the technical scope for improving energy efficiency is discussed.

ENERGY CONSERVATION MEASURES

60.   In the previous section it was shown that there is considerable variation in energy consumption between mills, even within the same product group, and that significant improvements in energy efficiency can be achieved by appropriate energy management. In this section, the technical measures which could be adopted are discussed in detail. In general, these measures involve substantial investment. Most are already commercially available overseas, even if they are not currently used in the UK. Some could be installed in existing plants, others would only be used in completely refurbished or new plants.

Water Removal from the Web

61.   About two tonnes of water normally are evaporated for every tonne of paper produced although modern pressing equipment permits a reduction in this to around 1½ tonnes. Drying by steam is almost seven times as costly as removing water in presses which in turn is about eight times as costly as removing water by vacuum.

62.   Wet-pressing The extent of water removal in the press section of the machine can be maximised by studying the characteristics of each machine and the particular grade of paper being produced, in order to get the best results from the energy expended in driving the press section of the machine. The following factors are particularly relevant:

- the length of time the paper is saturated during the contracting part of the press nip,

- the type of felt used,

- the resistance to the flow of water in the paper and the felt,

- the degree of rewetting in the expanding part of the press nip.

63.   Vacuum removal Modern paper and board machines make extensive use of vacuum to remove water, with consequently high power consumption.

64.   Drying The method of drying has a major effect on the properties of the paper produced and there are many reasons why the industry still relies on steam heated cylinders in spite of technical inefficiencies. The most significant improvement in drying by steam cylinder has been in ventilation.

Canopy hoods or totally enclosed hoods are now used in modern installations and ventilated by dry air to reduce the vapour pressure in the atmosphere surrounding the web. Many of these developments have been a requirement dictated by the need to maintain the pace of modern paper machines without introducing an excessive number of cylinders. In tissue production conventional multiple steam cylinders are not practicable on such light materials; so drying has to be done on one very large cylinder, with auxiliary drying by simultaneously blowing hot air at the sheet from an air cap.

Refining

65. Specific energy consumption in refining varies considerably because of the way the refiners are used. In part, therefore, this is energy management or good housekeeping, in part it relates to the capacity with which the refiners are run. Significant improvements in electrical energy could be achieved in the refining process.

Waste Heat Recovery

66. There are a number of opportunities within the paper-making process for recovery of waste heat by the introduction of appropriate heat exchangers. For example, water evaporated from the web is often discharged directly to the atmosphere as warm moist air. Recovery of the sensible and latent heat in this water vapour has been the subject of a demonstration project. The recovered energy is used to pre-heat the air intake to the drying section of the machine.

Radio Frequency Drying

67. The selective removal of moisture from the web by radio-frequency (RF) techniques has been used as a means of reducing variations in moisture content over the width of the web. The technique has the potential of saving energy because of the greater efficiency of heat transfer to the moisture, and because overdrying at the edges of the web, which is at present necessary to meet the specification for moisture content in the middle of the web, can be eliminated.

Dry-Forming Techniques

68. Alternative paper-making techniques avoid the use of wet pulp and hence the energy consumption required for drying. There are two variations:

    - the 'carding' techniques and

    - the 'air laid' techniques.

69. The carding technique meters the dry fibres onto the web by mechanical means. Fibre to fibre bonding is then achieved by spraying binders or by heat fusion. It is a well proven technology but is limited by the quality of the bond achieved and, more seriously, by the speed of production.

70. The air laid technique has the potential for a much higher rate of output. The fibres are suspended in an air stream and the moving web is

formed from this suspension. The characteristics of the air stream can be controlled and this has an effect on the characteristics on the paper web formed. Fibre to fibre bonding is achieved by the addition of binders or by heat fusion. The kind of board or paper produced in this fashion is suitable for filter, paper nappies; carpet backing components; tissues; and cushioning and insulation materials.

71.   The air layering process has greater potential for development than the carding system primarily because of the speed of output. There is also in the future the possibility for refining the process in order to produce finer quality paper. The advantages of the dry process are:

- the machinery can be much smaller and relatively inexpensive;

- it avoids the use of large water supplies;

- energy requirements would be much less (typically saving about 40%) because of the absence of drying;

- further investment to increase capacity would be possible in much smaller steps because the machine (and hence costs) is much smaller;

- the process should be capable of accomodating changes of grade more easily and less time therefore would be lost in machine start-up and shut-down;

- the process produces much less pollution;

- the location of the industry is not dependent on water supplies and can therefore be located either near the source of fibre or closer to markets;

- a new process usually offers the possibility of producing entirely new materials.

72.   At present, R&D work in the UK and elsewhere is not being actively pursued. The main use to date for the dry forming process has been for the manufacture of soft products such as towelling, tissue grades and nappy pads. Harder products have not achieved commercial status using the dry forming process mainly because the quality of the bond to give the appropriate strength to the paper has not been possible using a spray or heat treatment process.

73.   A hybrid wet-dry process which utilises the bonding strength of two outer wet laid webs pressed onto an inner layer of a dry laid web seems to offer major gains. The combined sandwich of wet and dry webs is brought into contact with a consolidating cylinder (a form of press dryer) where the base sheet is produced. The rest of the process is conventional water removal and drying. Interest in this process is high enough to induce a conference on recent test results and prospects in 1983 with some expectation of further development funding.

PROJECTIONS OF FUTURE SPECIFIC ENERGY CONSUMPTION

74.   The extent to which the opportunities for improving energy efficiency in the paper and board industry will be realised in the period up to 2000 depends very much on how much investment in new and refurbished plant occurs, which in turn depends on the output of the industry.

75.   Paper and board is an internationally traded commodity, and the trend of recent years has shown how vulnerable the UK industry is to foreign competition. The level of output depends not only on the UK demand for paper and board products, but on the future competitiveness of the UK producers. The extent to which UK producers can maintain or improve their competitive position involves many factors which are beyond the scope of this report to assess. However, it is reasonable to assume that the industry can only survive if it does invest in cost-cutting measures, including reduction of energy costs.

76.   In Volume I of the present study of energy efficiency, two scenarios of future output are described, based on the upper middle and lower middle scenarios of the Department of Energy's 1982 Energy Projections. Both scenarios assume a GDP growth rate of 1½%. The high output scenario is one which envisages relatively buoyant prospects for a traditional manufacturing industry, while the lower output scenario assumes a structural shift towards 'new' industries, especially those in the service sector.

77.   The projections of future specific energy consumption are shown in Table 6 for both scenarios, together with the projected fuel mix. The assumptions which have been made in arriving at these projections are discussed in the following paragraphs.

High Output Case

78.   It is assumed that the SEC in the high output case will decline to reach the current best practice level of 18 GJ/tonne by 2000. Although there may well be some structural change in the mix of product groups, the effect which this would have on the overall SEC is small and well within the uncertainty in identifying the best practice value. Hence it is assumed that there will be no change in product mix.

79.   Self-generation of electricity has been declining for many years. The remaining generating plant is relatively old steam turbine plant and there is as yet little evidence that the industry is replacing it with new gas turbine plant. Although this trend could conceivably be reversed, no such assumption is justified on the present evidence. It is therefore assumed that self-generation will continue to decline, but at a slower rate than previously, perhaps to 25% in 1990 and 20% in 2000.

80.   Coal is expected to increase its share of the steam-raising fuel requirement, mainly at the expense of oil; the economics favour this trend at present fuel prices, and should become more favourable if oil and gas prices rise relative to coal. The rate at which this trend will occur is difficult to determine, but considering the rapid changes in fuel mix which have occurred in the past (cf Table 4), it is not unreasonable to suggest that coal could supply 50% of the total energy requirement by 2000. Oil and gas would supply the balance equally, with perhaps a substantial proportion of

the remaining self-generation plant being gas turbine or diesel sets fired by gas or gas-oil. The corresponding values for SEC and fuel mix in 1990 are obtained by interpolation.

Low Scenario

81. In the low scenario case, similar trends are assumed, but at a somewhat slower rate. Thus the shift in average SEC towards current best practice may not be complete by 2000. However, considering the progress made between 1980 and 1982, and the estimated improvement through the application of management measures alone (5%), a value of 19 GJ/tonne by 2000 does not seem unreasonable. Similarly, a somewhat lower value for the penetration of coal-firing is assumed, with a value of 40% taken as a reasonable expectation.

Uncertainties in the Assumptions

82. It is clear that there are major uncertainties in projecting future specific energy requirements in the UK paper and board industry. For example, there could be a structural shift towards high added value products, such as printings and writings or tissues, which would affect the average SEC. However, this is a second-order effect compared with the considerable uncertainty in identifying the best-practice value for SEC within each product group. The range of SECs even within a product group is very large (cf paragraph 56), with individual mills in some groups reporting values below 10 GJ/tonnes.

83. No attempt has been made to identify in the projections the technological changes which will occur. In previous sections, some of the options were discussed. Some measures such as heat recovery, depend only on the economic prospects of the industry. Others, such as dry-forming, involve technological development. It is not possible to be definitive, however, about the extent to which these measures will contribute to the SEC in the period up to 2000.

84. Nevertheless, it is clear that there is ample scope for reducing specific energy consumption, and although the prospects for the industry in the UK remain uncertain, its survival depends on reducing costs, including energy costs. There is ample evidence that the projected SEC values are realistic.

Table 2. Trend of UK Production and Consumption of Paper and Board

	1971	1972	1973	1974	1975	1976	1977	1978	1979	1980	1981
UK Production* (10^3 tonnes)											
Newsprint	575	468	442	382	315	326	301	319	364	363	113
Other Paper & Board	3,792	3,907	4,265	4,245	3,343	3,862	3,844	3,877	3,889	3,454	3,295
Total Production	4,367	4,375	4,707	4,627	3,658	4,188	4,145	4,196	4,253	3,817	3,408
UK Consumption* (10^3 tonnes)											
Newsprint	1,369	1,447	1,639	1,493	1,254	1,280	1,271	1,329	1,396	1,374	1,349
Other Paper & Board	5,708	5,890	6,458	6,596	5,060	5,803	5,932	6,330	6,436	5,720	-
Total Consumption	7,077	7,337	8,097	8,089	6,314	7,083	7,203	7,659	7,832	7,094	-
Exports	216	226	278	387	229	302	311	337	366	394	
UK Producers' Domestic Market Share (%)	58.7	56.5	54.7	52.4	54.3	54.9	53.2	50.4	49.6	48.3	

Source: Annual Abstract of Statistics (ref 5).

Table 3. Trend in Number and Utilisation of UK Mills 1971-1982

	1971	1972	1973	1974	1975	1976	1977	1978	1979	1980	1981	1982
Number of UK Mills	168	153	149	147	146	142	141	141	136	133	124	115
Utilisation (%)	86	88	94	95	74	84	85	88	90	83	87	83

Source: BPBIF (ref 4).

Table 4. Energy Consumption by the Paper and Board Industry 1965-1982

Year	Production (10³ tonnes)	Energy Consumption (PJ)	SEC GJ/tonne	Fuel Mix (%) Coal & Coke	Oil	Nat Gas	Purchased Electricity	Self Generation Ratio*
1965	4,614	121.4	26.3	70.4	25.4	—	4.2	66.2
1966	4,600	118.0	25.7	67.5	28.0	—	4.5	64.7
1967	4,473	115.7	25.9	62.7	33.1	—	4.1	66.9
1968	4,702	125.5	26.7	60.2	35.1	—	4.6	65.1
1969	4,970	131.4	26.4	57.7	37.4	—	4.9	63.0
1970	4,941	119.9	24.3	56.0	38.4	—	5.4	67.6
1971	4,367	112.3	25.7	41.1	47.7	5.5	5.6	69.5
1972	4,375	112.7	25.8	34.2	45.6	14.9	5.3	62.4
1973	4,707	107.1	22.8	31.5	46.1	16.6	5.8	58.7
1974	4,627	103.4	22.3	29.6	44.3	20.2	5.9	59.6
1975	3,658	94.8	25.9	28.0	45.4	19.9	6.8	58.6
1976	4,188	108.7	26.0	26.0	42.0	25.2	6.7	52.7
1977	4,145	102.4	24.7	26.2	39.7	27.0	7.2	51.0
1978	4,196	105.4	25.1	25.3	40.7	24.7	9.3	43.4
1979	4,253	101.7	23.9	26.1	40.5	23.4	10.0	39.9
1980	3,817	93.3	24.4	26.2	38.0	24.4	11.4	36.3
1981	3,408	74.5	21.9	25.6	36.5	26.8	11.0	34.2
1982	3,186 (E)	69.4	21.8	26.5	32.0	29.5	12.0	32.0

* $\dfrac{\text{Self-generated electricity}}{\text{Self-generated + purchased electricity}}$

Source: Adapted from Table XII of reference 4.

Table 5.   Range of Specific Energy Consumption Data by Product Groups

Product Group	Wt Ave GJ/tonne	Lowest GJ/tonne	Highest GJ/tonne	'Best Practice'
Printings and Writings (uncoated paper)	29.7	17.7	49.2	20
Printings and Writings (coated paper)	38.3	22.3	55.2	-
Fine Paper Specialities	33.4	12.3	62.5	23
Wrapping and Packaging (corrugated case)	16.5	11.3	20.2	14
Wrapping and Packaging (specialities)	36.8	11.6	57.8	-
Household Tissues	19.7	9.4	28.9	16
Industrial and Special Purpose Papers	27.0	16.1	68.3	-
Packaging Boards	32.9	14.8	45.0	20
Board Specialities	23.0	9.6	44.0	20

Note: SEC is purchased energy per tonne of saleable product.   'Best practice' SEC is a value which is approximately representative of those mills which have a specific energy consumption around the 20th percentile of the distribution of SEC.   For some products groups, there are not enough mills to provide a 'best practice' value on this basis.

Table 6.   Projections of Future Specific Energy Consumption

	1980	1982	1990 High	1990 Low	2000 High	2000 Low
SEC (GJ/tonne)	24.4	21.8	19	20	18	19
Fuel Mix (%)						
Coal	26.2	26.5	35	30	50	40
Oil	38.0	32.0	26	29	18	23
Gas	24.4	29.5	26	28	18	23
Electricity*	11.4	12.0	13	13	14	14

* Purchased electricity only.

4.2   CONVERSION OF PAPER AND BOARD

INTRODUCTION

85.   The conversion of paper and board is identified in the 1980 Standard Industrial Classification as a separate group of activities (SIC 472) which is quite distinct from the manufacture of paper and board from pulp (SIC 471).   The products of the paper and board industry are inputs to the conversion industry.

86.   The products of the conversion industry are:

- decorative wallcoverings

- household and personal hygiene products

- stationery (notepaper, binders, etc)

- packaging products (bags, boxes, etc).

87.   There is little information on the subsector contained in reports in the IETS or Energy Audit series, and hence the analysis of energy use and energy conservation potential in this report is necessarily incomplete and speculative.   Nevertheless, it is an important energy using subsector, which according to reference 1 consumes about 25 PJ, ie 1.6% of the total energy used by manufacturing industry.   Some consideration of the future specific energy consumption is necessary in order to provide a complete picture of the sector.

ENERGY USE

88.   The sole source of information on energy use in the paper conversion industry is the BSO 1979 Purchases Inquiry (ref 1).   Table 7 shows the relevant data.   Oil is the main fuel used by the industry: coal is less than 4% of the total.   There is little difference in fuel mix between the different product groups.

89.   There is no information available on how the energy is used, although by analogy with other energy non-intensive industries (eg engineering and textiles) it might be assumed that space heating would account for a large proportion of the fossil fuel use, while motive power (for machinery) and lighting would account for electricity use.

90.   In order to obtain an estimate of energy use in 1980 (the base year for the present study) it is necessary to pro-rate the 1979 energy consumption by the value of the output of the industry.   Table 8 shows output data for 1979 and 1980 in both gross and net terms in 1980 money.   Note that the ratio between the 1979 and 1980 values differs for the two monetary measures. Gross output, which is a measure of the total value of sales, is a more appropriate measure of output volume than net output, which is a measure of added value.   Applying the ratio of gross output to the 1979 energy consumption gives an estimate of 23.4 PJ for the 1980 energy consumption. This assumes that the specific energy consumption has not changed between the two years.   The fuel mix is also assumed to be unchanged.

PROJECTIONS OF FUTURE SPECIFIC ENERGY CONSUMPTION

91.   In the absence of any information about the pattern of energy use and the potential for energy saving in the industry, the projections of future SEC can only be speculative. However, analogy with other energy non-intensive sectors would suggest that savings of around 15% by 2000 should be possible. These might be brought about through improvements in building insulation, boiler efficiency, and lighting.

92.   Table 9 shows the future SEC values which can be estimated from the above assumptions. The value for 1990 is obtained by interpolation between 1980 and 2000.

93.   Changes in fuel mix are also difficult to assess. It is reasonable to expect that the industry's reliance on oil will change. Coal is likely to be the most cost-effective alternative, and might take 20% of the market by 2000; some increase in gas use from the present low value might also occur. There is no reason to expect any change in electricity consumption.

Table 7.   Energy Consumption in the Conversion of Paper and Board
in 1979 (PJ)

Product Group	Solid Fuel	Oil + LPG	Nat Gas	Elec	Total
Wall coverings	0.3	1.7	0.4	0.4	2.9
Hygiene products	–	4.9	0.9	1.4	7.2
Stationery	0.4	1.4	0.3	0.4	2.5
Packaging products (paper)	–	1.4	1.5	0.6	3.5
Packaging products (board)	0.3	5.6	1.1	1.1	8.1
Miscellaneous products	–	0.7	0.3	0.2	1.2
Total	1.0	15.7	4.6	4.1	25.4
(%)	(3.9)	(61.8)	(18.1)	(16.1)	(100)

Source: BSO Purchases Inquiry (ref 1).

Table 8.   Conversion of Paper and Board: Monetary Output 1979-1980
(1980 £.10^6)

	1979	1980	1980/1979
Gross Output	3,901	3,589	0.920
Net Output	1,762	1,598	0.907

Source: Business Monitor (ref 8).

Table 9.   Projections of Future SEC for Conversion of Paper and Board

	1980	1990	2000
SEC (GJ/£k gross)	6.5	6.0	5.5
Fuel Mix (%): Solid	4	10	20
Oil	62	52	40
Gas	18	22	24
Electricity	16	16	16

4.3   PRINTING AND PUBLISHING

INTRODUCTION

94.   The printing and publishing industry, in contrast to paper and board manufacture, has one of the lowest energy intensities of all the subsectors in this study (see Volume I of this report for a comparison of energy intensities of all subsectors).

95.   Three reports written by PIRA in the IETS series cover most of the industry, viz:

- Stationery printing (ref 9)

- Newspaper and Magazine printing (ref 10)

- General printing and publishing (ref 11).

The only source of data on energy consumption for the subsector as a whole, however, is the 1979 Purchases Inquiry. This is used in the present report to obtain a quantitative estimate of energy use in the base year (1980) while the IETS reports are drawn upon for technical description of energy use and projections of future specific energy consumption.

ENERGY USE

96.   Total energy consumption, broken down by fuel type, for each of the main product groups of the Printing and Publishing industry in 1979 are shown in Table 10 taken from the BSO Purchases Inquiry (ref 1). Total energy consumption for the subsector is 19.1 PJ, which is approximately 1.2% of the total energy consumption in manufacturing industry. Energy use in 1980 is obtained by pro-rating according to gross output; as Table 11 shows, there was virtually no change. Consumption of oil and gas are about equal. Solid fuel is a very small proportion of the total. Electricity accounts for a relatively high proportion (18%) of total energy use.

97.   According to the IETS reports (ref 9, 10) the main use of energy is space heating; between 70 and 80% of the energy consumed by stationery printers visited was for space heating. Some of the larger sites printing newspaper and periodicals use some process heat for melting metal and drying.

98.   Electricity is used extensively for motive power for driving printing presses. At some sites, motive power accounts for 15-20% of total energy use. Compressed air systems and air-conditioning account for 2-3% of total energy use. Lighting accounts for 4-5%.

99.   At one site, which was the largest establishment visited, most of the electricity was self-generated using three 1.2 MW diesel generating sets with waste-heat recovery from exhaust gases and jacket cooling water. The waste-heat is used for space heating. The company estimated that the capital cost of the CHP installation would be recovered in two years from the fuel savings made.

ENERGY CONSERVATION POTENTIAL

100. Space Heating Both IETS Reports found that there was significant opportunity for reducing energy consumption in space heating. Specific measures identified include:

- better insulation and draught-proofing of buildings

- better use of existing boiler controls (thermostats and time-clocks)

- installation of thermostatic valves

- improvement of boiler efficiency through maintenance.

Savings of 10-15% could be achieved.

101. Process Power Process electricity consumption could be improved through better understanding of power factor correction. In some companies, managers were unaware of the existence of equipment for this purpose within their establishment; as a rule it was not periodically checked.

102. Lighting Few companies carried out a programme of cleaning lighting tubes and fittings, which were often covered by ink mist. The benefits to be gained from light coloured walls and ceilings and clean windows were not appreciated.

103. Overall potential energy savings of 10% in newspapers and 12% in stationery were estimated in the IETS Reports.

Barriers to Implementation of Energy Conservation

104. The reports found that the main reason quoted by companies for not implementing energy-saving measures was the relatively long payback period for recovery of the cost of improvements. However, the reports note that the payback periods quoted were seldom the result of careful analysis, and conclude that lack of knowledge is a major barrier to the implementation of measures.

PROJECTION OF FUTURE SPECIFIC ENERGY CONSUMPTION

105. The overall potential for energy saving is estimated to be 10-12%. This could be higher if the industry were to pursue energy conservation vigorously. For example, the widespread use of small scale CHP systems could reduce electricity demand considerably. However, the low energy intensity of this sector tends to suggest that energy conservation is less likely to be pursued vigorously than in other sectors. Hence, a 10% saving by 2000 is projected. Table 12 shows this corresponding value of specific energy consumption.

106. A modest penetration of the fuel mix by coal is assumed in Table 12, while electricity should increase its share, since most of the expected savings are in space heating.

107. Since energy consumption is not related to output, no differentiation in SEC between different output scenarios is made.

Table 10.   Energy Consumption in Printing and Publishing in 1979 (PJ)

	Solid Fuel	Oil + LPG	Nat Gas	Elec	Total
Newspapers	-	1.6	2.7	0.8	5.1
Periodicals	-	1.3	0.6	0.5	2.4
Books	-	1.3	1.0	0.6	2.9
Other	0.2	3.7	3.2	1.6	8.7
Total	0.2	7.9	7.5	3.5	19.1
(%)	1.0	41.4	39.3	18.3	100

Source: BSO Purchases Inquiry (ref 1).

Table 11.   Monetary Output 1979-80 (1980 £.10^6)

	1979	1980	1980/1979
Gross Output	6,755	6,798	1.006
Net Output	4,118	4,223	1.025

Source: Business Monitor (ref 12).

Table 12.   Projections of Future Specific Energy Consumption

		1980	1990	2000
SEC (GJ/£k)		2.8	2.7	2.5
Fuel Mix (%):	Solid Fuel	1.0	5	10
	Oil	41.4	38	35
	Gas	39.3	38	35
	Electricity	18.3	19	20

REFERENCES

1. '1979 Census of Production and Purchases Inquiry', Business Statistics Office, HMSO 1983.

2. 'Energy Use in the Paper Industry', IETS Report No 9, Department of Industry, 1979.

3. Energy Audit Series No 14, The Paper and Board Industry. Issued jointly by the Departments of Energy and Industry, 1982.

4. 'BPBIF Report on Energy Consumption and Cost for 1982', British Paper and Board Industry Federation, London 1983.

5. 'Annual Abstract of Statistics', Central Statistical Office, HMSO 1983.

6. 'Energy Management Demonstration Project', A report to the Department of Energy, BPBIF, 1983.

7. 'Inquiry into Private Generation of Electricity in Great Britain 1977', Department of Energy.

8. Business Monitor PA 472, HMSO 1980.

9. 'Energy Use in the Stationery Printing Industry', IETS Report No 7, Department of Industry, 1979.

10. 'Energy Use in the Newspaper and Magazine Printing Industries', IETS No 8, Department of Industry, 1979.

11. 'Energy Use in the General Printing and Publishing Industry', IETS No 39, Department of Industry, 1983.

12. Business Monitor PA 475, HMSO 1980.

ENERGY USE AND ENERGY EFFICIENCY IN UK MANUFACTURING INDUSTRY

UP TO THE YEAR 2000

SECTOR 5. FOOD, DRINK AND TOBACCO INDUSTRIES

R Hardcastle and A V Ward

Subsector	Paragraph No
Introduction	1
5.1 Dairy Industry	27
5.2 Processed Foods	106
5.3 Bakery Products	144
5.4 Sugar and Sugar Products	186
5.5 Drinks Industries	232
5.6 Oils and Fats	320
5.7 Miscellaneous Foods	350
5.8 Tobacco Industry	384
References	

SECTOR 5. FOOD, DRINK AND TOBACCO INDUSTRIES

INTRODUCTION

1.   In 1980 the food, drink and tobacco (FDT) industries accounted for 11.5% of total industrial consumption of energy on a heat supplied basis. No official statistics for this year are available to show energy consumption in the 21 sub-sectors which comprise this sector [1].

2.   The dairy, brewing and malting industries have been the subject of Energy Audit reports [2,3,4] and the Leatherhead Food Research Association [5] has published work on energy use in the food manufacturing industry. There is some information available on the Processed Foods sector and on the Tobacco industry from recent Industrial Energy Thrift Scheme Reports [6,7] but there is little published information on the remaining sub-sectors. Each sub-sector of the FDT industries for which any data is available is considered in turn. Unpublished material from the Leatherhead Food Research Association has been of inestimable value.

3.   The rest of this introduction sets out briefly why the FDT sub-sectors have been treated in some detail; those characteristics which are common to many (though not all) sub-sectors; and some of the conclusions of the exercise.

4.   It is apparent from the most cursory examination of data relating to energy use in the FDT industries that this sector is highly diverse both in the nature of the products and in the pattern of energy use. Crisps and rusks are highly energy intensive products (around 9 GJ/tonne), while tea and coffee packing use little energy (around 0.25 GJ/tonne or less). Overall, energy costs as a percentage of total purchases are low at around 2-4% in most sectors but in starch and brewing energy costs can reach 8-9%, while in sugar confectionery they may be about 1%. However, as a percentage of profit, energy costs are important throughout the entire industry.

5.   The question of what is meant by energy intensiveness is interesting. The 1979 Fuel Purchases Inquiry permits some answers. Table 1 shows both physical measures (GJ/tonne - liquid volumes of alcohol converted to weight measure at appropriate densities) and various financial measures. The question of what costs to use in the denominator is not easy: total purchases, total purchases less fuels, total purchases plus wages and salaries, net output and gross output all have some claim for inclusion.

6.   The measures shown suggest the most energy intensive subsectors are starch, spirits, brewing and malting but there is considerable variation in the least energy intensive sectors depending on which measure is selected.

7.   There is a diverse pattern of energy use throughout the various sub-sectors of the industry. While most sub-sectors use boilers to raise steam and hot water (around 70% to 80% of all energy is passed through the boiler) there are some sub-sectors which use little boiler output (ie little hot water or steam) and, those sectors which do use much steam and hot water, use it in very different processes. It is not enough to know the proportions of fuel passing through the boiler; it is important for energy conservation policy to identify where it is used in the industrial processes. Within the FDT industries the use of steam and hot water is associated with preparation

of food stuffs (blanching, peeling, mixing, blending, washing and cleaning); cooking or processing of the product which includes refining, mashing, distilling; and the evaporation, concentration and drying of the product to its final form. Pasteurisation and sterilisation which often occur at the end of the process are also important energy using processes within the FDT industries. It follows that it is important to understand how improvements in plant and equipment associated with these processes may influence energy use; it is also important to understand how heat may be recovered from these different processes as an element in conservation action.

8.   Several areas of FDT which are important energy users do not use steam or hot water based processes for their product. The most obvious example must be maltings, and bread and flour confectionery industries which use kilns and ovens which are typically directly or indirectly fired and which account for over half of total energy use within these establishments. Again, sectors such as tea and coffee packing and some areas of the tobacco industry use little energy in processing equipment – most energy is used for space heating. There are also the complex areas of energy use such as extraction and refining of vegetable oils, and distillation plant in whisky production, in both of which heat can be cascaded and used in associated downstream operations. Finally electricity used in chilled and frozen storage is important for some foodstuffs such as fruit, vegetables, dairy produce and margarine and fats.

9.   Another aspect of the variety within the FDT industries is the degree of concentration of production in a few large companies. 80% of beer is produced by seven brewers; two biscuit companies account for nearly 70% of all biscuit products and nearly half of snack foods; 70% of soft drinks are produced by two companies. At the same time over 5,000 companies employing less than 200 employees coexist in the FDT industries. It would be interesting to relate sector structure to types of establishment and pattern of energy use but this is beyond the scope of the present exercise.

10.   It is for these reasons therefore that the sub-sectors of the food, drink and tobacco industries have been treated in some detail. Because the pattern of energy is different, the opportunities for energy conservation are also varied in the different sectors. Furthermore, the Leatherhead Food Research Association (LFRA) possess details of energy use in these different sectors prepared as part of the IETS exercise. This information was made freely available making this detailed exercise possible.

11.   It is, however, important to recognise the limitations of the data and how it has been used to estimate energy conservation potential. The data represents a sample of establishments selected from the Business Statistics Office (BSO) listing of all establishments participating in the Annual Census of Production. The sample of firms agreed (and thus form a self-selected sample) to a one or two day visit from an energy consultant. The results of that visit together with data supplied by the establishments were analysed and subsequently published as part of the Industrial Energy Thrift Scheme (IETS) reports (see [6] or [7]).

12.   From the visit reports it is possible to assess the mean values of GJ/tonne or some similar measure of specific energy consumption (SEC) and observe the range of values around the mean. It is also possible from discussions with those who visited firms and wrote the Thrift Reports to obtain some feel for how representative were the firms (by different sizes, by

different product mix and other characteristics) in the sample, with respect to the industry as a whole.

13. Armed with this information - observed variance in GJ/tonne and some feel for the firms in the sample - it proved possible, in most cases, to identify 'best practice' figures for specific energy consumption (SEC) as well as 'average practice' SEC figures, usually in GJ/tonne.

14. Such data is necessary for this study for it is central to the methodology adopted here for estimating energy conservation potential. Volume I sets out in some detail why this methodology was used and the problems associated with definition and measurement across all industries in the study. The point to restate here is that energy conservation potential is defined and measured as the difference between energy consumption observed in 1980 and energy consumption which would have occurred had all firms been operating with the specific energy required of the 'best practice' firms.

15. In the FDT sector the data described above together with some other sources (such as surveys conducted in the dairy and brewing industry) provided evidence of mean and observed variance in GJ/tonne across a sample of firms, together with some feel for the nature, character and representativeness (with respect to the rest of the sector) of the firms in the sample. So it proved possible in most cases to identify 'best practice' as well as 'average practice' SEC figures.

16. In most cases the 'best practice' figure was found to lie around the lower quartile of the distribution. This was because the lowest specific energy figure was often associated with some special factors (size, location product mix) which made it unlikely that other firms in the industry could achieve such figures. Even in the case of firms producing apparently similar products, the pursuit of product quality by one or more firms may make comparisons of SEC very difficult.

17. Selecting 'best practice' figures was therefore neither an easy nor a precise exercise. The sample sizes varied considerably from several hundred in the dairy industry survey to only one establishment visited in the grain whisky industry as part of the IETS exercise. Most samples were around 20-30 firms but some sectors such as crisps or nuts might have only 3-5 firms. In such circumstances it is difficult to be precise about 'average' and 'best' practice.

18. Each of the following eight sectors of the FDT industry contains a note indicating the data on which the values of 'average' and 'best' practice figures were selected. In each case the choice of figures was discussed and agreed with officers at Leatherhead Food RA and others who would be expected to be familiar with energy use in the sector. The choice of best practice was arbitrary in the sense that adjacent figures of perhaps $\pm$ 10% could have been selected with equal plausibility; the choice was not arbitrary in that it was based on some, albeit imperfect, evidence and much discussion and consideration.

19. The second part of each study looks at the trends affecting markets, technology and other developments over the next two decades and several common strands affecting the FDT sectors may be identified. The first is the uncomfortable position in which many food manufacturers have found themselves

in the past decade. They feel they are being squeezed on the one hand between farmers whose prices are fixed largely by the CAP intervention prices or by sharply fluctuating commodity prices like cocoa, fats and oils, over which the manufacturer feels he has little control; and on the other hand, the growing power of the supermarket chains which now account in some food areas for over half of grocery sales. Food manufacturers have seen some brands effectively disappear in the face of generic and own label products.

20.   The squeeze factor is but one element in a difficult trading environment. Food sales have been moderately static for almost a decade growing slowly with population and with an income elasticity with respect to growth in the economy of around 0.2. Thus for a 1% growth in the economy only about 0.2% of the associated increase in income is spent on food products. Within this largely static picture of volume sales there are quite sharp changes of fortune in individual foodstuffs. Three trends may be noted.

- convenience foods - seen most clearly in hamburgers, cheese burgers and fast food chains but also seen in the growth of packaged, sliced, prepared and semi-prepared and snack foods;

- freshness, improved quality and the desire to prolong product life - seen most clearly in the growth of chilled and frozen foods and the growth of delicatessen and prepared side salads and cut meats;

- health/nutrition - this may account for the changing patterns of consumption of such products as bread, sugar and fats and also for the marked increase in sales of fruit juice, fresh fruit and vegetables and decaffeinated, desalted, and other diet conscious foods.

21.   If volume sales are broadly flat and margins are being squeezed between high commodity prices and powerful buyers then there is every incentive to food manufacturers to reduce costs, amongst other measures, in order to maintain profit margins. Recent efforts to become more energy efficient can be interpreted as such a measure.

22.   A significant proportion of EEDPS projects lie within the FDT sector (about 20%). EEDPS projects (apart from measures associated with good housekeeping and skilful management) have focused on boilers, boiler replacement and boiler control and on heat recovery from many of the industrial processes. These include exhaust gases from ovens; waste water from blanching, pasteurisation and other relatively hot processes; also, some attention has been paid to heat recovery from refrigeration processes. The targeting and monitoring exercises of the Brewers Society and the survey of energy use carried out by the Maltsters and Dairy Federation also indicate an interest.

23.   The contribution of innovation in new plant processes and equipment to energy conservation over the next 10-20 years is likely to make more than a marginal impact on efficient energy use. Innovations differ between the various industries but it may be noted here that:

- mechanical vapour recompression techniques may be quite widely used on evaporators as a pre-concentration technology (reducing liquids to a viscous or semi-dry state before spray or other drying technologies are applied);

- some developments have occurred in process plant which make blanching, sterilisation, pasteurisation and the traditional evaporation techniques more efficient;

- one major innovation may be thermo-compression techniques, which are now employed in sugar confectionery and biscuits manufacturing but may also be extended to other areas of the food industry over the next 10 years or so.

24. For the rest, processes and equipment are not likely to see radical changes although:

- more incremental innovation will make industrial processes a little faster, more reliable, subject to better quality control; and

- there will probably be a change to continuous processes from batch processes. This will require major input of control and instrumentation and it is this feature which may marginally increase energy intensiveness although at the same time it must certainly improve the monitoring of energy use at each stage of the food preparation chain. Another major facet of continuous processes is the opportunity to maximise the potential for heat recovery.

25. The main point is therefore that the food, drink and tobacco industries are highly diverse:

- in products;

- in the pattern of energy use; and

- in prospects for the next two decades;

- in opportunities for energy efficient actions.

It follows that each must be treated separately. This is why FDT industries are examined in some detail in the following pages.

26. However, it may be useful to start with a summary of the whole sector shown in Tables 1-3. The main points to note seem to be:

(1) Apart from Tobacco and 'Drink' (there is a case for splitting spirits and brewing) most of the sectors account for a similar proportion of the total energy consumption. It is not possible to argue that most energy is concentrated in a few subsectors.

(2) The potential for conservation varies between 20% and 40% but may be taken to be around 25%.

(3) This figure (25%) is higher than the 14% shown in Energy Paper 32 [8] but includes almost all sectors of the FDT industries.

(4) Ice Cream is the one subsector not included in the study.

(5)   Table 3 shows specific energy consumption data to 2000 for both high and low growth scenarios. Reference should be made to each section for the reasoning behind the figures.

(6)   The Main Report (Volume I) of the study of industrial energy conservation potential to 2000 contains estimates of actual energy consumption. It also contains a discussion of the high and low growth scenarios adopted in this study.

(7)   Changes in market structure between 1980 and 2000 are made according to the likely developments discussed in each section. These changes are partly reflected in the changing SECs in Table 3.

(8)   The assumptions and conclusions reached are entirely the responsibility of ETSU and may not necessarily accord with the opinions of the institutions and individuals contacted in the preparation of the Report.

TABLE 1: Energy Intensiveness in FDT Industries 1979

	Energy Consumption (less Derv & motor spirit), PJ	Energy Costs (less Derv & motor spirit), £000	Total Purchases (except merchanting), £000	Energy Costs as % of Total Purchases, %	Specific Energy Consumption, GJ/tonne
Marg. & compound cooking fats	2.46	5,487	202,743	2.7	3.9
Organic oils & fats	6.43	13,990	517,272	2.7	5.0-11.4
Bacon curing & meat processing	6.20	20,552	1,260,603	1.6	5.1-6.4
Poultry	4.15	13,719	420,315	3.3	3.1
Milk & milk prod.	22.07	52,715	2,534,657	2.1	1.5
Fruit & veg.	8.05	17,052	483,090	3.5	3.1-4.2 (9.5 crisps)
Fish	3.54	11,421	441,609	2.6	5.2
Grain milling	2.69	12,372	770,519	1.6	0.3 (4 Breakfast cereal)
Starch	7.23	12,414	134,298	9.2	7.9
Bread	15.34	50,545	744,767	6.8	3.5
Biscuits & crispbread	4.65	11,195	362,533	3.1	5.2-13.1
Sugar	21.65	27,510	494,467	5.6	5-10
Ice cream	1.27	4,724	102,422	4.6	
Cocoa, choc. & sugar confect.	9.89	10,409	866,389	1.2	6.2-11.8
Compound animal feed	4.49	18,225	1,496,362	1.2	0.6-3.2
Pet foods & manf. feeds	5.95	14,170	397,117	3.6	2.3
Misc. foods	13.69	31,401	1,324,675	2.4	2.2-7.1
Spirits distilling	26.57	52,791	647,707	8.2	28-60
Wines, cider, perry	0.96	2,562	56,150	4.6	1.4
Brewing & malting	25.22	54,727	692,078	7.9	3.2-4.5
Soft drinks	2.96	8,187	431,292	1.9	0.6
Tobacco	4.36	9,148	503,658	1.8	12.7-25.1

Energy Costs as % of Total Purchases		Specific Energy Consumption, GJ/tonne	
Top		**Top**	
1. Starch	9.2	Malt/spirits	28-60
2. Spirits	8.2	Tobacco	12-25
3. Brewing & malting	7.9	Starch	8
4. Bread & flour	6.8	Biscuits	5-13
Bottom		**Bottom**	
1. Compound animal feed	1.2	Soft drinks	0.5
2. Cocoa, chocolate	1.2	Compound feeds	0.5-3
3. Bacon curing	1.6	Grain milling	0.3
4. Tobacco	1.8	Some misc. foods; pet foods; milk	2.3

TABLE 2: Summary Table; Energy Consumption and Conservation Potential in Food, Drink and Tobacco Industries 1980, 10^6 GJ

	1980 Energy Consumption	Energy Conservation Potential[2]	
	10^6 GJ	10^6 GJ	%
Dairy industry	23.24	6.97	30
Processed foods	21.57	6.81	31
Bakery products	19.45	3.69	19
Sugar, etc[1]	29.69	5.9	20
Drinks industries	53.28	12.08	23
Oils and fats	13.71	4.9	35
Miscellaneous	19.93	5.88	29
Tobacco	2.07	0.85	41
TOTAL	182.94*	47.08	26
D.Energy Digest figure	197.81		
1979 Fuel Purchases Inquiry figure	199.84		
% total manufacturing industry	11.5		

* Ice cream (1.4 PJ); non-tonnage related products; exclusion of Derv; and output differences between 1980 and 1979 account for the discrepancies between estimated 1980 energy consumption, the 1980 Digest figure and the 1979 Purchases Inquiry figures.

1 Estimated.

2 Based on difference between energy consumption in 1980 using average practice SEC and best practice SEC.

TABLE 3: Summary Table; Specific Energy Consumptions in FDT, 1980-2000, 10^6 GJ (derived from weighted average calculations)

	1980	1990		2000	
		H	L	H	L
Dairy (MJ/litre)	1.54	1.45	1.40	1.33	1.37
Processed foods (GJ/tonne)	4.16	3.75	3.64	3.28	3.17
Bakery (GJ/tonne)*	1.81	1.68	1.65	1.58	1.54
Sugar & sugar products (GJ/tonne)	9.59	9.48	9.47	9.42	9.33
Drinks (MJ/litre)	3.37	2.86	2.88	2.53	2.63
Oils & fats (GJ/tonne)	6.37	6.00	5.98	5.63	5.59
Miscellaneous foods (GJ/tonne)	1.43	1.19	1.24	0.97	1.06
Tobacco (GJ/tonne)	13.82	13.09	13.09	12.21	12.21

* Excluding flour confectionery.

5.1 THE DAIRY INDUSTRY

27.   The UK dairy industry processed around 3,346 gallons (15,214 million litres) of milk in 1980. The milk sold off farms to wholesale producers is sent to either processing dairies for heat treatment and bottling or to manufacturing creameries.

28.   In the UK about half the milk sold off farm is sent for liquid consumption and nearly all of this is heat treated. The flow of milk from the farm to its various uses in 1980 is shown in Figure 1 (note that the data refers to England and Wales, not UK).

29.   It is clear from the diagram that the dairy industry may be separated into liquid milk processing and milk manufactured products. The two sides of the industry have very different characteristics and, even where organisations such as the Milk Marketing Board operate in both areas, they maintain separate divisions to serve the distinct markets.

30.   The main differences are:

  (a)   Liquid milk processing is in general a homogeneous activity where all firms carry out much the same operations in a similar fashion. By contrast, milk products vary considerably and there is little uniformity in the size and type of creamery.

  (b)   In energy terms, this implies that liquid processing is not too difficult to examine but in milk manufacturing the range of plant and equipment is much greater, the variety of production arrangements much greater, and examination is therefore more difficult.

The next few paragraphs (31-33) note some recent developments in the industry; then energy use in the liquid milk processing subsector is considered; this is followed by energy use in milk products.

31.   Although demand for liquid milk tends to be stable over the year the supply varies substantially between 370 million gallons per month in May and 265 million gallons per month in November, so that the average utilisation of capital plant in milk manufacturing varies accordingly. This has the effect of limiting the full utilisation of capital equipment and to some extent reduces the scope for energy conservation. However, the long term prospects are for an increase in the supply of raw milk which will reduce the seasonal variations in supply to manufacturing and, to some extent, will encourage investment in capital equipment.

32.   <u>Production Trends 1970-1980</u> Over the last decade milk production has increased steadily by around 2½% per annum with the market for milk changing gradually. Liquid milk consumption per head has been falling steadily since the end of the 1970s - consumption per head has fallen from 4.76 pints per person per week in 1970 to 4.40 pints per person per week in 1980. The proportion of total raw milk produced going to liquid consumption has therefore fallen gradually from over 60% in 1969/70 to less than 50% in 1980/81 with a corresponding rise in the proportion of milk going to manufacturing.

33.   The UK is self sufficient in liquid milk but meets only 70% of domestic demand for cheese and 50% for butter. The longer term prospects for the dairy trade are therefore for a modest expansion of milk supply and manufacturing subject to competition from imported dairy manufacturers.

Structure of the Dairy Industry

34.   Because the bottling of milk adds substantially to the volume and weight of product and the storage life is limited there is a tendency for liquid milk processing to be carried out near its market population. By contrast, manufacturing of milk reduces volume and weight and increases the storage life of the product. Location and manufacturing tends therefore to be close to the raw milk production areas in the north-west and the south-west.

35.   Processing of liquid milk in England and Wales is carried out in about 415 centres of which about 60 are large establishments (processing over 25,000 gallons a day), which process over 60% of the total milk. A further 120 are intermediate in size (4,500 to 25,000 gallons a day) and process around 32% of total output. The remaining milk is processed by the remaining 235 small dairies. Many are owned by Unigate, Express Dairies and Northern Foods which together account for around a half of total output. Many more belong to co-operatives which together account for a further 30% of total output.

36.   Milk manufacturing in the UK tends to be concentrated into a few large establishments. About 59 large centres produce over 80% of total milk manufactured products. Unigate have recently sold 16 creameries to the Milk Marketing Board which, together with the Express Dairy, now dominates the milk manufacturing industry. This structure in the industry seems to suggest that energy conservation is most likely to yield major results if concentrated on the small number of larger users.

37.   If one defines the dairy industry as including the production of raw milk through to final liquid processing for human consumption, the average costs of milk production are dominated by feedstuffs for the herd together with some labour and capital in the farming community and herd replacement. On this basis energy costs are very small - 0.25p per litre. Table 1 showed that energy costs across the milk and milk products sector represented 2.1% of total purchases and this may be taken as a more representative figure.

Liquid Milk Processing

38.   The energy use characteristics of liquid milk processing are significantly different from those in milk manufacturing. It is therefore useful to consider first energy use in the liquid milk processing sector of the industry. There have been no published official statistics on energy use in the industry since the Census of Production in 1968 although results from the Fuel Purchases Inquiry, 1979 [10] are now available. A survey of energy use in the liquid milk processing sub-sector of the industry was conducted by the Dairy Trade Federation (DTF) with assistance from ETSU in 1980 (for the year 1978/79) and forms the basis for some of the estimates made in this report [9]. The Fuel Purchases Inquiry [10] provides the basis for fuels used in the industry.

39.   This report is concerned with energy use in the <u>production</u> or <u>processing</u> of goods so no account is given of transport problems and energy use in distribution.

40.   Most milk sold (87%) in the UK is pasteurised, full cream milk. This means that the milk has not been separated (thus producing skim milk); nor has the fat been dispersed into the milk (homogenised); nor has it been ultra heat treated (UHT) or sterilised.

41.   These different types of milk use different quantities of energy. Sterilisation requires about five times as much energy per gallon as pasteurised milk but sales are small (and declining) amounting to around 6% of total milk sales in 1980/81. UHT milk and homogenised milk use more energy than pasteurised milk but less than sterilised milk. Sales of UHT (1%) and homogenised milk (6%) are currently small but UHT ('long life' milk) has been growing. The fastest growing area of liquid milk sales is in skim and semi skim milk which has doubled its sales in recent years. The cream/fats left by the separation process is used in butter making, so there is a problem of balancing requirements of skim milk and butter production.

Industrial Processes

42.   The industrial processes are shown in Figure 2. In pasteurising heat treatment of milk two main unit processes are pasteurisation itself and bottle washing. Pasteurisers at the time of the Audit (1978) used 85% regeneration and now (1982) approach 90%. Regeneration refers to the extraction of heat from hot milk which is being cooled in order to heat up the cool raw milk which is undergoing heat treatment. This is achieved through heat exchangers and because of the temperature gradients and the minimal losses which occur in the system it is generally thought in the industry that 90% represents the limit for regeneration of heat in this way. However, recent technical developments by the Alpha-Laval Company suggest that this might be raised still further.

43.   Bottle washers are of two main types – jet washers and soaker washers although the energy requirements of both types are very similar. Bottle washing machines use simple heat exchangers which transfer heat between the cooling and heating stages to provide an adequate supply of hot water and reduce losses of heat to the effluent.

Energy Consumption and Conservation Potential 1980

44.   The DTF survey reported that in 1978 almost all the fuels used in dairies was fuel oil and electricity with almost no gas. A separate account of fuel use in creameries is not available so the two sectors are combined in the figures given in Table 4. The fuel splits shown are taken from the 1979 Fuel Purchases Inquiry [10].

45.   It will be noted that the story is the same: fuel oil dominates fuel use in the industry (71%); there has been little movement to gas (10%); electricity accounts for 12.6%.

46.   The main process uses of energy in the liquid processing of milk are the heat treatment/pasteurising of milk and bottle washing processes which

together account for about 60% of energy use in operations although the pumping and control of large amounts of water used in the dairy are also important.

47.   Refrigeration is quite important in liquid processing of milk although taking only about 10% of energy use (on a heat supplied basis) compared with as much as 20% in dairies producing cheese and milk products. As with the latter, the electricity use is associated with motive power for the compressors. The other 40% of energy used in pasteurising dairies is used in a variety of ways - some space heating, washing of floors, tankers, canteen and other ancillary uses. Tables 4 and 5 summarise the utilisation of energy.

48.   The methodology used to assess conservation potential in this report is set out fully in Volume I. Briefly, conservation potential is assessed by comparing energy consumption in 1980 with energy consumption which would have occurred had all firms been operating at 'best practice'. This methodology clearly requires some evidence on 'average practice' and 'best practice' in the industry. In most sectors of this report the evidence has been derived from visits made under the Industrial Energy Thrift Scheme (IETS) operated by the Department of Industry. In the dairy industry there is evidence from the DTF survey of 1978. Both give figures for average energy consumption per unit throughput (MJ/litre processed) or SEC.

49.   The DTF survey of liquid milk processing showed that in 1978/9 the average energy consumption per litre of milk processed was 0.99 MJ/litre (comprising 0.89 MJ/fuel oil and 0.10 MJ/electricity). There were small variations by size of firms indicating that small firms were least efficient and medium sized firms most efficient. Full details are shown in Table 6.

50.   On close inspection of the raw data it turned out that establishments belonging to a particular organisation had consistently lower figures for MJ/litre processed. These establishments (and specific energy consumption figures) varied slightly across the size categories but a single best practice figure of 0.48 MJ/litre was taken. This figure was discussed with informed observers of the industry and agreed to represent an operating standard which all or most firms in the industry could achieve. Energy consumption in liquid milk processing in 1980 is shown as the product of output and SEC in Table 4.

51.   The calculations are shown in Table 7: energy consumption in liquid milk processing is 7.06 million GJ but on best practice figures could be 3.43 million GJ. So, conservation potential is 3.63 million GJ or 51% of the 1980 consumption.

Energy Conservation Measures

52.   What is perhaps surprising is the size of the energy conservation potential - 51% of energy consumption in 1980. This is rather higher than that suggested in the Audit and, more significantly, it derives entirely from measured differences in average and best practice in use in the industry. It is therefore all achievable by improving or increasing the rate of diffusion of best practice through the industry. It takes no account of further improvement in regeneration, nor does it take account of technical change in pasteurisers or heat pumps or heat recovery in dairies. All these may indeed improve energy consumption in the best practice firms (making best practice

even better) and this is considered later when likely changes in the industry to the year 2000 are dealt with.

53. (i) <u>Churn and tanker washing</u> There have been small savings which have accrued from tanker washing rather than churn washing but this has been a once and for all saving and no further savings can be expected.

(ii) <u>Increased regeneration in pasteurisers</u> As noted in paragraph 42 regeneration is now approaching 90% which is usually taken to be a limit in the industry. Further exploration of recent developments by one manufacturer which claims rather higher efficiencies in regeneration have still to be completed, but it may be that the ceiling could be raised towards 95% regeneration.

(iii) <u>Heat recovery via heat pumps</u> A major area for savings of process heat is the use of heat pumps. Two processes in particular might benefit. First, bottle washing currently uses heat exchangers to pass heat between the various stages to provide an adequate supply of clean hot water and reduce losses of heat to the effluent. Heat pumps or heat pipes would almost certainly increase the efficiency of heat transfer between the stages. However, detailed investigation of the possible savings have yet to be completed.

The second area is the general use of hot water for tanker washing, space heating and general hot water in canteens etc. For this heat pumps could certainly be used to extract the large amounts of heat from effluent water. Typically the ratio of the volumes of water to milk in most dairies is around 4-1 and the effluent temperature is usually less than 25°C. Hot water for bottle washing, tanker washing, space heating etc, is required at around 50°C.

The coefficient of performance of a heat pump at 25°C would be perhaps 4 or 5 (14.5-18.0 MJ moved per kWh supplied) which at current prices for electricity and steam may not be economic. It would however only take small changes in operating temperatures and fuel prices to make heat pumps in dairies a cost effective means of saving energy.

A demonstration project is currently underway to evaluate the use of heat pumps in a dairy [11]. Details of performance however have yet to be completed.

(iv) <u>Good housekeeping</u> Where the technology of the process is so uniform efficiency in energy use is largely due to the good management of general operations. As noted earlier the energy is actually used across a range of operations - canteen, tanker washing, space heating, as well as the process heat associated with pasteurising and bottle washing. So, the first step in good housekeeping is to identify and measure how much energy is used at each point. Whilst most firms have some measure of the total milk output and total energy and water use, few manage to measure energy consumption at each point in the manufacturing process. The fact that the relationship between water consumption and

energy consumption is so close offers the hope that a single
system of monitoring will be introduced for both water usage and
energy consumption. The use of chilled transportation suggest
some economies might be made if the total chain from farm to
doorstep is considered [12].

(v) Packaging The majority of the liquid milk in the UK is sold in
re-usable glass bottles. Any move to one trip containers
(typically waxed paper cartons) would save fuel in the dairy
because no bottle washing would be required. However, any
comparison of the energy requirement of the different containers
must take account of the full range of factors involved.

The few studies available on the issue show an overwhelming
balance in favour of returnable glass providing a high trippage
rate can be maintained. The cost advantage of glass has been
further enhanced by the introduction of the light weight glass
'pintie' and there is some prospect for further savings to be
made. Moves to introduce labelling on bottles rather than the
embossed identifying mark of the dairy and colour coding of the
milk top as a means of identifying the kind and type of milk has
so far been resisted. In energy terms it would be a retrograde
step as it would increase energy use in paper, printing, adhesives
and bottle washing.

The competition from carton based supplies sold through
supermarkets is potentially serious but the convenience factor at
current price relativities of door step supply of around 12 pints
of milk per week for the average household is strong. The cost
savings which may be available through supermarket suppliers would
need to be high in order to induce the necessary domestic
investment in cold space to accommodate this volume of milk
supply. Recent developments in PET (polyethylene terephthalte)
may lead to alternative, non reuseable, containers, which has
implications for the milk round. It may also be noted that most
continental milk containers are non-glass containers.

Manufactured Milk Products

54. Over half of milk production is used in manufacturing cheese and other
milk products such as butter, cream, condensed and evaporated milk and
yoghurt. Table 8 sets out some recent production figures for milk products
and shows clearly the very high proportions taken by cheese making and butter
making. Milk manufacturing is about twice as intensive as milk processing, so
there is particular interest in energy conservation in this sub-sector of the
industry.

55. The trend of sales in the industry is towards milk products and away
from liquid milk consumption. Future output in the industry as a whole is
therefore likely to be more, rather than less, energy intensive. As fuel
prices rise in real terms so there will be some incentive to ensure that the
necessary capital equipment is as fuel efficient as possible.

56. Industrial processes and conservation potential Whole milk contains
about 11-13% solids including fats. Premium milk products such as butter,

FIGURE 1: Milk Flow England and Wales 1980/1981, million litres

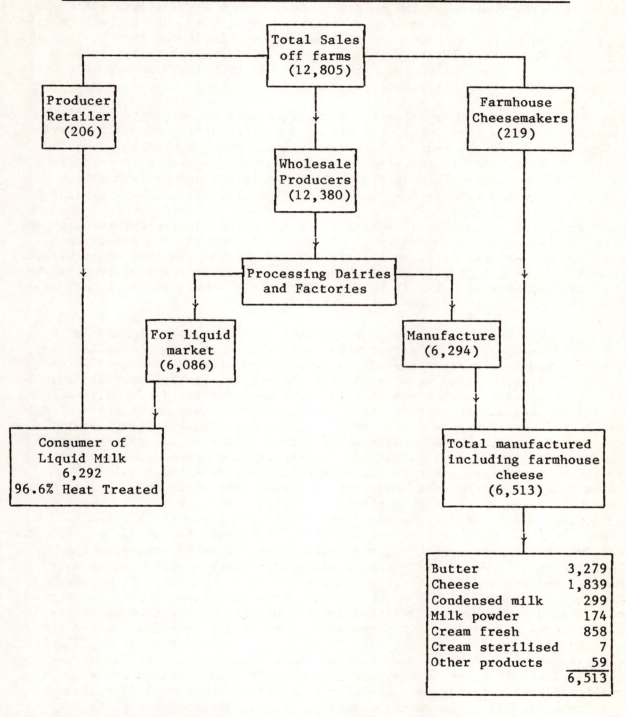

Note: Arrangements in Scotland and N.Ireland are slightly different from the
       illustration. Source [13].

cheese, cream etc, separate much of the solids into a more concentrated form, leaving a less concentrated residue which is, in turn, converted to manufactured products by evaporation and drying processes. Liquid skim milk, which is produced by separating cream from the milk before cream or butter making, is a by-product of the butter making process which accounts for about half of milk used in milk manufacturing processes. Over 60% of the skim milk is then dried to powder form for the food and catering industries as well as being exported. Some 13% is used in stock feeding and the remainder is produced as condensed milk, low fat cheeses and other liquid and powder forms. Figures 2 and 3 show the details.

57. Skim milk must be pre-concentrated by evaporation before drying for milk powders. Two main forms of drying are used – spray drying and roller drying although the latter is restricted to special products or used as peak load plant. Milk must be concentrated to around 27–30% solids for roller drying or for yoghurt and condensed milk production; and is pre-concentrated by evaporation to about 48 to 50% for spray drying.

58. In cheese making whole milk is curdled and the curds separated from the whey. Small cheese making centres then sell the whey in liquid form, but large creameries must first concentrate and/or dry the whey before sale. Whey is sometimes used in dry form and, because it has a lower solids content than whole milk or skimmed milk, it is sometimes pre-concentrated not by evaporation but by the process of reverse osmosis.

59. Evaporation The evaporator plant used in the industry is mainly of the multiple effect falling film type with between three and four effects, although some modern plant uses six or even seven effects. Typically the equipment is also fitted with vapour pre-heaters and thermal compressors. The plant operates by re-using the heat in the vapour generated by the first pre-heated session (the first effect) to concentrate subsequent flows of milk (second and other effects). The number of times the heat in the vapour can be used is referred to as the number of effects and this in large part depends upon the temperature gradient across the vapour, the pressure at which the vapour is held and the temperature of the incoming milk which requires concentration. The maximum temperature to which the milk can be heated is around 70°C because protein precipitates at that temperature. Below about 35–45°C the milk viscosity is such as to limit the flow rate through the evaporator. The practical number of effects available to old plant is about four but recent developments in evaporators now installed and operating provide for six or seven effects, so significant improvements can be made through replacement of plant.

60. Evaporators may be classified by the rate at which they concentrate the liquid being processed and by the manner in which the liquid is passed through the evaporator process. Typically in the dairy industry tubular falling film evaporators are used with capacities of around 3 kg per second but evaporators of up to 10 or 12 kg per second are currently in production in larger creameries. By contrast, evaporators for fruit juices and similar liquids which require lower evaporative capacity and concentration use plate climbing/falling film type evaporators with rather lower capacities.

61. The scale of energy saving through increasing the number of effects of evaporators is considerable. An increase of one effect per evaporator

(raising the average from 3/4 to 4/5 was estimated in the Audit [2] to save about 2-3 million gallons of oil (0.53 x 10^6 GJ) per year.

62. Mechanical Vapour Recompression (MVR) All evaporation processors are highly energy intensive since the separation of concentrates in the liquid is achieved by boiling. It is not surprising therefore that industries that require to concentrate or purify liquids have paid considerable attention to alternative ways of effecting concentration. Mechanical Vapour Recompression (MVR) employs a compressor to raise the temperature of vapour arising from boiling. The compressed vapour is returned as a heating source to sustain the boiling action. In primary energy terms the energy requirement of MVR is about 60% lower than the most efficient alternative process which, in the dairy industry, is taken to be a 6-7 effect traditional evaporator.

63. Energy savings through the use of MVR can be substantial because evaporation accounts for around 9,000 TJ per annum. If MVR were to replace multiple effect evaporators some 1,500 TJ would be saved [14] which is about 5-6% of milk manufacturing primary energy requirement. There are several problems associated with introducing MVR but the equipment is now in place and has proved effective. Another significant point concerning MVR is the relative importance of steam and electricity in the evaporation process. For a given output of concentrated liquor an MVR uses 3-8 times as much electricity and about a quarter of the steam and, overall, the MVR uses about a quarter of the energy on a heat supplied basis saving about 60% of energy on a primary fuel basis.

64. Reverse Osmosis This is another method of pre-concentration of milk which has the effect of saving energy. A reverse osmosis plant consists of a bank of several cylindrical membrane units called candles each of the order of about one metre long; so the plant is very compact and can be arranged to fit any available corner of an existing dairy and is thus a useful way to expand effective evaporation capacity where space is at a premium.

65. By concentrating the 9% total solid skim milk to 18% total solids the water content is halved. This requires quite modest pumping loads of around 10 kWh per tonne of water removed and, important in the food industry, involves no potential heat damage to the product. The cost of evaporation to 48% total solids also shows a saving because of the reduced load on the evaporator which can of course be smaller if the liquid milk is first treated by reverse osmosis. Reverse osmosis requires no heat input and uses mainly electricity. The long term interest in reverse osmosis or the related technology of ultra filtration is in extending the limit of solids concentration and in its application to other products such as skim milk. It is currently considered unlikely that reverse osmosis or ultra filtration will be developed to the point where the evaporation stage could be eliminated so the main application for this technology will always be at the pre-concentration stage.

66. Drying Spray drying is the dominant method of drying the concentrated milk. Some roller drying is still carried out but is limited to special products or peak load capacity. It will eventually be phased out. Drying accounts for nearly half the energy use but only about 10% of water removal when concentrating and drying one kg powdered milk.

67.   In spray drying the very high energy consumption is due to the rejection of large volumes of hot, water laden air. So, to minimise energy costs, best practice is to use as high a concentration level as possible in the feedstock (about 50% is the limit because of the viscosity of milk) and to exhaust the air at the lowest possible temperature (about 94°C is normal).

68.   The most widely used method to improve the efficiency of drying has been two stage drying. The total solids in the milk are raised to 88% total solids in a spray dryer and then to 96% in a second stage fluidised bed dryer. The savings achieved depend on the inlet temperature for the first stage - around 15% is a representative figure. Capital costs are higher than for a single stage dryer although the cost of the second stage is offset by the fact that the first stage can be smaller for the same capacity. The overall result is judged to be an increase in capital cost of about 30%. New investment in drying is expected to take the form of two stage drying equipment.

69.   About 25% of the total heat in the fuel of an indirect fired spray heater is lost in the flue gases. In principle no losses occur if the product is directly fired using the combustion gases as the drying medium. There is a hygiene problem when milk is directly fired: there is some chance that carcinogenic material in the combustion products of the gas may be absorbed in the milk granules which may then be passed to babies. Legal practice varies - it is illegal in some European countries but not in the UK or US. However, there are no examples in the UK of direct firing for milk drying.

70.   The other main method which has been considered for decreasing the energy used in drying is to recover heat from the exhaust gases. Although an obvious source of energy conservation the attempts to harness the exhaust heat have proved very difficult in practice. The main reason for this is the small proportion of milk powder (about half to 1% of total) present as fines in the dust. This quantity which amounts to up to two kilogrammes per hour in an efficient dryer is sufficient to present serious difficulties with either wet or dry recuperators. In principle there are three ways round the fines problem:-

      - re-design the atomiser to give a spray which does not produce fines

      - recover the fines before attempting to recover heat

      - use a wet technique such as scrubbing the exhaust gas.

Discussion of these three approaches in the Audit report suggest that the economics of recovery are marginal. However, if the value of the milk product recovered through collection and drying of fines were to increase then the combination of value of the milk products recovered as well as the re-use of the heat recovered may justify investment in heat recovery equipment. The energy savings available are fairly small - so much will depend upon the value of the milk powder.

71.   <u>Cheese Making</u> The processes in making cheese are very similar to those of other milk products. The whey may be part concentrated only or fully dried out. Table 9 sets out details of energy use in the sub-sector. Because cheese making is a low temperature process, substantial use could be made of heat pumps to re-cycle and re-use heat. The potential for heat pumps depends on the size of the factory and the extent to which a factory concentrates its whey.

Energy Consumption and Conservation Potential in Manufactured Milk Products

72.   As with the liquid processing sub-sector, energy conservation potential
is measured by calculating energy consumption in the industry on the basis
of average energy consumption per unit of output and subtracting from that
figure energy consumption in the industry if best practice were to be used by
all firms in the industry. In the absence of fuller information on the stocks
of assets it is difficult to identify a best practice but for the reasons
discussed below a figure of about 20% lower than the figure for average
practice has been estimated. The reasoning and information behind this
conclusion is as follows.

73.   Energy consumption in the milk products sector of the industry is
dominated by process energy use, see Table 9. Over 85% of all energy in
butter production and 10% in cheese making goes to evaporators, dryers and
refrigeration and only a small quantity falls into the category of
housekeeping. This means that the spread of observed practice will reflect
the possession and use of different vintages of capital assets. The observed
spread between best and worst based on one organisation's data is at least as
wide as in liquid processing but the range observed is difficult to
interpret.

74.   This is mainly because different creameries have a different range of
products and different utilisation rates. When added to the effect of
different vintages of equipment the range of observed figures for energy
consumption per unit throughput is very marked indeed (factors of 2-4 are not
unusual) but the difference between 'average' and 'best' is much less.
Because drying is the most energy intensive activity, butter/powder creameries
show the higher rates; but cheese making shows the greater variance because
not all cheese creameries dry whey.

75.   Having considered industry views and those of others, we have taken the
data in Table 7 as representative of the industry in 1980. Typically 'best'
is around 20% better than average.

76.   Data on average practice is quite robust and reasonably extensive. We
have data from the Audit [2] which, though old, is broadly confirmed by 1980
and 1982 data from the Milk Marketing Board which accounts for around one
third of the total output of milk products. Together these suggest figures of
around 1.52 MJ/litre for cheese and 2.2 MJ/litre for other milk products. A
small spread of something around 10% might be observed based on the size of
firms. Small firms are taken to be somewhat less efficient but, in cheese
making, they are less likely to dry the whey and hence will use less energy.
However, some small increase in MJ/litre has been assumed for small firms.
The very large firms which account for 80% of the output of the industry are
taken to be just a little (about 5%) more efficient because of the economies
of scale afforded by the large volume throughputs. It is also the case that
they may well have a faster capital replacement policy which would allow them
to install more efficient equipment more frequently. These details do not
affect significantly the use of single measures of best practice noted above
and used in Table 7.

77.   Energy consumption and conservation potential are shown in Table 10.
The main point to note is that conservation potential amounts to 3.34 million
GJ or c. 20% of energy consumption in milk manufacturing.

78.   Table 11 sets out where the potential savings estimated for 1980 might
arise. Both the liquid processing and the milk manufacturing sectors are
taken together. The position in the industry has not altered significantly
since the date of the Audit (1978) which identified rather similar categories
and proportions.

79.   The main point to note is that good housekeeping dominates energy
conservation in the liquid processing sub-sector while process plant is much
more important on the milk product side. These characteristics are important
in determining the course of developments in the industry to 2000 considered
in paragraphs 82 onwards.

80.   Table 11 can be used to infer which fuels are most likely to be
affected. It is clear that the scope for conservation in liquid processing
lies mainly with steam raising and steam use. Because most potential savings
are likely to be made in the use of steam any reductions in energy consumption
are likely to fall on oil (or coal if, sometime in the future, some switch to
coal occurs) for steam raising purposes. By contrast the milk products side
of the industry has a high use of electricity for refrigeration and motive
power. Indeed this proportion of electricity may increase if new technologies
such as mechanical vapour recompression (MVR) were to be used in the
concentration stage. As energy consumption is directly related to production
and economies are only feasible with replacement of capital plant the
potential for savings of oil will to some extent be reflected in a sharp
decline in the use of oil but some modest increase in the use of electricity.
The table therefore must be regarded as tentative in its apportionment of the
conservation potential between the various fuels.

81.   The main conclusion must be that energy conservation potential as
measured in this section of the paper amounts to some 3.34×10^6 GJ which
represents about 20% of total energy consumption in the industry. The
potential in the liquid processing side might be realised quickly but it will
be much slower although potentially more significant on the milk products
side. Future developments in the industry to around 2000 in both sectors of
the industry are considered in the following paragraphs.

Future Trends in the Dairy Industry

82.   Estimates of energy consumption in 2000 depend on both total output in
the industry and the energy consumption per unit output in 2000. Both
elements depend in turn on the likely course of technical and market
developments in the industry over the next five years. Here we consider both
and estimate future energy consumption per unit output and the structure of
output of the industry.

Market Developments: Liquid Milk

83.   As noted in paragraph 53, there is a question by how much the present
daily delivery of liquid milk will be curtailed over the next 15 years. Many
observers consider the system will survive to 2000 albeit with some reduction
of the frequency of rounds (seven days to five days) and some withdrawal from
particularly uneconomic rounds. This is because the system is large, long
established and seen by most households to be convenient and efficient at
current price relativities between doorstep and supermarket prices.

84. However, the long life characteristics of UHT milk and the fact that it does not need to be refrigerated may encourage sales of this and similar liquid milks. At present, the taste is not popular but technical change may make it more acceptable. Also, the move to skim and semi-skim milk may alter traditional 'milk' tastes but the judgement made here is that the delivery system remains broadly intact to 2000.

85. There is some pressure from farmers to process their own milk rather than sell to the Milk Marketing Board but such units would typically be smaller than those of the Board or of other large processor/retailers.

Technical Developments: Liquid Milk

86. It will be recalled that pasteurisation and bottlewashing account for around 60% of energy used in the liquid milk sector. The view of the industry is that neither process is likely to see significant changes. It is already well proven technology and there is nothing in the current R&D programme which would suggest major breakthroughs. Small marginal improvements in efficiency are of course possible, but no major change is foreseen.

87. Regeneration in the pasteurisation process may be an exception to this view. Alpha-Laval claim that their new pasteurisation plant is capable of regeneration of around 95%. Current understanding in the industry is that regeneration has a limit of around 90% (it is currently around 85%). The higher ceiling may lead to some further small energy savings but the claim has yet to be proven.

Market Developments: Manufactured Milk Products

88. Major changes may occur in the product and marketing area. Whey - more specifically the lactose in whey - is a versatile chemical. It can be processed to alcohol using fermentation; to methane by anaerobic digestion; to polymers by hydrolysis and can provide an array of sweeteners through lactose/sucrose variants. If such developments come to pass - and already orders have been placed with MMB - the effect would be a significant shift of MMB product activity and some small net savings in energy. This is because at present most of it (around 75%) is evaporated and dried in whole or part and this is energy intensive. Whey based products may be an example of innovations that ACARD have been seeking [14].

89. If the new array of products were to achieve a major penetration then much of the output would be as a syrup or similar heavy liquid. Although some of these processes are moderately energy intensive the absence of the drying stage saves energy. The net result is that around 2/3 to 3/4 of whey would be produced in as energy intensive form as is currently the case, providing a saving of around 1/4 to 1/3 on current energy use.

Technical Developments: Manufactured Milk Products

90. Technologies such as ultra-filtration and reverse osmosis are likely to be extended to hard cheeses as well as soft cheese. Further developments in reverse osmosis could achieve solids at the pre-concentration stage of around 30% which would significantly reduce transport costs of carrying bulk liquids.

91.   Evaporation will achieve some small further improvements: the major developments have already taken place in mechanical vapour recompression which has already been used for skimmed milk for some time. The improvements over the next 15 years are therefore likely to be in operational efficiency and improved utilisation rather than in the basic plant itself. In France MVR has seen significant sales (with financial encouragement) and might do so in the UK [see 15].

92.   It is unlikely that the traditional technology for evaporation (the multiple effect evaporation units) will exceed the current six or seven effect 'state of the art' plants. Improvements are likely to come through replacement of old three and four effect evaporators or perhaps via a move to MVR. However there has recently been some suggestion that a 'mimic' 12 effect evaporator could be produced. This point has not been followed up.

93.   Direct Firing Although not explicitly prohibited, direct firing of food products is unlikely to occur in this country. The practice is permitted and is used in the United States but not in many European countries. The risks associated with the combustion gases and other material as potential carcinogenic materials is probably too great for manufacturers to risk direct firing.

94.   Fuel Switching Sources in the industry suggest that some fuel switching from oil to coal for basic steam raising is quite likely over the next decade. However, it was not possible to form realistic estimates of the rate at which such switching might occur. Already there are two dairies which are currently installing coal fired boiler plant under the Department of Industry grant scheme. It is known that other plant is earmarked for conversion in the next five years but, because of the additional space and handling facilities needed for coal firing, some establishments have been effectively barred from potential switching. In some cases, it may be that whey will be processed in anaerobic digestion units and boilers converted to use methane.

95.   CHP Evaluation of the potential for CHP in dairies and creameries has previously suggested that CHP was not economic. The position is regularly reviewed and it may be that some time in the next 15 years it may become economic. This view would be strengthened if MVR (and the attendant switch to electricity) were to take place. However, in the absence of large scale MVR installations it would probably be the case that CHP was not worthwhile. This is mainly because plant utilisation would not justify the CHP investment. Plants, if built, would be small - in the range 1,500-2,000 kW - and, on balance, the view in the industry is that CHP will not be a significant factor in the next 15 years.

96.   Drying The technology for drying is already well established and no more than marginal improvements in efficiency of current plant is foreseen. There will be some extension into multiple stage drying - certainly to two stage drying and the possibility of introducing filter mats and scrubbers as a form of heat recovery from the drying process. But this is still current technology and improvements therefore over the next 15 years are likely to be incremental rather than significant breakthroughs.

97.   Low pressure hot water systems In many dairies there is no requirement for steam. Cheese making and other processes are low temperature processes (80°C) and even where higher temperatures are required as in pasteurisation,

this can be achieved by a low pressure hot water system. Details of the savings to be achieved are not available but there is no doubt that eliminating a steam requirement would significantly reduce the cost of installing and running a boiler.

Estimates of Specific Energy Consumption in 2000

98.   If the above developments do occur, then estimates by sources in the industry suggest that the consequent improvements in best and average practice could be as indicated in Table 12(b). Applying these changes to the estimates for 1980 gives specific energy consumption per litre of milk processed in 2000 as shown, while Table 12(c) outlines the assumed fuel split based on developments in the industry. The output structure estimate for the period to 1990/1 has been based on data provided by the industry. Thereafter the best advice was that milk supplies would stabilise as the European support system for dairy produce was held constant. This would lead to constant (or even rising) real costs of milk and output would no longer rise each year. Marginal producers would transfer to other areas of agriculture but the bulk producers would maintain or even increase output. Total supplies would therefore remain broadly constant. Two estimates have been produced (high and low) see Table 12(a).

99.   Within total milk supply, liquid milk consumption is assumed to fall away with the balance of milk supplies going to milk products. The general view is that cheese output will grow slowly over time and remaining supplies of milk for non liquid use will enter the butter/milk powder markets which has managed to absorb such supplies in the past decade.

100.   The data provided by the industry represents two spot points for 1980 and 2000. Estimates for 1990/1 were made by interpolation, using a simple model of the rate at which innovation diffuses into an industry. Volume I (the Main Report) contains a full discussion of the diffusion models adopted by this study.

101.   The rate of change in liquid milk processing was linear in nature as the necessary changes in techniques are mostly of the good housekeeping type and change is already underway although changes in cheese and butter making may be slower to occur (because investment in plant is needed). Specific energy consumptions were then graphed between 1980 and 2000, with an appropriate shaped curve fitted between the points. Estimates for 1990 were then interpolated.

Energy Conservation Potential in 2000

102.   As with estimates of the 1980 energy conservation potential given previously, the potential in 2000 is measured as the difference between energy consumption if the industry operates on 'average' practice and energy consumption when operating on 'best practice'. From the estimates of average and best practice in Table 12 it is concluded that the energy conservation potential in 2000 might be over 30% of estimated energy consumption in that year. Thus in 2000 the conservation potential should be about the same as in 1980.

103.   This is because the dairy industry will be moving to more energy intensive products (milk powder) and producing less low energy products

(liquid milk). Although the average and best practice levels improve over time, the differential narrows less in the area of milk powder, where improvements are contingent on investment in plant. Hence the differential between best and average practice will remain wide overall, and there will still be a large potential for energy conservation in 2000.

104. The main source of energy conservation in liquid milk processing will be good housekeeping. Perhaps 70% of all savings will come from this source. The remaining changes will be due to investment in pasteurisation and bottle washing equipment.

105. Milk products use about 85% of their energy in process use − pasteurisation, concentration, evaporation, drying and the basic butter making and cheese making equipment. Much of the conservation potential will be utilised from improved equipment particularly in evaporation plant and pre-concentration equipment. However, about half of all savings are estimated to be for utilisation. This forms part of process energy use but takes the form of non-production process energy − that is, scheduling, cleaning and maintenance which minimises the disruption of production. It is also important to maximise continuous production runs and avoid a current problem of operating only on one or two scattered days in the week. Increased supplies of milk which are expected to occur over the next decade or so should assist in a better utilisation of plant. Table 13 summarises the breakdown of energy conservation potential.

FIGURE 2: Liquid Milk Processes: Milk, Butter, Milk Powder

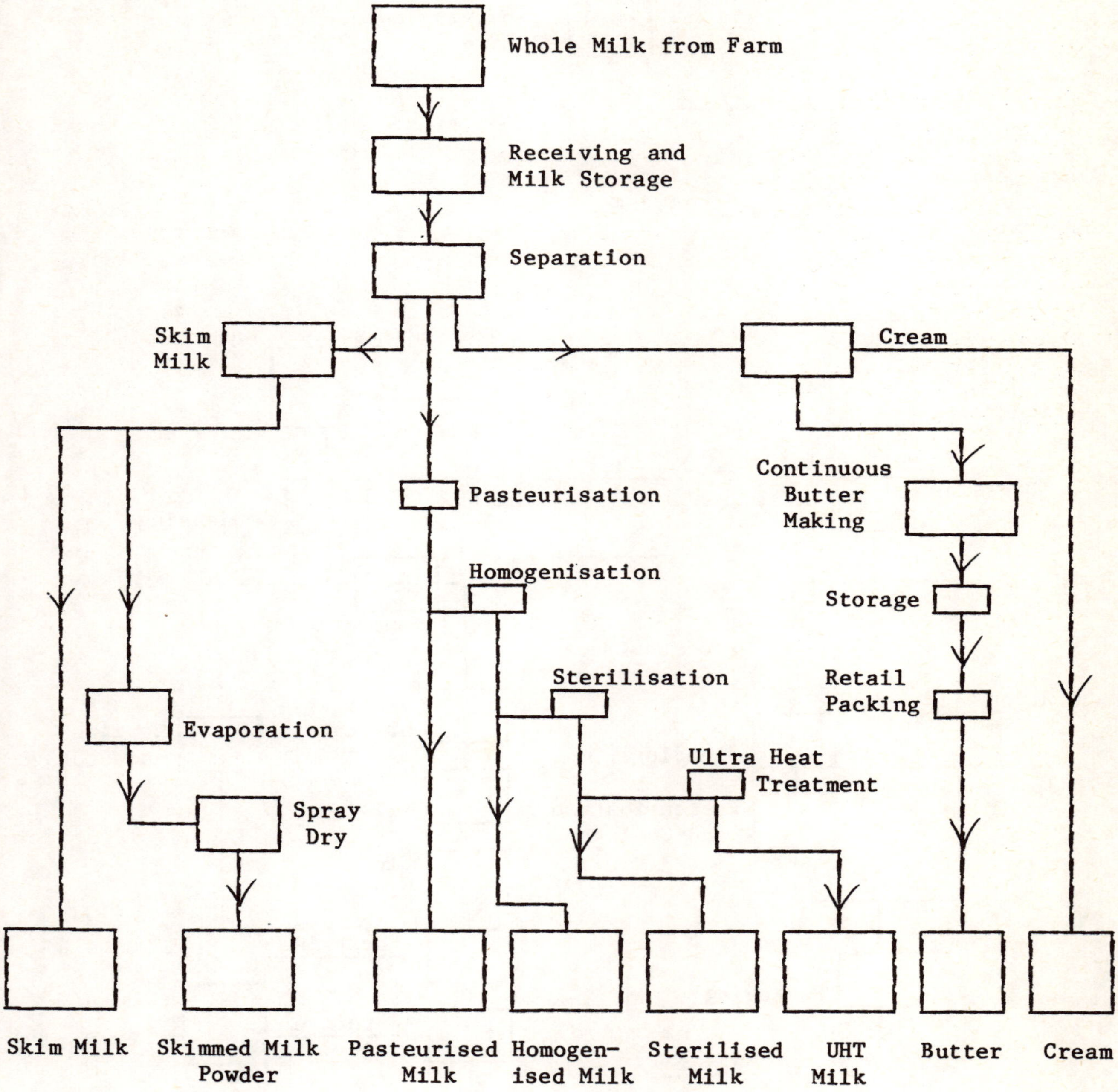

Whole Milk from Farm

Receiving and Milk Storage

Separation

Skim Milk

Cream

Pasteurisation

Continuous Butter Making

Homogenisation

Storage

Sterilisation

Retail Packing

Ultra Heat Treatment

Evaporation

Spray Dry

Skim Milk

Skimmed Milk Powder

Pasteurised Milk

Homogen- ised Milk

Sterilised Milk

UHT Milk

Butter

Cream

FIGURE 3: Cheesemaking Processes

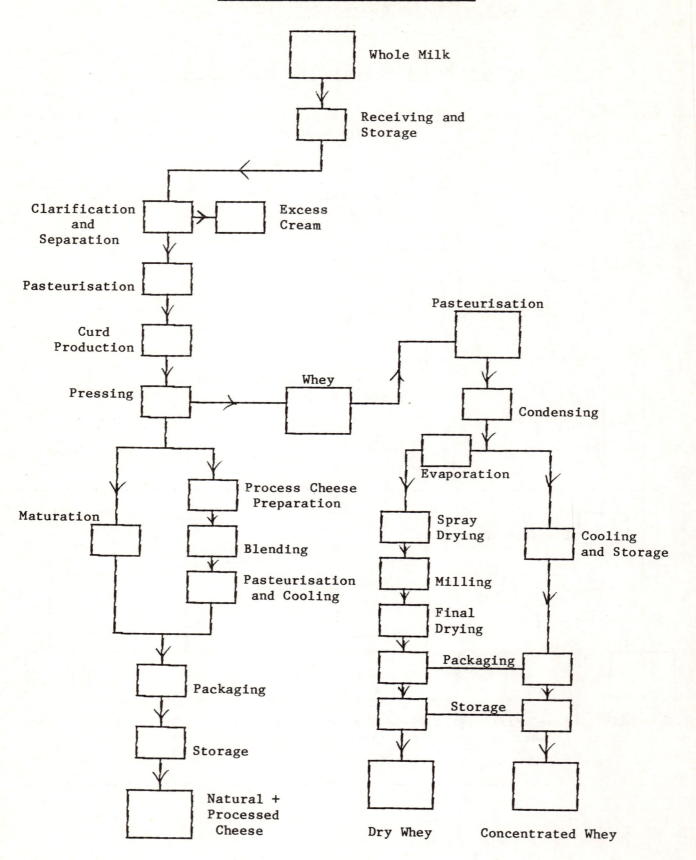

TABLE 4: Output and Energy Consumption and Fuels Used in Milk and Milk Manufacturing Industry 1980/1

	Output, mill. litres	Average Practice, MJ/litre	Energy Consumption, 10^6 GJ	Coal	Oil	Gas	Elec.	Total
						10^6 GJ		
Liquid milk processing	7,136	0.99	7.06					
Cheese	2,335	1.52	3.55	1.35	16.58	2.30	2.93	23.24
Butter, milk powder, cream, crumb etc.	5,743	2.20	12.63					
	15,214		23.24					

TABLE 5: Utilisation of Energy in Liquid Milk Processing Industry 1980

	Process %	Space Heating %	Total
Liquid milk processing	60	40	100
of which			
pasteurisation	30		
bottle washing	30		
other washing (tankers, equipment)	20		
refrigeration and lighting	20 / 100		

% fuels through boiler about 80 - 85%

TABLE 6: Dairy Trade Federation Survey Data
 - Liquid Milk Processing 1978/9

Size of Firm 10^3 litre	Fuel Oil MJ/litre	Electricity MJ/litre	Total Fuel MJ/litre	Output 10^3 litre	% Output	Total Energy Consumption 10^6 GJ
0-9,990	1.24	0.13	1.37	183,947	4	0.252
10-24,999	0.95	0.1	1.05	601,893	13	0.631
25-49,999	0.65	0.08	0.73	1,180,113	26	0.861
50-99,999	1.06	0.11	1.17	1,783,677	39	2.086
1,000,000 +	0.78	0.1	0.88	838,280	18	0.737
				4,587,910		4.570

Energy consumption if all firms operated at 'best practice' in the size range.

Size of Firm 10^3 litre	Fuel Oil MJ/litre	Electricity MJ/litre	Total Fuel MJ/litre	Output 10^3 litre		Total Energy Consumption 10^6 GJ
0-9,999	0.66	0.09	0.75	182,947		0.137
10-24,999			0.75	601,893		0.451
25-49,999	0.41	0.07	0.48	1,180,113		0.566
50-99,999			0.48	1,783,677		0.856
10,000 +	0.41	0.07	0.48	838,289		0.402
						2.414

∴ energy conservation potential given by difference
    4.570 - 2.414 = 2.155 x 10^6 GJ.
    As percentage of total energy consumption in 1978/9: 47.2%.

TABLE 7: Energy Consumption and Conservation Potential in Milk
and Milk Manufacturing, 1980

	Output mill. litres milk processed	Average Practice, MJ/litre	Best Practice, MJ/litre	Energy Consumption on Average Practice, 10^6 GJ	Energy Consumption on Best Practice, 10^6 GJ
Liquid milk processing	7,136	0.99	0.48	7.06	3.43
Cheese	2,335	1.52	1.22	3.55	2.85
Butter, powder, cream, crumb etc.	5,743	2.20	1.74	12.63	9.99
	15,214			23.24	16.27

TABLE 8: Production Figures for Milk Products

Year	Sales of Liquid Milk Daily Average (10^3 litres)
1978/9	20,265
1980/1	19,549

Year	Consumption per Head UK, pints/week
1964/5	4.89
1969/70	4.76
1975/6	4.95
1979/80	4.48
1980/1	4.40

Milk Product	Production of Milk Products in UK, 10^3 tonnes				
	1976	1977	1978	1979	1980
Butter	89.3	134.0	163.3	160.5	168.4
Cheese	203.6	206.7	215.6	234.2	237.1
Cream (total fresh and sterilised)	110.9	113.1	117.1	120.9	122.5
Condensed milk	186.8	229.5	185.9	174.4	142.7
Milk powder	218.8	297.2	326.8	287.1	301.8

Source: [13].

TABLE 9: Energy Utilisation in Milk Manufacturing 1980

	Process %	Space Heating %	Total %
Cheese	70	30	100
of which electricity for refrigeration	30		
Butter and other products	85	15	100
of which electricity	30		
Spray drying	25		
Evaporation	25		

% fuel passed through boiler 80% in both subsectors

TABLE 10: Summary of Energy Conservation Potential in Milk Manufacturing Sector, 1980

Energy conservation potential in liquid milk processing given by:

$$7.06 - 3.43 = 3.63 \times 10^6 \text{ GJ}$$
or 48% of 1980 consumption

Energy conservation potential in manufactured milk products given by:

$$16.18 - 12.84 = 3.34 \times 10^6 \text{ GJ}$$
or 21% of 1980 consumption

Total energy conservation potential in dairy industry is:

$$23.24 - 16.27 = 6.97 \times 10^6 \text{ GJ}$$
or 30% of 1980 consumption

TABLE 11: Milk Processing and Milk Manufacturing: Categories of Energy
Conservation Potential 1980

Type of Conservation Measure	Conservation Potential 10^6 GJ	% of Total
Milk Processing		
Housekeeping	2.54	70
Add-on Equipment	0.18	5
Replacement Equipment	0.73	20
New Process Plant	0.18	5
	3.63	100
Milk Manufacture		
Housekeeping	0.33	10
Add-on Equipment	0.17	5
Replacement Equipment	2.0	60
New Process Plant	0.84	25
	3.34	100
Total Housekeeping	2.87	33
Total Add-on Equipment	0.35	5
Total Replacement Equipment	2.73	45
Total New Process Plant	1.02	17
	6.97	100

TABLE 12: Milk: Output Structure, Specific Energy Consumptions
and Fuel Split 1980-2000

(a) Structure of Output (percentage)	1980	1990 H	1990 L	2000 H	2000 L
Liquid milk	47	36	39	32	29
Cheese	15	13	14	15	15
Butter, powder, crumb, cream, etc.	38	51	47	53	56
	100	100	100	100	100

(b) SEC, MJ/litre	1980 Average	1980 Best	1990 Average	2000 Average	2000 Best
Liquid milk	0.99	0.48	0.72	0.69	0.39
Cheese	1.52	1.22	1.39	1.22	1.04
Butter, powder, crumb, cream etc.	2.20	1.74	1.98	1.76	1.40

(c) Fuel Split Assumptions	1980	2000
Coal	6	15
Oil	71	50
Gas	10	15
Electricity	13	20
	100	100

TABLE 13: Sources of Energy Conservation in 2000, Expressed as a
Proportion of Total Estimated Potential

Type	Liquid Milk Processing	Cheese Making	Butter/Milk Powder
Good housekeeping incl utilisation	70	40	40
Bolt on components			
Replacement plant and equipment	30	60	60
New process plant			
TOTAL	100	100	100

5.2   THE PROCESSED FOODS SECTOR

106.   It is convenient to distinguish frozen fruit and vegetables from other processes such as canning and bottling and preserves such as jams and marmalade.   All establishments contain two or three energy using processes most of which are supplied by steam or hot water from gas or oil fired boilers.

107.   In the food industry process energy usually accounts for 60-80% of total energy and around 70% of total energy is supplied through boilers.   But, electricity, of course, is much higher, as a proportion of process energy, in the frozen sector.   Space heating is generally in the order of 15% of total energy requirements.   The individual proportions vary according to the processes involved and are shown across the different sectors in the processed foods industry in Tables 14, 15, 17 and 18.   A note of each of the main product group explains why these proportions vary.

FROZEN FRUIT AND VEGETABLES

Industrial Processes and Energy Utilisation

108.   The fruit and vegetables are brought in, cleaned, peeled, blanched (if necessary heating to 90-95°C to inactivate enzymes), cooled, graded and frozen.   The produce is then packaged and stored.   Figures 4-6 show the details taken from [16] and amended for UK practice.

109.   Frozen potato chips require different treatment.   The potatoes are steam peeled, cut, water blanched, part fried by passing through hot cooking oil, cooled, packaged and frozen.   This is more energy intensive than most other forms of fruit and vegetables but the tonnages are large and the main supplier very efficient, so the specific energy consumption is about the same as for most other fruit and vegetable processes.

Other Fruit and Vegetables

110.   Crisps   The main processes are as described above - cleaning, steam peeling, cutting, blanching and frying and packaging, the main difference (and the main energy intensive element) is the fryer for frying the crisps.

111.   These are usually directly or indirectly fired gas ovens and account for about one half of total process energy use.   Other processes will require hot water for cleaning and steam peeling.   The packaging hall requires to be heated so boiler fuel accounts for something over half of all energy use and space heating for around 20% total energy.

112.   Canned Products   The main processes are lye or steam peeling, cleaning, blanching, brine and syrup production, filling and closing, followed by pasteurisation (fruits) or sterilisation (vegetables) in steam or water heated retorts at 100-125°C.   The main energy carrier therefore is steam and hot water which is supplied by boilers - 80-90% of energy is process energy and about 90% of this is supplied through the boiler.

113.   Pickles and Sauces   The processes are similar to those described above; peeling, blanching and cooking.   Typically pickles and sauces are bottled so there will also be a bottling, packing and labelling facility.   After bottling

the product is pasteurised at about 70-80°C. About 90% of all energy is
process energy and about 90% of all energy is passed through boilers. Space
heating is small - perhaps 5% of total energy.

114. <u>Jams</u> Some produce is held in frozen storage so the electricity/ refrigeration element in energy use is higher than might otherwise be expected. The
main processes are blending and boiling of the fruit and syrup to a required
consistency. The main energy using process is evaporation (boiling at about
100-112°C) but bottle washing and bottling is moderately energy intensive.
Space heating is more important than in other sectors - perhaps the 15% shown
in Table 15 is an under estimate. Indeed the Food RA in an earlier study
showed that space heating in the whole of this sub-sector is about 22% so the
figures shown in this work are possibly under estimates. Total process heat
in the sector is about 80% and most of this is supplied by boilers.

Energy Consumption and Conservation Potential 1980

115. Energy consumption is estimated as the product of 'average practice' and
output for 1980 taken from Business Monitor [17]. The output of some
products is only available as gross sales (£M) so average costs (£/tonne)
advised by Leatherhead Food RA were used to convert the figures to tonnes.
Energy conservation potential is assessed as the difference between 1980
energy consumption and the consumption that would have occurred had all firms
operated at 'best practice'. Evidence on best practice and average practice
is required.

116. The data is available from visits made by consultants from the
Leatherhead Food RA and Campden Food Preservation RA as part of IETS visits
[6]. Some 101 sites were visited in meat and fish processing industries and
23 sites in the fruit and vegetable industry, so the sample sizes are
respectable accounting for around 20% of total employment in the industry and
being representative of the size distribution of firms in the industry (see
Tables 4 and 5 in [6]). The sample is not ideal - it contains few firms in
crisps and nuts - but it is enough to base some judgements about average
practice and best practice in the industries.

117. Discussion with officers at Leatherhead Food RA and Campden Food
Preservation RA led to judgements on average practice and best practice
figures that are shown in Table 14.

118. The product of these figures suggests that energy consumption in 1980
was 10.9 x 10^6 GJ. If all establishments operated at best practice then
energy consumption in 1980 would have been 8.3 x 10^6 GJ. So energy
conservation potential is estimated at 2.6 x 10^6 GJ or 24% of energy
consumption in 1980, see Table 16.

MEAT AND FISH PRODUCTS

Industrial Processes and Energy Utilisation

119. The division between frozen, chilled, cooked and canned products
reflects not simply the different products but the different patterns of
energy use. Each is considered in turn and illustrated in Table 17.

120. **Poultry** The birds are delivered live and then hung, slaughtered, plucked and singed, eviscerated, cooled, inspected and frozen. None of these activities is particularly energy intensive apart, of course, from refrigeration. There is some hot water for general cleaning purposes but as it is normal in frozen food factories to accept low working temperatures there is little energy demand for space heating. About 90% of energy is for process operations of which around 40% is refrigeration. Only 50% of energy used is obtained via boilers as steam or hot water mostly for cleaning and space heating.

121. **Fish** Most fish is delivered from the quay side as plate frozen fish so it is part thawed using steam chambers; filleted; and refrozen. As with poultry plants there is little space heating. Several fish plants and meat factories use liquid nitrogen as a refrigerant rather than electricity, so 'other fuels' in Table 17 refers mainly to refrigerants. The fish sector is complicated by the presence of several integrated works handling frozen fish, fish products such as fish fingers, canning of fish and manufacture of fish pastes. They are usually classified to the frozen sector as that is usually the principal product.

122. **Meat** At the abattoir the animals are stunned, slaughtered, skinned or singed, eviscerated and butchered. In most cases the body is partly sectioned and the flesh for human consumption is sent to refrigeration for eventual dispatch to retail or wholesale butchers. The remaining meat and carcass is sent for animal foodstuff preparation and meat rendering for the production of animal fats. Chilled meat may well be processed using gas packaging. Table 18 shows the details of energy use in this area.

123. **Cooked/Smoked Meats and Fish** The industrial processes are summarised in Figures 7 and 8. Bacon is injected with brine, stored and then passed through a low temperature oven (about 80°C), cooled and chilled. Ham, tongue and cooked sausages are cooked or boiled in hot water vessels before chilling. Some meat (ingredients in recipe packs) receives two processes.

124. Meat pies require a baking operation; roast pork and beef sold for slicing also undergo the usual oven preparation and cooking (which are usually directly or indirectly fired gas ovens) before chilling.

125. There is a general point to be made about energy use in the frozen-chilled sector of fruit and vegetables and meat and fish products. The point is that energy use occurs throughout the chain from field to retail store because chilled transport and special packaging are important aspects of meeting customers' requirements for fresh, chilled food. A systems approach is required but this study looks only at energy use in the manufacturing process.

126. Space heating is still typically less than 15% of total energy but process heat is water and steam based (apart from the ovens) so a high proportion of fuel (over 60%) passes through boilers. Electricity for refrigeration accounts for perhaps 20% of overall energy use and about 1/3 of process energy.

127. **Canned Meat and Fish** Meat and fish are prepared, canned and then cooked in steam jacketed retorts at about 80-130°C as with canned vegetables. It is a steam and hot water based system. Table 18 provides details of energy use.

'Heat processing' may be a more appropriate term than canning because glass jars, retortable pouches and other rigid or semi-rigid containers are also processed.

128. Energy Conservation Potential in 1980   Using output data from Business Monitors and specific energy figures from the Leatherhead Food RA, energy consumption in 1980 was estimated as 10.8×10^6 GJ which is close to the Leatherhead Food RA figures (9.6×10^6 GJ) but less than the 1979 fuel price enquiry data (11.68×10^6 GJ). Output differences will account for some of the discrepancy. If all firms operated at best practice levels, energy consumption might be 6.5×10^6 GJ which, as a percentage of the 1980 energy consumption figures, is nearly 40%. Table 19 sets out the details.

Energy Conservation Measures in the Processed Foods Sector

129. The Leatherhead Food RA make recommendations which include the following:

- good housekeeping measures such as, improved insulation, monitoring of steam use, etc;

- some heat recovery from industrial processes (blanching, fryers, sterilisation and refrigeration plant);

- improved burner controls on boilers;

- improving the utilisation rate of plant if possible by continuous rather than batch production particularly for canning and sauce making;

- improve recovery of condensate and use of flash steam.

130. Developments to 2000 in the Processed Foods Sector   The market and technological forces likely to act on the fruit and vegetable sector and the meat and fish products sector over the next 15-20 years are very similar. These two sub-sectors are therefore considered together.

Market Conditions: Three Main Trends

131. Three important trends which are generally expected to increase in importance over the next two decades may be observed in current markets. The first trend is the notion of freshness. There has been some small increase in the quantity of fresh food purchased; there is a move away from wrapped and packaged goods towards 'fresh' goods which may be covered but are not processed in any way. Thus sales of chilled produce have increased. This trend has been reinforced by the requirements for date stamping which places a premium upon fast distribution between farm manufacturer and retailer because of the short shelf life. This trend is well established in dairy produce and some other areas of the food industry but is now observable in fruit, vegetable, meat and fish products.

132. The second trend is the move towards convenience foods. Fast food chains have extended to a wide variety of foods and an enormous variety of snacks (eg pizzas). This has been one of the major growth areas in food industry in the last decade. The field is expected to grow further. This is

because the single person householder is expected to grow in numbers amongst the young and old in the next two decades and this is expected to maintain and increase the demand for ready prepared, pre-packaged single portions. Frozen or chilled meat, fish and vegetables will be important components of such meals and snacks.

133. The third observable trend is towards those foods which increase health and have a nutrition value, or at least are not supposed to impair health. Fruit and vegetable products are expected to gain in market share compared with bread, flour, potatoes, sugar products etc; meat and fish also have high protein and relatively low fat content and they would expect to fare well from this trend.

134. Other Factors The balance between domestically produced frozen and canned products has not stabilised and the broad proportions (roughly one to two) are expected to change further over the next two decades. However, both shares may well be affected by the growth of chilled and fresh produce.

135. Output Structure and Specific Energy Consumption in 2000 No estimates of output developments or market forecasts over the next one to two decades are available within the industry, so that Table 20(a) therefore presents output structure forecasts based on informed opinion and the trends described in paragraphs 131-134. The structure of output within fruit and vegetables assumes a fall in canned, further growth in frozen and faster growth in chilled products. Table 20(b) presents estimated specific energy consumptions for each product group, based on the following analysis of technological developments.

Technological Developments in the Industry

136. It is convenient to distinguish between canning, chilling and freezing. In future, considerable emphasis will be placed upon chilling products, rather than deep freezing them. The technology required lies mainly in more sensitive control and monitoring equipment because the difference between deep freezing and chilling is largely one of temperature and temperature control. More fundamental R&D on the preserving qualities of chilling, is being carried out at a number of research centres. The results will permit finer control and more efficient use of chilling equipment than current practice allows.

137. A similar development is underway with frozen storage of some products. The main benefit is to spread the load between harvest, product processing and final sale. There are thought to be significant cost and production advantages in frozen storage of semi-processed foodstuffs because the savings from continous production offset any additional costs associated with frozen storage. A full account of effects on food quality of deep frozen storage for long periods of time is not known in detail (so this too is a subject for R&D) but the benefits of this development are such as to push the industry in this direction unless R&D proves otherwise.

138. Canning is unlikely to see any significant developments. The main energy using processes are sterilisation and blanching and while investment in new plant and equipment is likely to yield savings in energy use, the rate of increase in investment will inhibit the rate of energy savings. A significant proportion of energy use may be conserved by good housekeeping measures and these may be expected to be introduced fairly rapidly in those companies which

have not already taken the measures. Bolt-on equipment, where modest
expenditure might be expected to yield significant energy savings, includes
sensing equipment and monitoring equipment for boilers and (rather more
expensive) heat recovery systems attached to either exhaust gases, to hot
water systems used in blanching or sterilisation. Introduction of these
measures may be expected to occur in a steady fashion over the next two
decades. There is currently some R&D work in the process equipment industries
in the area of sterilisation equipment (retorts, autoclaves, sterilisers)
which may yield savings through speed, labour and the opportunity for heat
recovery. Automatic control systems for blanching and sterilisation should
increase process efficiency in canning.

139. Other developments which may affect the industry include:

. Aseptic packaging

Here sterile food is placed in sterile containers which are most
likely to be paper-based laminates or plastic containers. The
advantages are the saving of high energy cost in sterilisation and
savings in cost of the package. At present aseptic processing is
largely limited to dairy products and fruit juices and further R&D is
necessary before the process will be commercially applied to
particulate or low acid foods.

. Retortable Pouches

These are laminated foil or similar bags capable of withstanding
sterilisation temperature (up to 125°C). Although the process energy
requirements are similar there are savings in packaging and transport
costs.

. Wrapping and Packaging

At present, although most cans are packed in board containers, a
significant proportion of cans are packed in trays with a shrink wrap
polythene liner. The shrink wrapping is applied by heat passing
through 12 kW electric ovens which, in a controlled fashion, reduce
the polythene and mould it to the half container and cans. This is a
moderately expensive form of packaging and developments are under way
to install a cold stretch wrap. This at present is only available
for large pallet loads but this will no doubt be extended to smaller
loads.

. Dehydration

After active marketing some years ago food companies have not pursued
this area. Soups and vegetables (pot snacks) continue to be
dehydrated and specialist foods for travel, camping and other cases
where light weight and space is important continue to be marketed but
the main thrust has moved to fresh foods. However, convenience and
fast preparation times (5 minute meals) are also important marketing
points so this area may expand. Dehydration is an energy intensive
activity.

140. Energy cost represents only some 2-3% of manufacturing costs, so
conservation, although a useful addition to profits, is not seen yet as a
major priority for management attention. Although static markets and pressure
on margins will influence the industry over the next two decades, there is
little reason to suppose that the introduction of these measures will be other
than a steady, incremental, development. The diffusion of these various
conservation measures and technological change is assumed therefore to be
linear. Informed opinion suggests that average practice might fall gradually
by around 20-25% by 2000, while best practice might still fall further by
around 15% implying some narrowing of the current differentials between best
and average practice.

141. Developments in Meat and Fish The same factors are likely to affect
meat and fish as those noted above although no visits or other direct
information is available to support this view. Past trends and informed
expectations suggest that demand for poultry will almost certainly increase
significantly, while fish consumption may remain broadly unchanged.
Consumption in total is likely to remain stable, though within this total
there is likely to be a fall in lamb consumption and a rise in pork and
poultry consumption. These assumptions lie behind the estimates in Table
20(a) but must remain uncertain.

142. One element in this uncertainty is the possible development of
restructured protein and the ability to reconstitute fish and some meat
products. This, at present, is in the R&D state but, if successful, could
have some impact on catering trades and prepared foods. RHM have a fungal
protein cleared for human consumption but have yet to market foods containing
this protein. However, developments in this area are so uncertain that they
have been assumed to have an insignificant impact upon output or specific
energy requirements in the sector through to the end of the century. If they
were to take off the main impact would be on refrigeration.

143. No information is available on how specific energy consumption might
change over the period to 2000. The gap between average practice and best
practice in 1980 is wide (c. 30% in most sectors); so it has been assumed that
average practice falls by 20-30% to 2000 in most cases and that best practice
falls a further 10% in the same period. This is very uncertain but it is not
implausible and completes the account of this sector. Table 20 sets out the
assumptions for the whole subsector.

FIGURE 4: Industrial Processes in Frozen Fruit and Vegetables

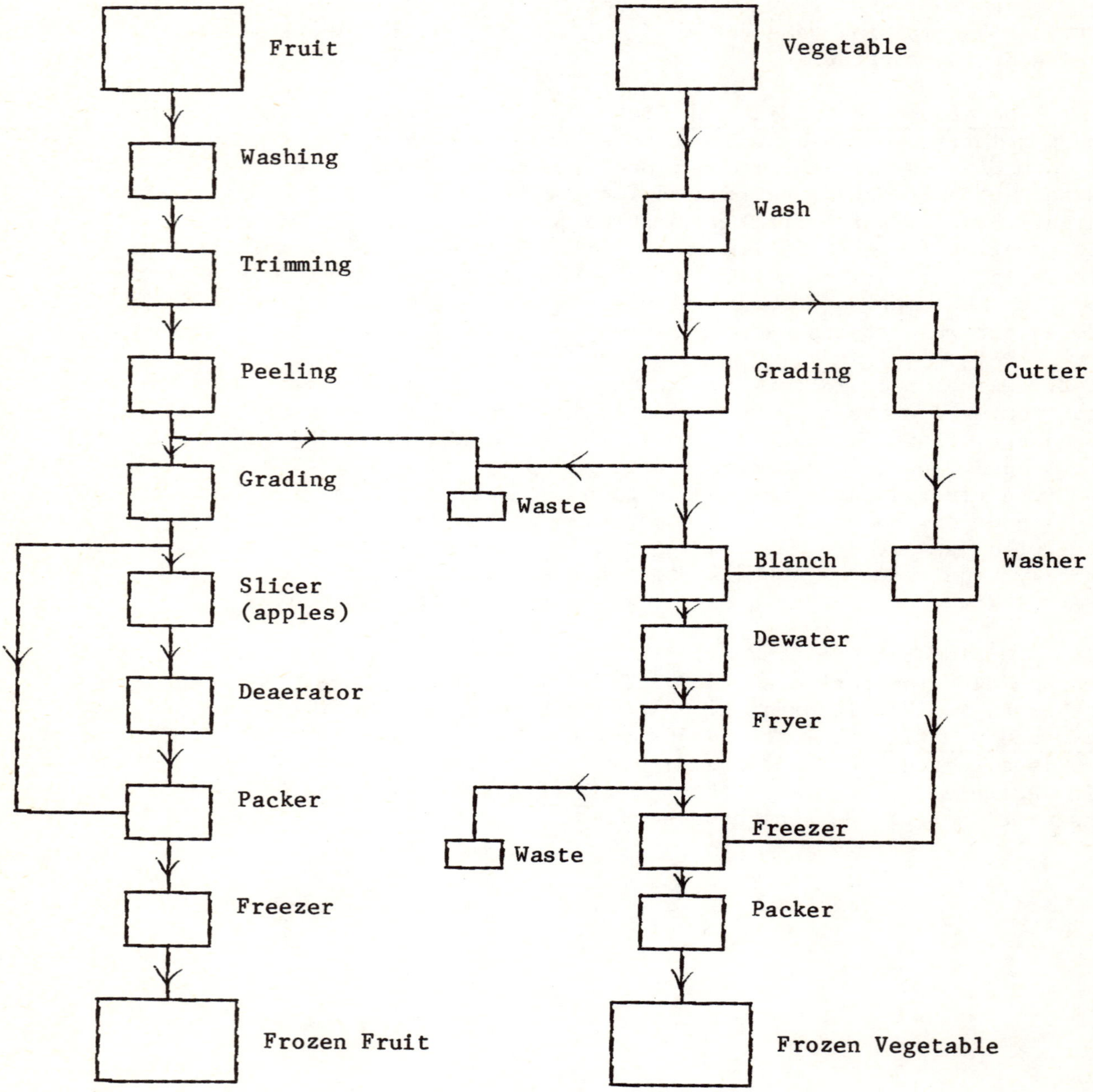

FIGURE 5: Industrial Processes in Vegetable Canning

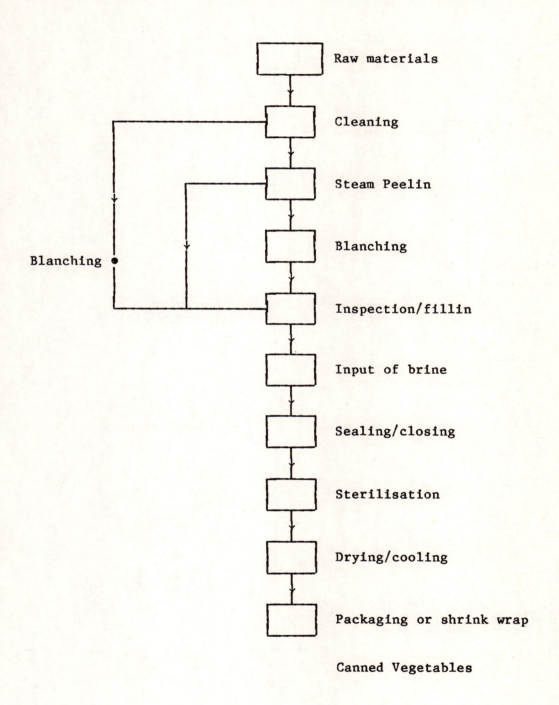

FIGURE 6: Industrial Processes in Fruit Canning

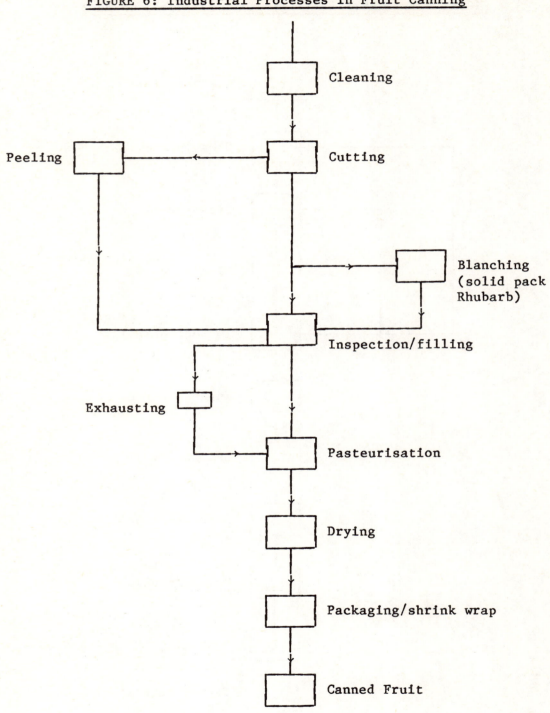

FIGURE 7: An Example of Industrial Processes for Meat Products

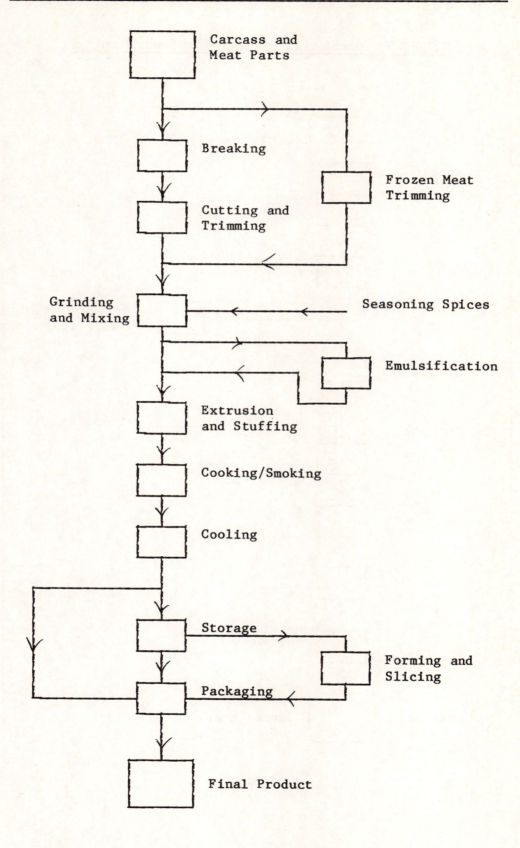

FIGURE 8: Industrial Processes in Meat Canning

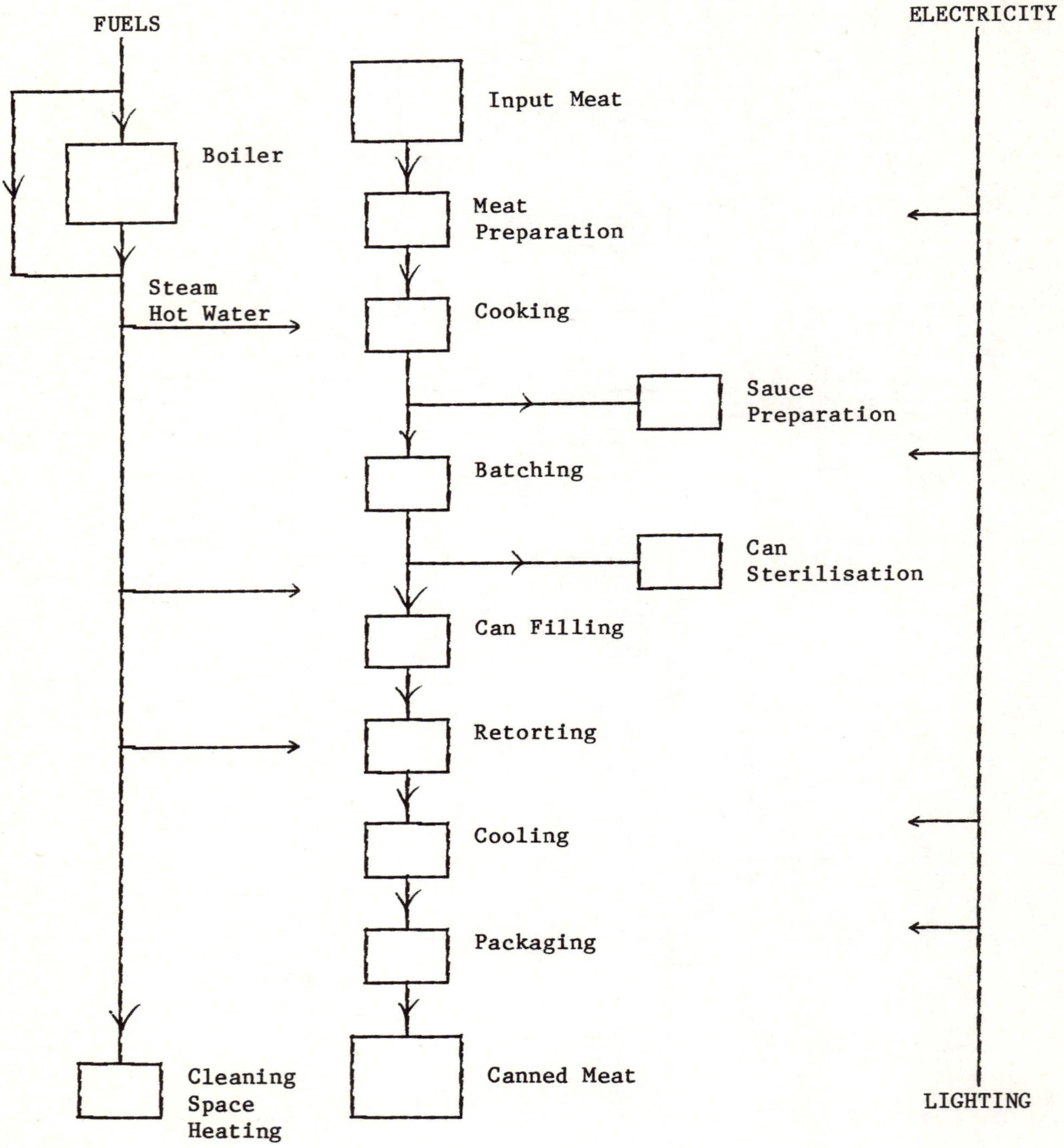

TABLE 14:   Output and Energy Consumption in the Fruit and Vegetable Sector 1980

	Output, 10^3 tonne	Average Practice,	Best Practice,	Oil,	Coal,	Gas,	Elec,	Total,
		GJ/tonne				10^6 GJ		
Frozen	485.4	3.5	2.9	0.05		1.30	0.35	1.70
Other F & V								
Crisps & savouries	196.6	9.5	8.2	0.13		1.57	0.17	1.87
Canned	1,229.3	3.1	2.3	2.40	0.26	0.95	0.20	3.81
Pickles & sauces	709.4	3.1	2.1	1.37		0.59	0.24	2.20
Jams & jellies	303.3	4.2	3.1	1.06	0.05	0.06	0.10	1.27
	2,924.0			5.01	0.31	4.47	1.06	10.85

TABLE 15:   Energy Use in Fruit and Vegetable Sector 1980 (%)

Energy Use	Frozen %	Crisps %	Canned %	Pickles %	Jams %
Process:	85	75	90	90	80
of which Electricity	20	10	5	10	5
of which Gas Ovens		50			
Space Heating	10	20	5	5	15
Other	5	5	5	5	5
	100	100	100	100	100
% through boilers	60/80	50/60	90	80	90

TABLE 16:   Energy Conservation Potential in Fruit and Vegetable Sector 1980

Energy consumption if all establishments operate at best practice = 8.28×10^6
∴ conservation potential = 10.85 − 8.28 = 2.57×10^6 GJ
Expressed as % of total energy consumption = 24.0%.

TABLE 17: Output and Energy Consumption in Meat and Fish Sector 1980

	Output, 10^3 tonne	Average Practice,	Best Practice, GJ/tonne	Oil,	Other*,	Gas,	Elec,	Total, 10^6 GJ
Frozen								
Meat	160	5.1	3.5	0.08	–	0.49	0.25	0.82
Fish	262	5.2	3.1	0.75	0.02	0.09	0.50	1.36
Poultry	610	3.1	1.6	0.90	0.15	0.15	0.69	1.89
TOTAL	1,032			1.73	0.17	0.73	1.44	4.07
Processed other than frozen								
Chilled, cooked inc. bacon	1,032	5.3	3.1	3.37	0.15	0.84	1.19	5.47
Canned meat & fish inc. pastes	185	6.4	5.0	0.82	0.11	0.13	0.12	1.18
TOTAL	1,217			4.19	0.26	0.97	1.23	6.65

* Mainly refrigerants in frozen fish and meat industries.

TABLE 18: Pattern of Energy Use in Meat and Fish Products 1980

Energy Use	Frozen Meat, Fish and Poultry	Chilled, Cooked, Cured Meat and Fish	Canned Meat and Fish and Pastes
Process Energy	90	90	80
Space Heating	5	5	5
Other	5	5	15
	100	100	100
% through boiler	50	60	75
Refrigeration	40	20	5

TABLE 19: Energy Conservation Potential in the Meat and Fish Products Sector 1980

Energy consumption if all firms operated at best practice = 6.48 x 10^6 GJ
Conservation potential = 10.72 – 6.48 = 4.24 x 10^6 GJ
As percentage of energy consumption in 1980 = 39%.

TABLE 20: Fruit and Vegetables, Meat and Fish; Output Structure and Specific Energy Consumption, 1980-2000

(a) Structure of Output

	1980	1990 H	1990 L	2000 H	2000 L
Frozen fruit & veg	9	9	10	9	11
Crisps & savouries	4	5	4	6	4
Canned	24	20	21	17	18
Pickles & sauces	14	14	15	14	16
Jams & jellies	6	5	5	4	4
Frozen meat	3	3	3	3	3
Fish	5	5	5	4	5
Poultry	12	13	13	15	13
Bacon, chilled meats	20	23	21	25	23
Canned meat	3	3	3	3	3
Total	100	100	100	100	100

(b) Specific Energy Consumption
(GJ/tonne)

	1980 Average	1980 Best	1990 Average	2000 Average	2000 Best
Frozen fruit & veg	3.5	2.9	3.15	2.80	2.61
Crisps & savouries	9.5	8.2	9.00	8.50	7.38
Canned	3.1	2.3	2.79	2.48	2.07
Pickles & sauces	3.1	2.1	2.70	2.30	1.89
Jams & jellies	4.2	3.1	3.78	3.36	2.79
Frozen meat	5.1	3.5	4.59	4.08	3.15
Fish	5.2	3.1	4.42	3.64	2.79
Poultry	3.1	1.6	2.60	2.10	1.44
Bacon, chilled meats	5.3	3.1	4.50	3.70	2.79
Canned meat	6.4	5.0	5.76	5.12	4.50

5.3 BAKERY PRODUCTS

144. The products covered in this sub-sector include: grain, bread, morning goods (rolls, croissants, buns and scones), cakes, pastries, pies and puddings. Biscuits, including rusks, crispbread and wafers are also considered here as the processes are similar.

145. Flour is the main raw material. Bread and morning goods are yeast raised flour and water mixes which are proved, divided and baked. Flour confectionery and biscuits have different mixes (they include eggs and sugar, fat, currants etc) but follow the same processes of mixing, rolling, dividing and baking. Coatings, often chocolate, and fillings of jam and or cream may be added after the baking process. Biscuits typically are baked to a lower final moisture content than the other products. Figures 9-11 show the processes.

BREAD, FLOUR CONFECTIONERY AND BISCUITS

Energy Consumption and Conservation Potential in 1980

146. Energy consumption has been estimated from the product of specific energy consumption and output data for bread, biscuits (in tonnes) and by value of sales (000£) for flour confectionery. It will be appreciated that a volume measure of cakes, pastries and puddings is not available.

147. The data on specific energy consumption has been taken from the Leatherhead Food Research Association (Food RA) visits and made on behalf of the Department of Industry Energy Thrift Scheme. Most of these visits were carried out in 1980 or 1981 so data refers usually to 1980. A total of 43 bakeries were visited accounting for 8-26% of total production in the sectors. The range of establishments visited was broadly representative of other sites in the industry.

148. The results accord closely with the Food RA's own estimates calculated by grossing up to the industry level from their sample. The results also accord closely with the Fuel Purchase Inquiry of 1979.

149. The specific energy consumption figures for bread which were derived from the Leatherhead RA data show little variance ranging from just under 3 GJ/tonne to something over 5 GJ/tonne. This was a little surprising as the sample included both small area bakers and large national bakers. It seems that, providing the ovens are utilised to their designed capacity the variance in energy consumption per unit throughput remains fairly small despite large differences in the actual throughput. After some discussion about other factors affecting variance in specific energy consumption such as ovens it was agreed with Leatherhead Food RA that average practice was 3.5 GJ/tonne and best practice was 3.0 GJ/tonne.

150. Energy consumption in flour confectionery proved very variable: Leatherhead Food RA data showed variations in energy use (GJ/000£ sales value) in excess of nine times. This is not too surprising because both the products and the premises vary greatly in nature. It is the oven which accounts for most energy: so hard baked cakes use more energy than lightly baked pies whose fillings must remain moist. For the same reasons (variety of product) it is

not possible to establish a volume measure of flour confectionery: so energy use is expressed per unit (000£) sales.

151.   It proved very difficult to establish an average and best practice figure: the average represents the arithmetic mean of RA estimates of energy use and known total sales for 1980 which gives a figure of 6.34 GJ/000£ sales. Best practice was taken to be 4 GJ/000£ because this figure was about 1/3 less than average (which accords with many other findings in the study) and two companies visited already operated below that level. Nevertheless, the results are arbitrary in the sense that figures of 3.6-4.4 GJ/000£ could have been used with equal plausibility, so there is some uncertainty in the estimates of conservation potential in this sub-sector.

152.   Energy consumption in <u>biscuits</u> depends in large part on the nature of the product. Rusks and crispbread require higher temperatures and longer baking times compared with standard biscuits. Chocolate and some sugar coated or filled sweet biscuits require cooling. The proportion of various kinds of biscuits/crispbread in total output are shown in the Business Monitors and the output figures have been multiplied by specific energy figures derived from the IETS visits. Most biscuits are produced at about 5 GJ/tonne but rusks require much drying so take around 13 GJ/tonne.

153.   The final figures on energy consumption in this sector in 1980 are consistent with the 1979 Fuel Purchase Inquiry data. Table 21 summarises the results.

154.   The fuel breakdown in Table 21 is taken from IETS data. The larger and continuously used bread ovens have switched to gas in the last decade and this fuel now accounts for 67% of fuel use in the sector. Oil takes 23%. By contrast the typically smaller cake and pastry ovens have stayed with oil (52%) although gas has taken 33% of the market. The higher electricity usage in flour confectionery (15%) is largely due to the milling, melting and cooling of chocolate and other fillings. Taking bread and flour confectionery together oil takes half the energy use, gas a third and electricity the remainder.

155.   Fuel use in biscuits is dominated by gas (72%) used in ovens, with the remainder split between electricity and oil with a very small quantity of coal.

156.   Most bread is baked in indirectly fired ovens heated mainly by gas. The ovens were once direct fired and were heated by coal. The prospects for switching back to coal will be discussed later, but it seems possible that ovens will be fired by electricity.

157.   Biscuit manufacturers have traditionally used directly fired long travelling ovens fired by gas. For a mixture of reasons these ovens are being replaced by indirectly fired ovens but continue to use gas.

158.   Steam raising using coal, gas or oil is a modest proportion of energy use, perhaps a half or less. The steam is used for proving bread, drying and a significant amount of hot water for cleaning equipment. Most of the boiler heat therefore, together with all of the oven heat, may be classified to

process energy use. Space heating accounts for around 10-20% of total energy, seldom more. In some factories the oven heat is so intense that buildings are built to a minimum insulation standard to assist heat dissipation.

159. Electricity is mainly used for process use: power mixing machines, refrigeration, lighting and conveyors. Some small bakeries use electrical heated ovens. In some biscuit factories dielectric heaters are used at the end of the baking oven for final drying of the biscuit. Refrigeration is the most important use of electricity particularly in flour confectionery as more items are sold chilled or frozen such as gateaux, trifles and pies. Refrigeration is also important for chilling fillings and coatings, particularly for chocolate and cream and for storage of the finished product. Table 22 shows the pattern of energy use.

160. Table 23 shows energy conservation potential calculations. The potential is measured by the difference between energy consumption in 1980 as calculated and energy consumption in 1980 which would have occurred had all firms operated at best practice. The calculations show a conservation potential of 3.12×10^6 GJ (16.17-13.05) in the total industries of which 0.6×10^6 GJ occurs in biscuits and 2.52×10^6 GJ occurs in bread and flour confectionery. This represents 13.3 and 21.6% respectively of their energy consumption in 1980.

Energy Conservation Measures in Bread and Flour Confectionery and Biscuits Sector

161. Bread and flour confectionery and biscuit manufacture have a high content of process energy, the production technology dominated by ovens but including other processes. This high proportion of energy going through ovens (about half of process energy and about 40% overall) is the focus of energy conservation but other processes (fermentation, chilling) and the use of steam/hot water for space heating are also important.

162. Energy conservation measures therefore are best focused on replacement equipment because the opportunities for good housekeeping savings must be limited if non-process energy amounts only to 15%-20% of total energy use. Energy management however must also affect the efficiency of process energy use.

163. The main conservation measures noted in the Thrift Scheme report are noted below.

Process Equipment

(1)   Oven burner replacement and separate metering of this most important energy user.

(2)   Insulation and thermostatically controlled temperature of ovens equipment and buildings.

(3)   Gases vented from the oven at 200°C so significant heat recovery possibilities exist. In none of the factories visited was this heat recovered. The heat could be used for recycling for refiring; heating of water or air.

(4) Replacement of electric ovens by gas or oil where this is still to be done.

(5) Better utilisation of the oven – reducing idle time and preheating time.

Space Heating

(6) Separate electric fires and water heaters were noted despite the heat dissipation problems in production areas. There is opportunity for heat recovery for use in space heating.

Boilers

(7) Because of the limited use of boilers for small quantities of steam and some space and water heating they have a low utilisation rate. They are therefore often badly neglected and need tuning. Lagging of pipework was poor.

Refrigeration Plant

(8) Roughly 2/3 of power input to refrigeration is converted to heat which could be recovered.

. Refrigeration coil defrosting should be by hot refrigerant gas rather than by electricity.

164. **Developments to 2000** The major trends in trading conditions for <u>bread</u> foreseen over the next 10-15 years are:-

. fairly static bread sales

. increase in share of brown bread (although share fell back from 12% to 10% in 1982)

. continuing growth in morning goods, rolls, croissants etc.

. increase in catering trade market particularly by fast food chains to meet demand for burger buns

. decline in 'hand made' small batch production of cakes

. stability (perhaps) in large volume cake production

. polarisation of production of bread between large volume bakeries for sale to supermarkets; small on-site bakery in corner shop/bread shop; and the in-store bakeries typically within the supermarkets.

The last source may prove vulnerable to low margins and the desire by supermarkets to use floor space for more profitable merchandise.

165. Three important factors underlie the whole sub-sector of this area of the food industry.

(i)   Health consciousness

There has been a recognisable trend in the last 5 or 10 years
towards starch reduced bread and biscuits; non sweetened food; a
move towards purer products with fewer additives; and, of course,
the recurring problem of sugar content, and the propensity for
dental caries. The cake industry has suffered a steady decline in
sales in the last 5-10 years. The bread making sector also
suffered although sales of wholemeal flour have increased.

(ii) Convenience

The demand for convenience foods arises from a number of factors:

. the increased demand for leisure time being traded for
  increased expenditure on prepared foods and snack foods, eg
  hamburgers, craquottes (which now take £23 million sales)

. the increase in the number of one and two person households
  both amongst the relatively young and the elderly.

(iii) Low margins

Margins are low in the bread market because the industry is
fiercely competitive with two companies controlling three
quarters of the market. This is because demand has been slack for
the past few years and capacity and output has not adjusted. The
biscuit market has also suffered recent contraction, eg the
closure of the United Biscuits plant in Liverpool.

In short then, the trading conditions in which the industry has operated and
is likely to operate over the next 10-15 years are those of stability but with
major product and marketing changes taking place below this static surface.
With low margins there is some incentive to improve the cost of the production
through technical change.

166. **Technological Developments** Over 55% of process energy is absorbed by
the ovens for baking bread and biscuits so much of the technical change will
be focused on oven design. A combination of improved oven design and analysis
of the travelling oven and ancillary equipment associated with a bread line
can save something in the order of 20% of energy costs. This is associated as
much with improved utilisation and operation of the line (pre-heating down
time burner controls etc) as with the insulation and design of the oven
itself. Spooners Ltd and Simon Vicars Ltd which produce ovens for bread and
biscuits have made some efforts to reduce energy by improving the directly
fired oven, which uses much less fuel than the normal indirectly fired oven
[18, 19].

167. On the biscuit manufacturing side there appears to be much less concern
with operational costs. The trends in process plant are to make it less
substantial so that the design life and physical life are both around 15
years; to improve speeds; process controls; and to maintain reliability.
Energy costs seem much less important to process plant manufacturers which may
reflect the buyers' (biscuit manufacturers') priorities. This is because
biscuit manufacturers consider energy costs represent only around 3% or less

of total manufacturing costs (see Table 1) so capacity, speed and reliability are far more important attributes of their plant and equipment. However, given roughly equal specifications between two or three biscuit manufacturers' suppliers, energy costs will be an important factor in deciding between rival manufacturers' equipment. This has led to some improvement in the insulation of ovens and to improved process control designs to reduce operational running costs.

168.   Other conservation activity already on-going is associated with heat recovery from the ovens, the heat recovered being used to pre-heat the air for the ovens and space heating and pre-heating water for boiler and other uses of hot water in the plant.

169.   Boilers are much less important in bread and biscuit manufacture than in other sectors of the food industries. In many establishments less than half (perhaps a third) of energy passes through the boiler, the hot water being used primarily for space heating and some cleaning. In some establishments the replacements of boilers by direct forms of water heating are being considered. Heat recovery from the ovens would be sufficient to maintain a base-load and separately (directly) heated water would serve for many cleanin operations. These could, over a 10 or 15 year period, significantly affect the quantity of fuels bought in for boiler use.

170.   The main technical changes in the industry are likely to be marginal - further improvements in the ovens, marginal improvements in track and ancillary equipment, better control systems. Some recent papers [18, 19] suggest that electric fired ovens may be preferred by 2000.

171.   One factor which might change this outlook is the extension of thermo extrusion techniques used in biscuit manufacture. This technology has been introduced by Nabisco Foods and other snack/biscuit manufacturers and essentially comprises a small unit (about a quarter of the size of a typical biscuit making plant) in which a prepared mix is subjected to intense pressures and is extruded through different shaped nozzles. The compression is sufficient to cook the mixture with the nozzle performing forming and cutting processes. Thus, in a physically small machine several processes have been combined. The product is highly aerated, similar to wafers, and depending upon the nozzle used, can be produced in a wide variety of shapes and sizes. Some traditional biscuit manufacturers like United Biscuits regar them as not directly competitive and certainly at present, dense materials associated with shortcake, rich tea and digestive biscuits could not be produced by thermo-compression techniques. However, aeration is not an inherent part of the thermo-compression technology so some pasta products and pet food products can be, and are, produced using this technology. If product ranges based upon this technology were to be successfully implemented (and craquottes have taken £23 million sales in less than five years) it could be a major substitute or an enhancement of the biscuits/snacks market.

172.   <u>Implications for Energy Consumption</u> Pulling all these factors together suggests that if bread and biscuit manufacturers are being squeezed by high cost raw materials and the buying power of retailing chains then they have every incentive to be low cost producers. Even when energy costs are relatively low (c. 7% of total purchases) energy expenditure represents a very much higher percentage of profit in what will be essentially a static volume

sales industry. It follows that they have every incentive to adopt energy conservation measures.

173. Nevertheless, energy efficiency is associated in large part with replacement of ovens and other process plant and the investment cycle is relatively long at around 15-20 years. Thus improvements will come only gradually. Total savings in specific energy consumption would therefore be of the order of some 15-20% compared with 1980 average practice figures. Best practice might improve by some 10 or 15% on 1980 levels. These judgements have been based on discussions with some firms in the industry and with Leatherhead Food RA. They form the basis of the forecast SECs shown in Table 27(b). The move towards a 20% reduction in specific energy requirements has been assumed to be linear over the period 1980 to 2000 rather than a curve. If thermo-extrusion techniques were to penetrate at a faster rate then specific energy consumption might well fall in a non-linear fashion. Good housekeeping and other low cost energy saving measures help to achieve all the 20% potential within 20 years.

GRAIN MILLING SECTOR

174. It is usual to distinguish wheat milling from other grains such as barley, oats etc mainly because barley and other grains require drying before storage but also because of the different markets to which grains are sent. Breakfast cereals are another obvious candidate for separate treatment. Energy use in these sectors is described in Tables 24 and 25. The industrial processes in the grain milling sector are shown in Figure 12.

Wheat Milling

175. About 70% of bread flour is now produced from UK wheat but whatever the source the wheat is taken from store and ground through electrically driven rollers. There is often some adjustment of the moisture content depending upon storage conditions. There is also a continuous sieving operation and flour, as it passes through the sieve, moves to a hopper from which it is weighed and bagged and passed to storage and packing.

176. A major feature of a grain mill is that it must be kept dry so there is little requirement for hot water. It follows that boiler systems for hot water and steam are required only for space heating and office/canteen uses.

177. Process energy therefore accounts for around one half of all energy, most of which is electricity. Boilers supply around half of the total energy most of which is used for space heating.

Other Milling

178. Barley and other grains often require drying usually in steam heated chambers. Once the moisture content has been corrected the milling, sieving and bagging operations are similar to those used in wheat milling.

179. The pattern of energy use is therefore similar: a little more process energy (around 65%) because of the drying operations and a similar high proportion of electricity in process energy. Some heat recovery is practised for space heating which accounts for no more than around 20% of the total

energy use. Boilers are sometimes not installed at all but, where they are, they are small and no more than 40% of fuels is passed through them.

Cereal Breakfast Food

180.   This is a very energy intensive product group. Wheat, maize, and other cereal and rice grains are milled, mixed with water and formed by extrusion and cooked or toasted to a totally dry state.

181.   These operations require water for preparation and cooking and there is a little cleaning in place. Where toasting or baking are required as with cornflakes, Weetabix, etc, these use indirectly fired ovens. There is some heat recovery from these ovens for space heating and hot water.

182.   Process energy therefore accounts for around 75% of the total with about 20% going to space heating. Perhaps 60% of all fuel passes through the boiler although it will be less in those establishments using ovens.

183.   Energy Conservation Potential in the Grain Milling Sector Using Business Monitor data for output and Leatherhead Food Research Association specific energy figures, energy consumption in 1980 for the industry as a whole is estimated as 3.27×10^6 GJ. If the industry operated at best practice levels then energy consumption in 1980 might have been 2.70×10^6 GJ, giving a conservation potential target of some 0.57×10^6 GJ, which represents about 17% of total energy consumption. The potential is described in Table 26.

Energy Conservation Measures in the Grain Milling Sector

184.   . Good housekeeping measures such as insulation, lighting levels, etc.

   . Some heat recovery is possible from ovens and from cooking processes

   . Boiler sensing equipment.

Developments in Grain Milling to 2000

185. No visits or other sources of information on grain milling were available. Certain assumptions were made about the structure of output and SEC over the period and these are incorporated in Table 27.

FIGURE 9: Main Industrial Processes In Bread Manufacture

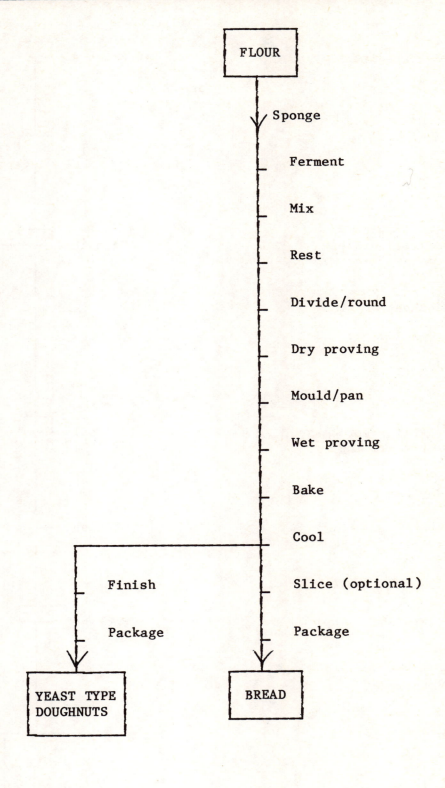

FIGURE 10: Industrial Processes in Flour Confectionery

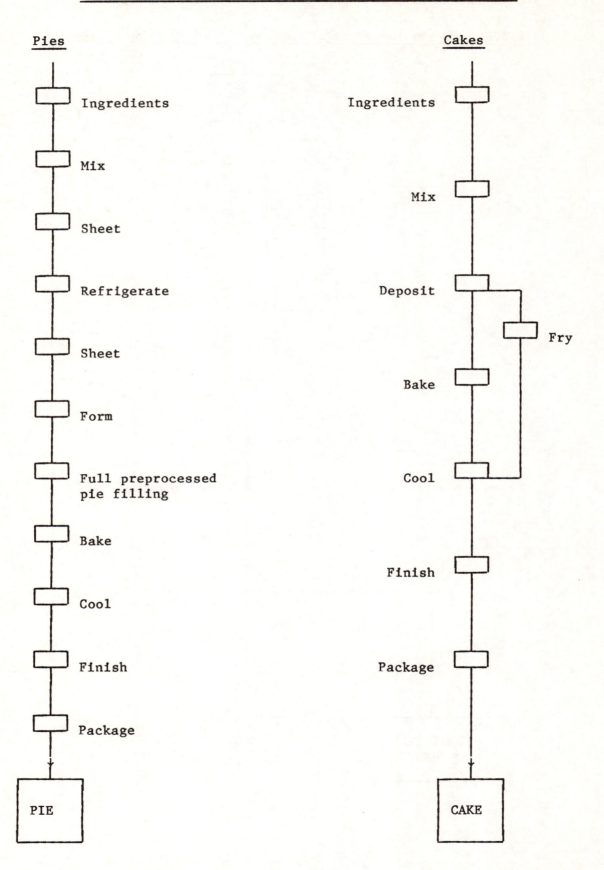

FIGURE 11: Industrial Processes in Biscuit Manufacture

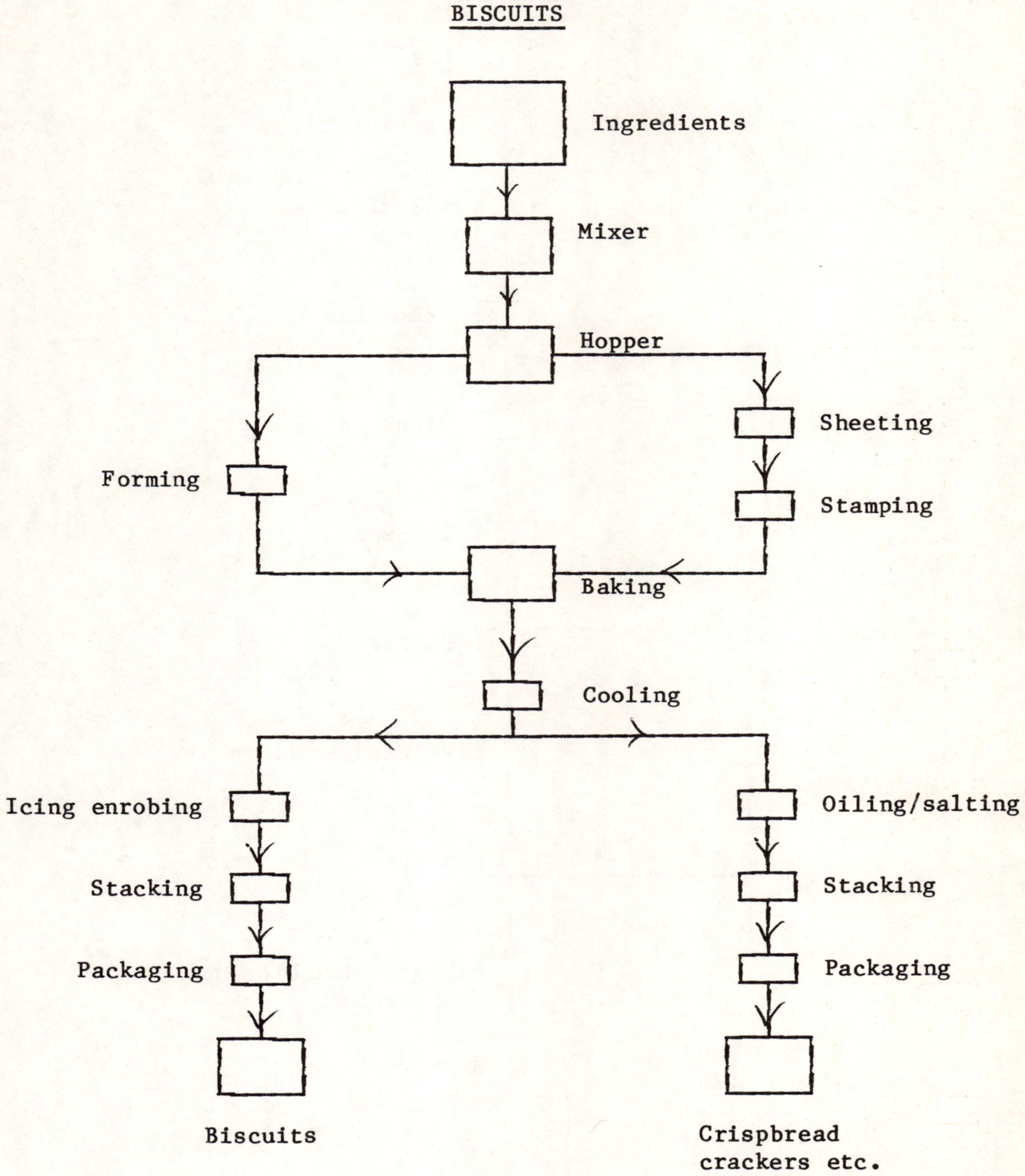

BISCUITS

FIGURE 12: Industrial Processes in Grain Milling

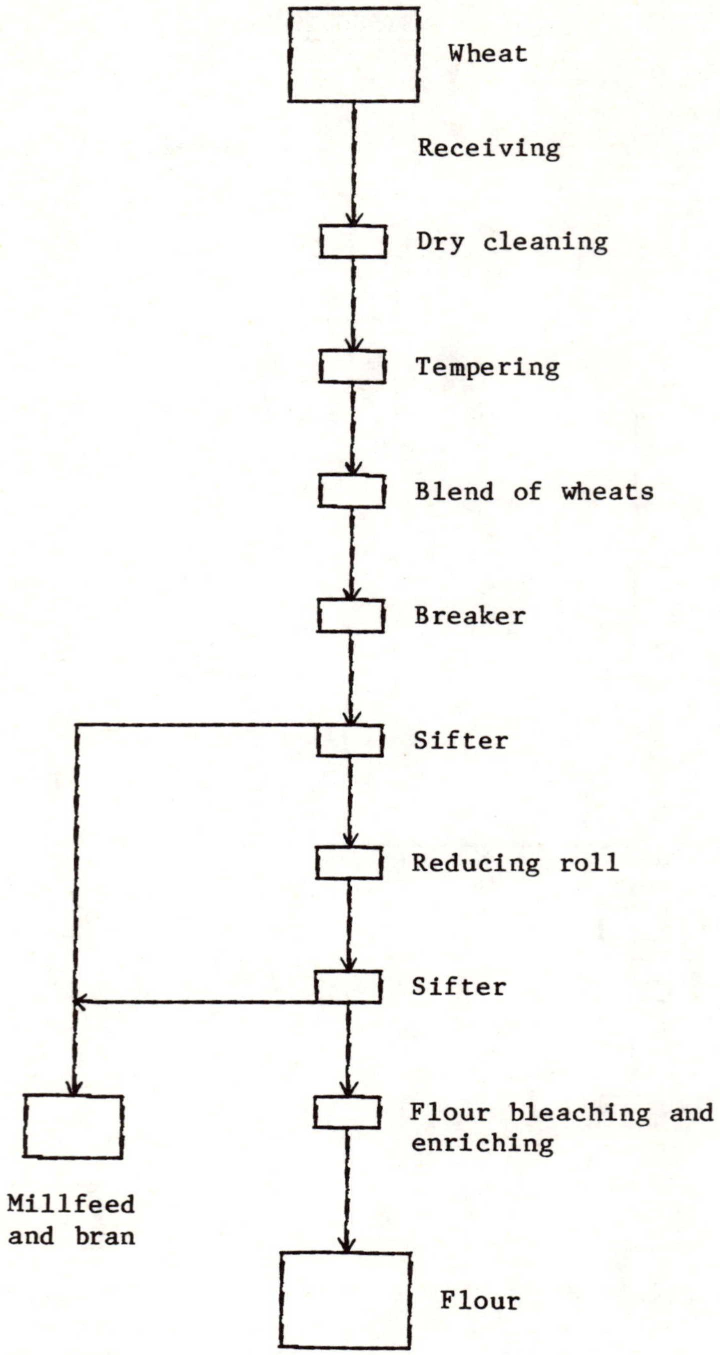

```
                    ┌─────────┐
                    │         │   Wheat
                    │         │
                    └────┬────┘
                         │
                      Receiving
                         │
                         ▼
                    ┌─────────┐
                    │         │   Dry cleaning
                    └────┬────┘
                         │
                         ▼
                    ┌─────────┐
                    │         │   Tempering
                    └────┬────┘
                         │
                         ▼
                    ┌─────────┐
                    │         │   Blend of wheats
                    └────┬────┘
                         │
                         ▼
                    ┌─────────┐
                    │         │   Breaker
                    └────┬────┘
                         │
         ┌───────────────┤
         │          ┌────▼────┐
         │          │         │   Sifter
         │          └────┬────┘
         │               │
         │               ▼
         │          ┌─────────┐
         │          │         │   Reducing roll
         │          └────┬────┘
         │               │
         │               ▼
         │          ┌─────────┐
         ├──────────┤         │   Sifter
         │          └────┬────┘
         │               │
         ▼               ▼
    ┌─────────┐     ┌─────────┐
    │         │     │         │   Flour bleaching and
    └─────────┘     └────┬────┘   enriching
                         │
    Millfeed             ▼
    and bran        ┌─────────┐
                    │         │   Flour
                    │         │
                    └─────────┘
```

TABLE 21: Output and Energy Consumption in Bread, Flour and
Biscuits Sector, 1980

	Output 10⁶ tonne	Average Practice GJ/tonne	Gas	Oil	Elec	Coal	Energy Consumption 10⁶ GJ
				10⁶ GJ			
Bread	2,255.3	3.50	5.28	1.82	0.79	–	7.89
Flour con-fectionery	594.6*	6.34*	1.22	1.96	0.59	–	3.77
							11.66
Biscuits							
Rusks	45.18	13.1				)	0.59
Chocolate/ sweet biscuits	227.16	9.1	3.25	0.41	0.66	0.19)	2.07
						)	
Other	355.72	5.2				)	1.85
							4.51

* 000£ sales.

TABLE 22: Utilisation of Energy in Bread, Flour and Biscuits Sector, 1980

	Elec	Gas	Oil 10⁶ GJ	Coal	Total
Bread & Flour Confectionery					
Process:	1.10	5.69	1.96	–	8.75
Of which ovens	0.34	5.07	1.30	–	6.71
Other process	0.76	0.62	0.66	–	2.04
Space Heating	0.23	0.76	1.30	–	2.29
Other	0.05	0.05	0.52	–	0.62
	1.38	6.50	3.78	–	11.66
Biscuits					
Process:	0.61	2.46	0.31	0.15	3.53
Of which ovens		1.63	0.20	–	
Other process	0.61	0.83	0.11	0.15	
Space Heating	–	0.65	0.08	0.04	0.77
Other	0.05	0.14	0.02		0.21
	0.66	3.25	0.41	0.19	4.51

TABLE 23: Energy Conservation Potential in Bread, Flour and
Biscuits Sector, 1980

	Best Practice	Energy Consumption if Operating at Best Practice, 10^6 GJ
Bread	3.0 GJ/tonne	6.76
Flour confectionery	4.0 GJ/10^3 sales	2.38
		9.14
Biscuits	4.2–12.0	3.91

Energy Conservation Potential as percentage of energy consumption 1980:
Bread and Flour Confectionery: 21.6%.   Biscuits = 13.3%.

TABLE 24: Output and Energy Consumption in Grain Milling, 1980

	Output 10^3 tonne	Average Practice GJ/tonne	Oil	Gas	Elec	Total
				10^6 GJ		
Wheat milling	4,940.3	0.36	0.07	0.40	1.31	1.78
Other milling	507.8	0.25	0.03	0.04	0.06	0.13
Cereal break-fast food	331.9	4.10	0.41	0.61	0.34	1.36
	5,780.0		0.51	1.05	1.71	3.27

TABLE 25: Energy Use in Grain Milling (%), 1980

	Wheat Milling	Other Milling	Cereal Breakfast Food
Process Energy of which:	50	65	75
mostly electricity for milling, drying, sieving	80	80	50
Space Heating	45	30	20
Other	5	5	5
	100	100	100
Fuel through boiler %	50	40	60

TABLE 26: Energy Conservation Potential in the Grain Milling Sector, 1980

	Output, tonne	Best Practice SEC, GJ/tonne	Total Energy Consumption, 10^6 GJ
Wheat milling	4,940.3	0.27	1.33
Other milling	507.8	0.22	0.11
Cereal breakfast	331.9	3.80	1.26
Energy consumption in operating at best practice:			2.70

Expressed as percentage of total energy consumption in 1980

$$= \frac{3.27 - 2.70}{3.27} \times 100 = 17\%.$$

TABLE 27: Bread, Biscuits, Grain Milling: Output Structure, Specific
Energy Consumption, 1980-2000

(a) Structure of Output (percentage)

	1980	1990 H	1990 L	2000 H	2000 L
Bread	26	25	25	24	24
Flour confectionery*	–	–	–	–	–
Rusks	1	1	1	1	1
Chocolate/sweet biscuits	3	3	3	3	3
Other biscuits	4	4	4	5	5
Wheat milling	56	56	57	55	56
Other milling	6	6	6	6	6
Cereal breakfast foods	4	5	4	6	5
	100	100	100	100	100

* 000£ sales (not included in output structure, which is on a volume output basis).

(b) Specific Energy Consumption (GJ/tonne)

	Average	Best	Average	Average	Best
Bread	3.50	3.00	3.15	2.80	2.55
Flour confectionery*	6.34	4.00	5.17	5.07	3.40
Rusks	13.10	12.00	12.25	11.40	10.80
Chocolate biscuits	9.10	7.50	8.51	7.92	6.75
Other biscuits	5.20	4.20	4.86	4.52	3.78
Wheat milling	0.36	0.27	0.33	0.29	0.25
Other milling	0.25	0.22	0.23	0.20	0.18
Cereal breakfast foods	4.10	3.80	3.98	3.85	3.27

* GJ/000£

5.4 THE SUGAR AND SUGAR PRODUCTS INDUSTRIES

186. The sugar and sugar product industries subsector includes a range of activities which are discussed under two separate headings:

- sugar manufacture,

- cocoa, chocolate, and sugar confectionery.

SUGAR MANUFACTURE

187. There are only two firms in the UK which are involved in the manufacture of sugar:

- British Sugar plc, which extracts sugar from domestically-grown sugar beet, produced 1.24 million tonnes in 1980,

- Tate & Lyle Ltd, which refines imported cane sugar, produced about 1.4 million tonnes in 1980.

188. It is important to note that there are fundamental differences between the activities of the two companies in the UK which affect their energy consumption. British Sugar carries out the complete process of extracting and refining sugar, starting from a raw material containing 16% sugar, whereas Tate & Lyle imports and refines an already crystallised raw cane sugar.

189. For reasons of commercial confidentiality, the BSO Purchases Inquiry does not report detailed data on the sugar industry. However, energy consumption data has been provided by the respective companies for the purposes of this study.

Beet Sugar Manufacture

190. Figure 13 shows the various process steps involved in the manufacture of sugar from sugar beet. The 1980 energy consumption is shown in Table 28.

191. The beets are washed and sliced and passed to a diffuser which extracts the sugar from the sliced beet. Exhausted sliced beet, or pulp as it is known, is then pressed and with added molasses dried and pelletised for feed for animal livestock. The process juice is passed to lime carbonation where the sugar solution is treated with milk of lime and carbon dioxide which produces a precipitate of calcium carbonate carrying down with it many impurities, including gums, waxes and resins. Filtration and further purification then takes place. The purified juice is then evaporated, passed to vacuum pans, crystallised, centrifuged, granulated and dried.

192. Beet sugar production uses heat several times through the process, first carrying out multiple effect evaporation using the resultant vapours for crystallisation and heating process juice and incoming beet. The resultant low grade heat, i.e. low temperature vapours and condensate are further used for juice heating, space heating, etc.

193. Continuous operation 24 hours a day and 7 days per week during the processing season provides excellent opportunities for heat recovery. This

low grade heat can be used both directly in the sugar production and also in boiler house and pulp drying plant. However, certain of the beet factories also operate outside of the general processing season, producing sugar from stored high density juice. In these circumstances the scope for heat recovery is reduced. In recent years, four of the smaller beet factories have been closed, the beet grown for these factories being shared between the larger and more efficient factories.

194.   CHP accounts for most of the electricity used in the beet sugar industry. All factories have turbines installed and, as these or the boilers feeding them require replacement, they are being renewed with bigger, more efficient sets. Steam produced at a pressure of 45 bars is passed through the turbine for electricity generation; the exhaust steam which is utilised as process steam for the heating of evaporators, gives total energy efficiencies of around 75-85% compared with the 33% efficiency of central electricity generation.

195.   In this report, we are concerned only with energy consumption by manufacturing industry in the UK, and not the energy consumption required in the overseas production and transport of imported feedstock. It follows that the UK energy consumption of the beet industry is substantially higher than that of the cane-sugar refining industry, since the latter imports a semi-manufactured feedstock. Furthermore, the beet industry produces substantial quantities of animal feedstuff as a byproduct. The energy used in the drying of this byproduct is included in the energy consumption of the sugar manufacture, because of the integrated nature of the energy flows involved.

196.   The specific energy consumption for beet sugar manufacture in 1980 was about 10.5 GJ/tonne. Since 1980, however, considerable efforts have been made to reduce energy consumption; by the end of 1983 the SEC was 7.5 GJ/tonne, with the best practice SEC estimated to be around 6 GJ/tonne.

Cane Sugar Refining

197.   Figure 14 shows the main processes involved in refining raw cane sugar. The main difference from beet sugar production is that the raw cane sugar has already undergone a cycle of separation, purification and crystallisation before importation. The refining steps are similar to those applied to the process juice extracted from beet, viz lime carbonation, evaporation, vacuum crystallisation, granulation and drying.

198.   Raw cane sugar consists of impure sugar crystals with an adhering film of syrup. It is mixed with recycled syrup to soften the adhering film of syrup; the mixture is then centrifuged to separate as much of the syrup from the sugar as possible. The separated sugar is dissolved in reclaimed liquors from the subsequent separation process, and enters the lime carbonation process. Surplus syrup is further treated to recover additional sugar, and the final syrups which cannot be further refined are passed out of the process as byproduct molasses.

199.   As with beet sugar production, energy in cane sugar refinining is cascaded from one process to another. The energy consumption at the largest cane refinery was approximately 3.5 GJ/tonne in 1980, and has fallen by about 4% per annum in the period 1978-1982. The specific energy consumption for

the whole industry was 4.0 GJ/tonne in 1980. Since then, however, closures of less efficient refineries have occurred. There are now only two cane refineries in the UK.

Energy Conservation Potential in Sugar Manufacture

200. It has already been noted that significant improvements have been made in energy efficiency in both beet sugar production and cane sugar refining. In both cases, the improvements have largely been brought about by improved energy management. This involves regular monitoring of energy use at each phase of the sugar refining process; closer control of boiler maintenance; regular tuning of temperature and time controls in the high steam using processes such as vacuum pans and drying chambers. Maximum use from heat recovery was designed into the plant and with all space heating met with recovered heat there is little further scope for significant gains in this direction.

201. Further improvements can be expected in the future. The beet sugar industry is continuing its drive for improved efficiency, with the current best practice SEC (6 GJ/tonne) as a target for average practice. Similar improvements in energy consumption in cane sugar refining are expected.

202. Since 1980, the closure of certain plants which were predominantly coal-fired has reduced the proportion of coal used by both British Sugar and Tate & Lyle. However, both companies expect to increase their coal consumption substantially in the future.

Market Trends to 2000

203. Consumption of sugar products by household, catering trades, food manufacturers and animal feed manufacturers has fallen by about 10% over the past nine years. It is generally thought that the downward trend will continue.

204. Both the beet sugar industry and the cane sugar refining industry are subject to European Community controls on their output.

205. In the case of the beet sugar industry, output is restricted by quotas under the Common Agricultural Policy. These quotas currently allow the UK to produce 1.114 million tonnes for sale within the European Community at the intervention price, and allow some excess production for sale outside the Community.

206. The cane sugar refining industry is controlled by quotas for the import of raw cane sugar from certain African, Caribbean and Pacific region countries under the Lomé agreement. Under this agreement, the producer countries are paid a fixed price for their raw sugar by the EEC, through an arrangement with the refiners. The UK refining industry refines almost all European imports of raw cane sugar.

COCOA, CHOCOLATE AND SUGAR CONFECTIONERY

207. Data on the cocoa, chocolate and sugar confectionery industries have been made available from Leatherhead Food RA from an IETS study which is not yet published. Some 21 visits were made involving most of the major firms in chocolate and sugar confectionery as well as many smaller companies.

Cocoa and Chocolate Confectionery

208. Cocoa beans form the basis of the chocolate section of industry. The beans are roasted in directly fired rotary ovens, ground to a mass and then pressed to produce cocoa powder. Some cocoa mass is sent to dairies to produce chocolate crumb. This product is made by mixing milk, sugar and chocolate to a given consistency and the mixture is then totally dried out into a hard cake. The blocks of chocolate crumb are then stored until required for the manufacture of milk chocolate.

209. Chocolate manufacture consists of mixing cocoa and sugar, refining and 'conching' for 24 hours or more until the viscosity and taste is judged correct. Chocolate is then thinned for enrobing (chocolate coating for biscuits and assorted centres) and a thicker chocolate is used for depositing (moulds for slab chocolate). Chocolate confectionery is usually chilled so that it sets quickly.

210. The main energy intensive processes are cocoa roasting and the mixing/melting vessels because, although the temperatures are not high, the processes are continuous and continue for some period of time. The ambient temperatures in the plant are usually quite high at all times because the products remain warm and pliable until chilled in order to set the chocolate. Once set, the blocks and chocolates are passed for packaging. Figure 15 shows the processes.

Other Sugar Confectionery

211. This covers an enormous range of products - toffee and toffee based sweets; boiled sweets; coated mixtures; gums and jellies; and liquorice based products. The basic requirement is the mixing of the ingredients; cooking; evaporation and, in many cases, the formation of a rope of the product which is then stamped, cooled and passed for packaging.

212. The main point at which energy is used is the mixing/cooking/ evaporation stage which takes place in large jacketed vessels. But at all stages of the manufacture the process plant is kept warm to keep the product pliable. Once a rope has been formed, sweets units are stamped out and coated (or left uncoated) and the product is then allowed to cool or is chilled to hasten setting before packaging. Figure 16 shows the processes involved and Table 28 shows energy use in 1980.

213. There is great variety in the range of products, the kind of industrial processes and the associated specific energy use in the sector. The range of specific energy requirements recorded by the Food RA lies between 2.6 GJ/tonne and 16.8 GJ/tonne, so clearly there is a need to group together like products in some way. Observation of the data and discussion with those who made the visits suggests that there are broadly two main groups:

(i)   cocoa, chocolate and chocolate based confectionery

(ii)   other sugar confectionery, a large part of which is manufactured by boiling (boiled sweets but also toffee, fudge etc).

Energy Consumption and Conservation, 1980

214. Most energy is used in the form of boiler fuels. Conversion to gas from oil and coal has occurred at a fast rate since 1970 and now takes about 30% of the total; oil retains 18% and coal still takes 34%. Electricity is important in this industry for refrigerated storage and chilling; it accounts for 16% of total energy.

215. Establishments mainly use steam or hot water based systems fired by a boiler using gas or oil. Most of the steam/hot water is used for two or three main industrial processes which account for around 70% of total energy. The industrial processes typically involve mixing and melting; cooking; forming and chilling (so electricity can be an important element in process energy); and some concentration or evaporation process. Finally the products are packaged and stored.

216. In general, the chocolate and chocolate based confectionery products use twice as much energy (around 10-13 GJ/tonne) as the boiled sweets, sugar and other confectionery which use around 6-7 GJ/tonne. Best practice SECs are generally about 20% lower than these values.

Energy Conservation Measures in the Chocolate and Sugar Confectionery Sector

217. The following measures have been identified by the Leatherhead Food RA:

- Good housekeeping:

  . draught proofing and insulation

  . reduction in additional office heating.

- Bolt on equipment:

  . heat recovery from evaporation and cooking vessels

- Boiler:

  . sensing equipment

  . regular checks on efficiency, burner controls.

Developments to 2000

218. Market Developments The market conditions which set the framework for productive investment are unlikely to show significant changes from the generally static pattern observed in the last decade.

219. Markets were characterised by informed opinion in the industry as:

  . growth in chocolate sales rising fast enough to offset falls in sugar confectionery;

  . sharp movements between lines due to fashions, advertising and competitors' responses;

- constant pressure to maintain market shares;

- pressure on profit margins which are being squeezed by the relatively
high costs of raw materials (cocoa, milk and sugar) over which
manufacturers have little control, the increasing power of retail
chains, and flagging consumer demand. The chocolate confectionery
manufacturers Cadbury, Mars, Rowntree etc protect their brands
vigorously and have been less affected by generic and own brand lines
of the big supermarkets. Nevertheless, as so much of their sales are
to the 5-15 age group they have been, and will be, affected by
population age distribution changes;

- hence they recognise the need to become, and remain, low cost
producers in order to maintain profit which implies some interest and
incentive to minimise energy requirements.

Given this trading environment there is some interest in those technical
developments which may affect the cost of production over the next 10-15
years.

220. <u>Technological Developments</u> Two developments may influence industrial
processes and have implications for energy requirements:

- continuous production;

- thermocompression techniques.

221. There is a recognisable move towards continuous production and the
installation of much improved microprocessor based control and
instrumentation. The main incentive to introduce this equipment is more
efficient utilisation of capital assets, some minor improvements in manning
arrangements, but above all better quality control of the product. The
improved control and instrumentation gear should give much more precise
measures of both mix, proportions, ingredients, reduce if not eliminate waste,
finer control over the weights for packaging and allow, within the margins of
taste and quality, far more instant response to the quite rapid changes in the
cost of materials. Manufacturers might thus be able to minimise the cost of
ingredients for a given quality and taste of product at fairly short notice as
one or more of the ingredients changes in relative cost.

222. The implications for energy are a little uncertain. Assuming output
remains broadly unchanged but equipment is now being utilised more
effectively, then energy costs per unit output might well fall. However,
the increased use of control gear and instrumentation gear together with a
requirement for faster chilling because of the increased speed of throughput
will increase the demand for electricity.

223. One informed opinion was that energy costs per unit output might
increase by some 10% or 15% as a result of continuous production with
sophisticated control gear. This was largely due to the increased requirement
for rapid chilling in order to maintain throughput.

224. The second technical change which is observed much less clearly in
the industry is the opportunities for using thermo-compression techniques. It
is reported that some sugar confectioners already are experimenting or have

used such technology but this has to be confirmed. It will be recalled that the main advantages are the small physical space used by the machines; the increased control over throughput; and the ability in one machine to combine the processes of cooking, forming and cutting. This clearly has significant implications for energy use and a culmination of reduced process energy and heating requirements in a much smaller building may well save something of the order of 1/3 of energy requirements per unit output for a given throughput. However, these figures are derived largely from the biscuit and flour confectionery industry and data from sugar confectionery is not available.

225. Another point which is of general significance is that nearly 40% of energy use in the cocoa, chocolate and sugar confectionery sector is in non process energy, i.e. space heating and 'other'. This implies that good housekeeping measures might have a much more significant effect than in those industries where process uses dominate energy use. It is not too surprising that there are quite significant differences in specific energy requirements between average practice and best practice, nor that in the period to 2000 significant improvements in average practice might reasonably be expected. The range of conservation measures were noted earlier and there is no reason to suppose that the majority of those should not diffuse quite rapidly through the industry in the next ten years. Thereafter, subsequent improvements must be dependent upon process plant and equipment.

226. Improvements in standard process plant and equipment in the industry - mixing vessels; steam jacketed heating vessels for cooking; machines for forming and cutting and the arrangements for chilling seem to offer only marginal improvements over existing equipment. Improvements such as the process plant manufacturers appear to have in hand relate to speed process controls, compactness and capacity measures. The impression gained was that little attention had been paid to operating costs and indeed with the increased use of electronic controls and the requirements for faster and higher quantities of chilling these are more likely to increase than decrease specific energy requirements.

PROJECTIONS OF FUTURE SPECIFIC ENERGY CONSUMPTION

227. The assumptions which have been made to project future specific energy consumption (SEC) in the sugar and sugar products subsector are shown in Table 29. For each of the four product categories we specify an individual SEC value which is a judgement based on the various trends discussed previously. The proportion of each product category in the total output of the subsector is also estimated.

228. For the sugar manufacturing industries, the fall in SEC by 1990 largely reflects changes which have already taken place. Some further improvement in SEC by 2000 is expected. The shift in the proportion of output between beet and cane sugar reflects changes which have already occurred; no further changes in the structure of output are assumed.

229. The SECs for cocoa and chocolate and sugar confectionery manufacture are assumed to increase by 5% by 2000. This reflects the view of sources within the industry, that improvements in energy efficiency will be more than offset by the introduction of more energy intensive processes. A continuing shift in the structure of output from sugar confectionery to chocolate products is assumed.

230.   The aggregate effect of these assumptions causes the overall average SEC for the subsector to fall from 7.5 GJ/tonne in 1980 to 6.5 GJ/tonne in 1990 and 6.0 GJ/tonne in 2000.   This implies an improvement in energy efficiency of 13% by 1990 and 20% by 2000.

231.   A shift in the overall fuel mix away from petroleum and, to a lesser extent natural gas, towards coal is assumed, based on the views of the various industry sources.

FIGURE 13: Processes in Beet Sugar Manufacture

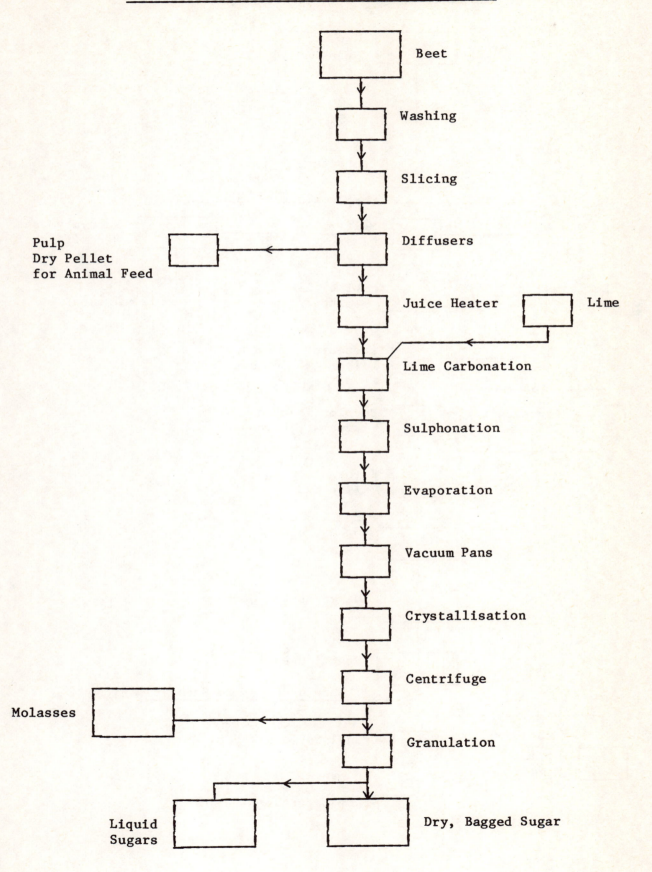

FIGURE 14: Processes in Cane Sugar Refining

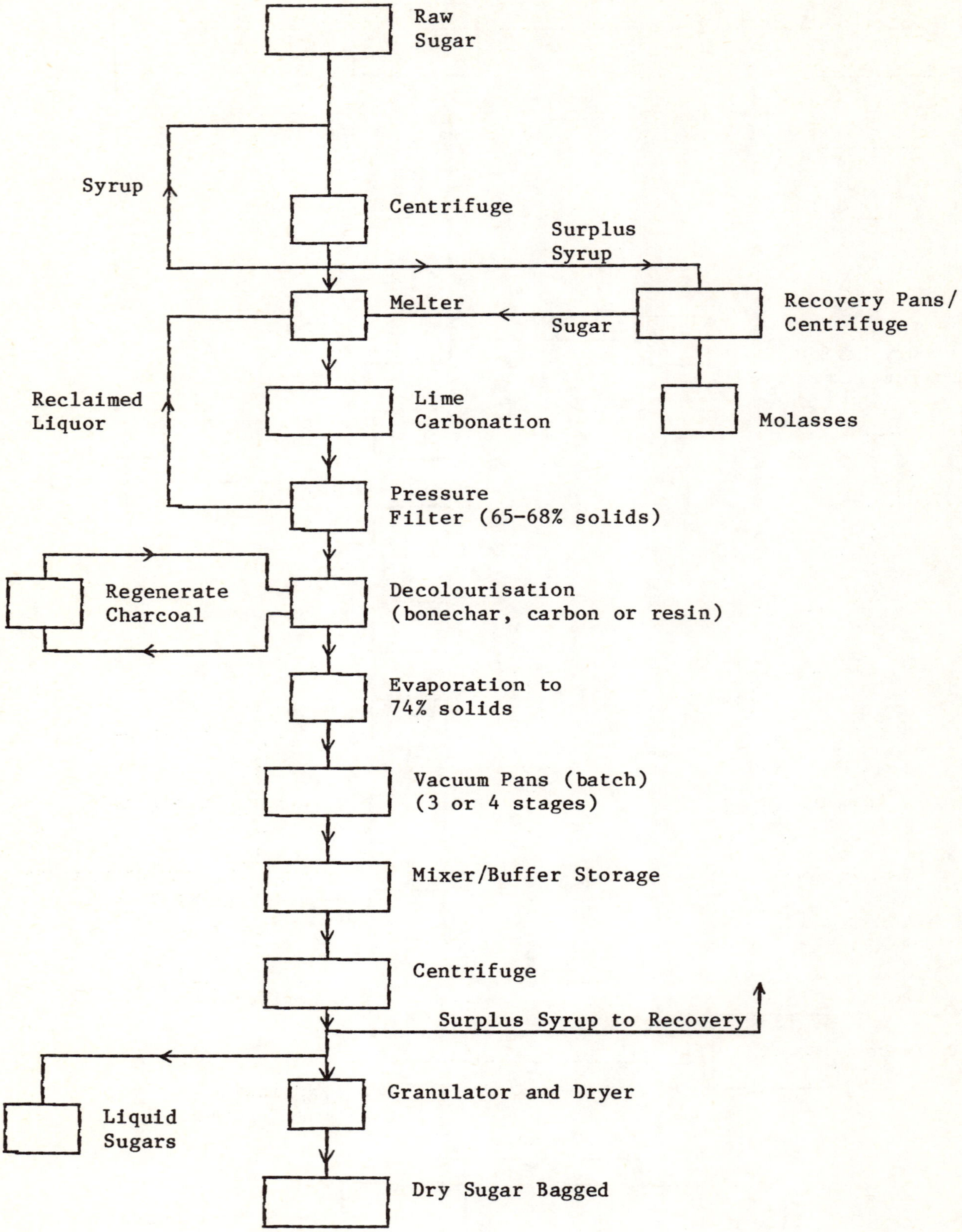

Raw Sugar

Syrup

Centrifuge

Surplus Syrup

Melter ← Sugar ← Recovery Pans/ Centrifuge

Molasses

Reclaimed Liquor

Lime Carbonation

Pressure Filter (65-68% solids)

Regenerate Charcoal

Decolourisation (bonechar, carbon or resin)

Evaporation to 74% solids

Vacuum Pans (batch) (3 or 4 stages)

Mixer/Buffer Storage

Centrifuge

Surplus Syrup to Recovery

Liquid Sugars

Granulator and Dryer

Dry Sugar Bagged

FIGURE 15: Processes in Cocoa and Chocolate Industry

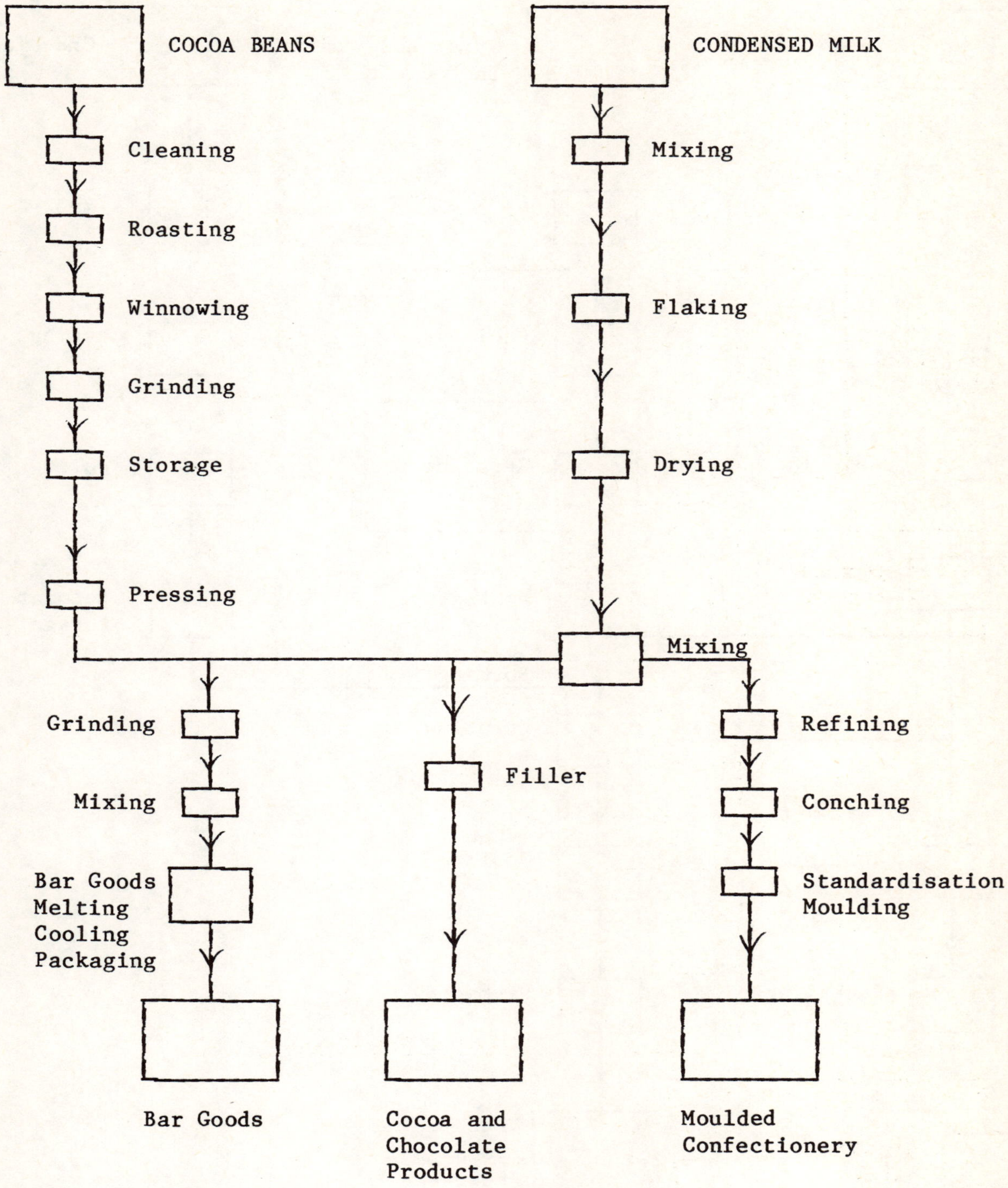

COCOA BEANS

CONDENSED MILK

Cleaning

Roasting

Winnowing

Grinding

Storage

Pressing

Mixing

Flaking

Drying

Mixing

Grinding

Mixing

Bar Goods
Melting
Cooling
Packaging

Filler

Refining

Conching

Standardisation
Moulding

Bar Goods

Cocoa and
Chocolate
Products

Moulded
Confectionery

FIGURE 16: Processes in Sweet Confectionery Industry

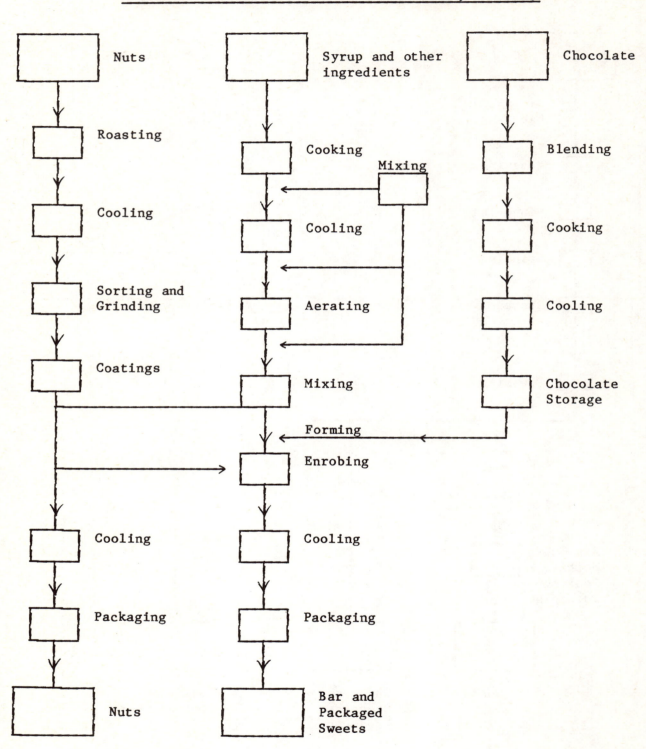

TABLE 28: 1980 Energy Consumption

	Output k tonnes	Total Energy Consumption PJ	Fuel Mix				Average SEC GJ/tonne
			Coal %	Petroleum %	Nat Gas %	Elec %	
Beet sugar	1,240	13.0	24	51	25	–	10.5
Cane sugar	1,400	5.6	30	10	60	–	4.0
Cocoa and chocolate	500	6.0	38	17	29	16	12.0
Sugar confectionery	335	2.1	26	19	37	18	6.2
Total	3,475	26.7	29	33	33	5	7.7

Sources: British Sugar, Tate & Lyle, Leatherhead Food RA.

TABLE 29: Projections of Future Specific Energy Consumption

		1980	1990	2000
Beet Sugar	% of Total Output	35.7	40.0	42.0
	SEC (GJ/t)	10.5	7.0	6.0
Cane Sugar	% of Total Output	40.3	35.0	33.0
	SEC (GJ/t)	4.0	3.0	3.0
Cocoa and Chocolate	% of Total Output	14.4	16.0	17.0
	SEC (GJ/t)	12.0	12.3	12.6
Sugar Confectionery	% of Total Output	9.6	9.0	8.0
	SEC (GJ/t)	6.2	6.3	6.5
Overall Average SEC (GJ/t)		7.7	6.4	6.2
Fuel Mix: (%)	Coal	29	35	45
	Petroleum	33	28	20
	Natural gas	33	32	30
	Electricity	5	5	5

5.5 THE DRINKS INDUSTRIES

232. In most of the establishments in this sector, other than those in the spirits industry, process energy accounts for 60-80% of total energy, most (around 80%) fuel passes through the boiler for steam raising and/or hot water which is used in processes (typically mashing, blending, bottle washing and cleaning); and space heating might account for 10-40% of total energy.

233. Spirits distilling is a very different industrial process - and malt whisky making is a rather different activity from grain distilling. Process energy in spirits distilling is a much higher proportion of total energy (about 90%) and the specific energy consumption is much higher than other products because of the distillation stage of whisky manufacture. It must also be noted that in grain distilling the volume of spent grain is large enough to be a useful by-product and is subsequently evaporated and dried for sale as animal feed. The distillation process is very important in malt whisky but grain distilling has two major energy use processes and it is often accompanied by evaporation and drying of the spent grains.

SPIRITS DISTILLING

234. This sub-sector is the largest energy user and one of the largest energy users in the food, drink and tobacco industry. The sector is naturally divided both by structure and by product between malt whisky distilling and all other spirits which are distilled from grain.

Malt Whisky

235. Malt whisky production is a batch process carried out by over 100 small to medium sized distilleries. The processes are fairly uniform between distilleries and are:

· Mashing and fermentation
  Malt is ground and mixed with hot water in a large circular vessel called a mash tun. The hot water causes the enzymes in the malt to convert the soluble starches to fermentable sugars.

236. Once the mashing is completed the solution (wort) is decanted, cooled, dosed with yeast and allowed to ferment for about two days. The resulting liquor is called wash and contains alcohol at about 4-7% by weight. The main energy requirement is hot water which is supplied from boilers.

237. · Distillation
  The wash is distilled in a large copper still - usually twice - although two distillers carry out a third distillation. The wash in the still is boiled and the vapour is led by the neck of the still to a water cooled condenser.

The energy requirements at this stage are heat for the still - which is usually a steam coil via the boiler - but some directly fired coal or gas stills also exist. Energy levels are also influenced by the efficiency of the condenser.

238. The first distillation separates crude alcohol from the wash and produces what is called low wines. All low wines then pass to the second,

smaller spirit stills. The first flow of spirits (foreshots) contain
impurities so although collected they are recycled in the next batch. The
main run of pure spirit is collected but after some time the distillate
weakens in strength which causes the temperature of the still to rise and oil
vapours cloud the distillate (these distillates are called feints). At this
stage the main distillate passes to storage and the 'foreshots' and 'feints'
added to the next batch of low wines for distillation.

239. . Maturation
       Whisky must by law be matured for at least three years. Malt
       whiskys are matured for typically 7, 10 or 12 years at around 68.5%
       alcohol (120°) in oak casks.

240. . Malting
       Most of the malt used in the process is bought in from maltsters.
       However, a proportion is still made by distillers. The process is as
       described in the report on the maltings industry, although the
       kilning operation for malt whisky grains includes the burning of peat
       which imparts an essential flavour to the malt and gives the malt
       whisky its distinctive characteristic smell and taste.

Grain Whisky Distillation

241. The processes outlined in Figure 17 are somewhat different and are
carried out on a much larger scale by about a dozen large producers.

       . Wort Preparation and Fermentation
         Grain whisky is made from a mixture of raw grain (usually maize or
         unmalted barley) and malt. The grain is ground and cooked by
         injection of steam under pressure. The cooked cereal is then mixed
         with ground malt. The enzymes in the malt convert the starch to
         soluble sugars to produce a wort. The wort is fermented by addition
         of yeast and the resultant wash is distilled. The wort is fermented
         in batches to a wash containing 5.5%-8.5% alcohol by weight.

242. . Distillation
       The still used in grain whisky products is a continuous still
       called a Coffey still (patented by Coffey in 1831). It consists of a
       pair of interconnected rectangular close copper columns 12-15 metres
       high containing a series of horizontal perforated plates. One column
       is called the analyser, the other a rectifier.

243. Hot wash is run into the top of the analyser column and steam is fed
into the bottom. Alcohol vapours and steam rise to the top of the column,
spent wash runs out at the bottom. The vapour from the top of the analyser
column is then passed to the bottom of the rectifier column. As the vapour
rises up the column it is cooled and less volatile elements condensed by cold
undiluted wash passing through a pipe in the centre of the column. The
vapours reaching the top are almost pure spirits and are finally condensed out
in a water cooled coil.

244. . Maturation
       After distillation grain whisky is matured for at least three years
       in oak casks.

· Blending and Bottling
  These processes use significant quantities of hot water.

· Other Grains and Neutral Spirits
  The spirit for other beverages such as gin and vodka is produced in
  the same Coffey still method but is rectified either singly or doubly
  to produce a spirit free of constituents other than ethanol as a
  flavourless neutral base spirit. A big tonnage of ethanol is used in
  perfumery, chemical and methylated spirit manufacture.

Energy Use and Conservation in Spirits Manufacture

245. The main energy consuming processes in malt whisky manufacture are the
first distillation process (over 60% of process energy) and the second
distillation process (15%) and mashing (13%) which makes up the total process
energy consumption. Table 31 provides the details.

246. There is a small amount of space heating and of course a certain amount
of lighting and other energy uses. Typically process energy accounts for 85%
of the total. Most distilleries use steam heated stills raised via a central
boiler but some distilleries still use directly fired coal, oil or butane
stills. The latter incur large flue losses. The second distillation may
use heat recovered from the first distillation which is one reason why the
second distillation process uses so much less energy than the first. Ther
is, however, a limit to how far the heat can be cascaded although it may be
that there are further opportunities for heat recovery for re-use for
sterilisation and cleaning.

247. Detailed figures on the pattern of energy use in grain whisky production
are not available. The main energy using processes however, are obvious
enough. The cereal cooking operation to prepare the wort is carried out at
4.4 bar and 165°C and is a very energy intensive process. The major energy
using process however is the first distillation operation although it does
require only low pressure steam. The by-products (dark grain) are, in most
grain whisky operations, of a large enough volume to justify the expensive
evaporation and drying processes in which the spent grains are prepared as an
animal feed. Malt whisky producers also produce dark grains.

248. Table 30 provides the details of fuel usage in the spirits distillation
sector. A little coal is still used (2%); the main fuels are oil (23%) and
gas (64%); while electricity accounts for around 10% of total energy.

249. Average specific energy data is available for malt whisky distilleries
from the Leatherhead Food RA visits carried out as part of the IETS visits.
Over 30 sites provided adequate data on SEC and other points across the drinks
sectors. Some six malt whisky sites but only one grain whisky site was
visited. Average practice SEC for malt whisky might be around 60 GJ/10^3
litres. Best practice might perhaps be around 55 GJ/10^3 litres. However,
as only one grain whisky plant was visited the figure of 20 GJ/10^3 litres
may be unrepresentative. Other sources suggest a figure of around 28 GJ/10^3
litres [20]. The consulting engineers who prepared the second estimate based
this on a number of investigations which they had carried out around grain
distillers and is taken to be a more generally representative figure than the
single visit made by the Leatherhead Food RA. So, 28 GJ/10^3 litres alcohol

is taken to be average practice while 20 GJ/10^3 litres may represent a best practice establishment.

250. Energy consumption in 1980 therefore is estimated as the product of output and specific energy consumption. The Business Monitor does not show malt whisky production separately from grain whisky and other spirits. Using Scotch Whisky Association figures in order to assess the proportion of malt whisky produced provides the estimates for Table 30. Fuel allocations are based upon Leatherhead Food RA estimates.

251. There is a complication in the output and SEC figures which should be made clear. Ethyl alcohol is a product of further fractionation of spirits within the grain whisky distillery. Business Monitor PQ 231 classes ethyl alcohol with 'other spirits' (gin, vodka, rum, etc) in the output figures so it is not possible to separate the various drinks.

252. Business Monitor data represents <u>sales</u>. While this may usually be taken to be a reasonable proxy for output, the production of spirits, particularly whisky, may vary due to export sales and production for stocks. Since 1979–80 the whisky industry has reduced production in order to reduce stocks so there may be some error in using sales based output data.

253. Another complication is the production of ethyl alcohol from petrochemical sources. From the Custom and Excise data (all alcohol production must be declared) it is not obvious how to disentangle the effects of production, sales, and two sources of ethyl alcohol in order to determine the correct quantity of ethyl alcohol derived from grain distilling. However, it is considered the Business Monitor figures do represent only ethyl alcohol from grain distilling. If so, then it seems reasonable to regard the SEC for grain distilling as appropriate for all spirits. If this view is not correct and ethyl alcohol is produced as a specialist product, in bulk, then it may be produced more efficiently than the 28 GJ/tonne used here.

254. Energy consumption in 1980 is shown in Table 30 estimated as 21.94 x 10^6 GJ; if all establishments operated at best practice energy consumption might be 16.84 x 10^6 GJ suggesting a conservation potential of 5.10 x 10^6 GJ, about 24% of 1980 consumption. The energy conservation potential of this sector is combined with that of the soft drinks etc sector in Table 32.

<u>Energy Conservation Measures in the Spirits Sector</u>

255. As most energy is absorbed by the distillation process conservation measures must focus on this area. There is a need to closely evaluate the apparent gains from steam heating via a boiler rather than direct firing and where appropriate encourage conversion to steam heating for the stills rather than direct heating.

256. There may also be opportunities to install heat recovery equipment in the stills and particularly in the cereal cooking vessels used in grain distilling which offers the opportunity for securing high grade waste heat.

257. Other measures include:

- improving boiler efficiency, through installation of sensing controls

- examination of how far processes may be carried out at lower temperatures and with a lower water content

- establishing faster and more continuous processes

- heat recovery from hot stills to preheat wastes

- heat recovery from wort to preheat liquor used for mashing

- in grain and malt distilling there are opportunities for mechanical water removal from the spent grain which reduces significantly the cost of final evaporation and drying

- it is always useful to recover condensate and flash steam for re-use elsewhere.

WINES, CIDERS AND PERRYS AND SOFT DRINKS SECTOR

258. All establishments in these sub-sectors are typical type two establishments: most fuels (around 80-90%) are passed through boilers for steam and hot water raising for two or three industrial processes (mostly pasteurisation); space heating; and hot water for cleaning. Space heating is typically a higher proportion (about 30%) in these sectors compared with whisky distilling. Tables 30 and 31 provide an account of energy consumption and the pattern of utilisation of energy in each of these sectors. The following paragraphs describe the industrial processes and explain why the proportions in Tables 30 and 31 vary in the way that they do.

Ciders and Perrys

259. Cider and perry (which is made from pears) is produced by crushing the fruit to extract juice and fermenting the juice to generate the alcohol content in the cider or perry. After fermentation is complete the liquor is racked off, allowed to settle and clear to a bright liquid, then matured for two months or so before bottling with or without carbonation. Some apple juice is concentrated and stored for use in other processes.

260. In general, the energy requirements within the industry are for hot water/steam typically at less than 100-120°C for mashing, pasteurisation, bottle washing and cleaning. Most of the energy (around 80%) passes through the boiler which is gas or oil fired. The Bulmer's plant in Hereford (which extracts pectin and dries it) uses waste heat from the coal fired CHP plant established by the Midland Electricity Board.

261. Because process energy is less than 60% of total energy, space heating, although not large in absolute terms, is a much higher proportion of total energy (around 30-50%) of the total compared with whisky distilling. Electricity use is slightly higher in cider making than in wine making because the product is often chilled before carbonation of the bottles.

Wine Making

262. British wines are made from imported grape concentrate or dried fruit. After soaking or mixing the mash is fermented, racked off and allowed to clear. Most British wines are sweet so, to avoid secondary fermentation, wines are pasteurised to deactivate the enzymes. Dryer wines do not require pasteurisation.

263. Wine making is not an energy intensive matter. Bottle washing is the main energy user (and pasteurisation where necessary). Hot water and steam, typically at less than 120°C, is supplied via a boiler which accounts for about 90% of all energy used. Process energy is not a major element, perhaps around half of the total, so space heating is proportionally a larger total than in spirits distilling. Electricity is not a major fuel in wine making.

Soft Drinks

264. These comprise three kinds of drink:

- carbonated, eg Corona and Pepsi

- squashes, concentrates (natural fruit juice, or synthetic)

- fruit juices, but produced in dairies or canneries.

The industrial processes are very similar to liquid milk processing and the later stages of cider and wine making. Water is mixed with the essences and flavours to the required consistency; pasteurised; bottled; chilled (where necessary) and carbonated before packaging. The energy intensive area is bottle washing although refrigeration for chilling may be significant.

265. Process energy is around half of the total mostly supplied through a boiler (80% of all energy is passed through a boiler) and the major element perhaps accounting for 2/3 of process energy, is bottle washing and cleaning. Space heating accounts for perhaps 1/3 to 1/2 of the total. Details are given in Table 31.

Energy Consumption and Conservation Potential in the Soft Drinks, Wines, Ciders and Perrys Sector, 1980

266. The data from IETS to some 30 sites in the soft drinks, wines and ciders sector provided the evidence on which to assess 'average practice' and 'best practice' in GJ/000 litres of output. The details of the sample were not available but if like most other samples within the IETS data the sample should be reasonably representative.

267. Discussion with officers at Leatherhead Food RA suggested the figures used in Table 30. It will be seen that energy consumption in this sector is small at 4.61×10^6 GJ. If all establishments operated at best practice then energy consumption might be around 3.44×10^6 GJ suggesting a conservation potential of around 1.17×10^6 GJ which as a percentage of the 1980 energy consumption of wine, ciders, perrys and soft drinks is around 25%. The energy conservation potential of this sector is combined with that of the spirits sector in Table 32.

Energy Conservation Measures in the Soft Drinks, Wines, Ciders and
Perrys Sector

268. In the wines, ciders and soft drinks sub-sector where around 1/2 of
energy is used for space heating and cleaning there is scope for significant
savings of the good housekeeping type such as draught proofing, better
building insulation, careful use or non use of auxiliary electric fires in
offices, etc. Perhaps the most significant measure on the industrial process
side is to establish a medium pressure hot water system rather than steam
raising. Such a change saves as a minimum the latent heat of steam. It also
focuses attention on the points of use of hot water and so encourages good
housekeeping measures.

269. As around 80-90% of all energy is passed through a boiler it is clear
that regular checks of boiler efficiency, installation of oxygen and other
sensing equipment, and other controls, whether or not already in operation,
together with careful assessment of boiler utilisation are an essential
element in energy conservation measures.

BREWING AND MALTING SECTORS

270. Although these industries are traditionally considered together they are
very different in energy use terms. Brewing requires two or three processes
all of which use steam or hot water. It follows that most (80%) fuels pass
through the boiler and service both process operations and space heating and
cleaning.

271. By contrast the malting industry is dominated by a single kilning
operation which absorbs about 80% of all process energy.

272. Tables 33 and 34 indicate the size and pattern of energy consumption in
these industries. The following paragraphs provide an account of the
industrial processes and examine how energy is used. Each industry is
considered in turn.

The Brewing Industry

273. Beer is a beverage made by alcoholic fermentation of carbohydrates
derived mainly from cereals but which are not distilled. Although many
different cereals may be used, most beers in the UK are made from malted
barley with added hops for flavour. Other grains and sugars are added as
blends for taste or as malt substitutes. The industrial processes in the
brewing industry are outlined in Figure 18.

274. Malt is milled and mixed with hot liquor to produce a mash. Mashing
yields a solution of sugars known as sweet wort. This is run off into another
vessel (the copper) where the hops are added and the liquor boiled for one or
two hours. After cooling the boiled wort is transferred to another vessel
where yeast is added and fermentation (where the sugars are converted to
alcohol and carbon dioxide) takes place. After fermentation most beer is
clarified and pasteurised and packaged into large or small containers.

275. Broadly three types of beer may be distinguished: ales, lagers and
stouts. Ales and stouts are top fermented beers, which are consumed soon

after production. They form about 70% of total sales. Lagers are bottom fermented beers which require conditioning by chilled storage for some weeks before sale. They are the fastest growing area of the beer market currently taking about 30% and expected by many to grow steadily to about 40-45% of the market by the late 1980s. Stouts are brewed with roasted barley and, in the UK, are dominated by Guinness, although other brewers also produce a range of stouts.

276. The majority of beer is bulk conditioned, that is matured in the brewery and filtered. Beer of this type is usually pasteurised (before kegging or after bottling or canning). Some bottled and all cask-conditioned beer contains yeast and is unpasteurised, secondary fermentation taking place in the container.

277. The UK is the third largest brewing nation in the world with an annual production of about 40 million bulk barrels (6.6 x 10^9 litres) in 1980, about the same as in 1978. Exports and imports of beers are small at about 5% of total sales and much of this represents the import of Guinness from Ireland. There were about 140 wholesale brewers (130 in 1983) but the industry is dominated by 22 breweries which produce over one million hectolitres per annum, most of which belong to the big six brewers. There is a long 'tail' of small brewers (50) among the 102 respondents to the Brewers Society Survey 1977 and 1979 which produced less than 3 x 10^7 litres.

Energy Use and Conservation Potential in the Brewing Industry, 1980

278. Almost all fuels (about 80%) are passed through the boiler to provide steam. Electricity as a proportion of total energy, see Table 33, is around 10% (on a heat supplied basis) and is used for refrigeration/chilling (30%), lighting (10%) and motive power for process plant (60%). As noted earlier lager production uses more electricity than other beers.

279. Oil, gas and coal account for boiler fuels in roughly equal parts although the switch from fuel oil is by no means ending and some 25% of total boiler fuels is accounted for by fuel oil. Details of fuel use in the industry are given in Table 34. The main processes in which the steam is used are as follows:

- mashing - where the milled malt is mixed with hot liquor

- wort separation

- wort boiling - hops are added and the wort boiled for 1-2 hours at an evaporation rate of about 5% an hour using steam in a jacketed copper

- hops separation

- wort cooling - where the wort is passed through a heat exchanger and is cooled by incoming liquor

- adjustment - the strength of the wort is adjusted by adding to or diluting the wort

- fermentation - the yeast is added to the wort, fermented for 3-7 days at 10-20°.

In addition there is a requirement for cleaning of tanks, vessels, pipework and the heat treatment of liquor for adjustment of wort strength.

280. Analysis of energy use in brewing therefore is all about the distribution of steam and where it might be saved. No detailed study exists of the use of the distribution of steam in breweries but the scattered evidence suggests that the following allocation is about the right order of magnitude:

	% of Steam
Mash mixing	3
Brewing	20
Liquor heating (wort boiling)	24
Fermentation	3
Bottling	10
Kegging	20
Total metered steam	80

The balance of 20% is typically 'lost' around the buildings for space and water heating, particularly hot water for cleaning purposes.

281. It is clear that brewers may differ from each other in their energy consumption because:

- they produce different proportions of ales, stouts and lagers

- produce different proportions in casks, bottles and cans (including canning and bottling from other breweries)

- of the size of the throughput

- single or multi-shift working which is important because heat recovery is an important element in energy conservation as heat from cooling the boiled wort may be used to pre-heat the mash liquor of the next brew. The effectiveness of heat recycling depends on the brewing schedule

- they use either borehole water or towns water

- boiling time and total evaporation from the copper varies between brewers.

282. Establishing an average practice and a best practice between brewers is a matter therefore of deciding on like groups and this is not an easy task. The main information derives from a survey conducted by the Brewers Society in 1977 and repeated in 1979. It is clear from the results that energy consumption varies directly with the size of the brewery; the proportion of output in four main types of beer (chilled filtered and pasteurised; cask conditioned and lagers) and the quantity and type of packaging carried out (bulk tankers, casks, cans and bottles) [21].

283. Lager producers use more electricity because it is conditioned and chilled for several weeks before sale and it is also clear that brewers with bottling plants use more energy. However, the difference between the different types of beer seem not to be a major factor although clearly an

additional process such as pasteurising or chilling and filtering must make some difference to energy consumption per unit output.

284. Estimates of average and best practice are also available from Leatherhead Food RA data derived from their visits to some 40 breweries under the IETS. The data is consistent with the Brewers Society survey results which give lowest values around 2.0 MJ/litre and high values around 4.5 MJ/litre with most breweries clustered around 3.25-3.75 MJ/litre. The Leatherhead Food RA data was taken because it is recent (1980) and it suggests average practice around 3.2 MJ/litre with best practice at 2.5 MJ/litre.

285. As noted above a more detailed analysis could be used specifying three or four categories of brewers each with an average and best practice but this broad picture seems adequate for the present purpose. A detailed study could be mounted using Brewers Society data.

286. Energy consumption is assessed as the product of average practice figures and output in 1980 obtained from the Business Monitor. This yields 20.74×10^6 GJ which is consistent with other data.

287. If all breweries operated at best practice energy consumption might have been 16.2×10^6 GJ suggesting a conservation potential of 4.54×10^6 GJ which expressed as a percentage of consumption in 1980 is 22% (see Table 35 where the malting sector has been included).

Energy Conservation Measures in the Brewing Sector

288. If this description of energy use in brewing is broadly right then we may apply the kinds of conservation measures appropriate for type two industries:

- energy auditing and energy check-listing is important because energy use will be spread across many points in the buildings and component elements in the process and the first step is to identify where the energy goes in some detail

- water use and energy use is clearly associated (bottle washing, pasteurisation, cleaning, etc.), so efforts to control water are a direct form of energy conservation. Use of large heat exchangers which maximise the heat transfer between hot and cool liquors often reduces overall demand for water and thereby reduces energy requrements.

- better electricity use is largely a matter of proper management of the refrigeration and chilling systems.

- it has been established experimentally that the wort-boiling time could be substantially reduced without significantly affecting flavour. There is some question as to whether this could be achieved in large-scale production. Recent experiments at Greenall Whitley Breweries have shown that high temperature continuous wort-boiling can save in excess of 50% of the energy with no detrimental effect on quality.

. other energy conservation measures are clearly laid out in the Audit
  Report and evidence to date suggests that although the industry is
  active in pursuing many of these measures there is still scope for
  some further application.

. high gravity brewing (beer production from wort of higher than normal
  original gravity and diluted to normal strength after fermentation)
  gives substantial economies in heating and cooling of wort and beer,
  reducing the specific energy per unit output.

. this may also be linked with continuous brewing production as this
  has additional benefits from heat recovery but can give serious
  quality problems.

The Malting Sector

289. Malt is cereal grain, usually barley, which has been steeped in water
and allowed to germinate. During the germination process the grain sprouts
and produces enzymes which have the property of transforming starch into
soluble sugars. The barley is germinated until the enzymes reach a certain
state (when the hard starch endosperm in the grain has been transformed to a
friable state); then germination is stopped and the grain stabilised by
kilning. Kilning is also essential for the development of colour and flavour
in the malt which may be varied by controlling temperature and air flow during
the kilning process.

290. The maltsters task is to produce a malted grain which gives the brewer
or distiller the particular characteristics of enzyme activity, colour,
extract (solubilisable content) and flavour required.

291. The Malting Process (see Figure 19)

. Steeping
  Barley as bought is typically 16-18% moisture and must be dried to
  about 11-12% before storage. This is often undertaken by the
  maltster. It is taken from storage and steeped in water to gain a
  moisture content of around 42-46% by weight.

. Germination
  The steeped barley is then allowed to germinate. Most malt is
  produced by pneumatic germination but some special malts are made by
  the older method of floor malting. In the former method the grain is
  warmed or cooled as necessary by the forced package of humidified and
  temperature controlled air using a range of box drum or hybrid
  vessels containing the malted grain.

. Kilning
  The kiln is a shallow bed drier usually of brick though recent
  vessels are steel in which the grain is laid to a depth of around 1-2
  metres. It is dried by passing hot air (direct firing via oil or
  gas) through the bed. The air may be drawn through or pushed through
  by fans.

292. The moisture content is reduced from around 45% to 2½-4% according to
the type of malt required. The rate of kilning is important and is determined

by the bed depth, rate of air flow, air temperature and humidity. For whisky malts peat is burnt in the kiln so that peat condensate settles on the mould which contributes to the final flavour of the whisky.

293. The malt is then passed to storage after the rootlets have been removed. These rootlets are known as 'culm' and are used as animal feed. Storage for at least one month is regarded as part of the malting process.

294. UK production is about 1.3×10^6 tonnes (1.9×10^6 tonnes if brewers grains are included as an output) (1980) of which about one million is produced by special maltsters for sale. Of the 1.3×10^6 tonnes, 35% is used in distilling; about 45% in brewing; 16% is exported and the remaining 4% is used in vinegar and as malt extract for a range of foodstuffs such as malt loafs, home brewing, biscuits and malted milk beverages.

Energy Consumption and Conservation in Malting, 1980

295. Information about energy consumption in the industry is derived mainly from a survey carried out by the Maltsters Association of Great Britain (MAGB). The survey covered 207 kilns producing 73% of total malt in the UK in 1980. After scaling up, the survey suggests a total energy consumption of 7.16×10^6 GJ on a heat supplied basis. A series of seven visits to maltsters by the Leatherhead Food RA as part of the IETS operation also provides evidence on SEC.

296. The weighted average energy consumption was 4.78 GJ/tonne and of this 82% was for kilning with the remaining energy going for barley drying (which will be dependent on the moisture content in the supply from the farmer or grain storage).

297. Energy use in the malting industry is clearly dominated by the kilning operation. Variations in other activities may be larger but even small variations in efficiency and practice associated with kilning have significant energy implications.

298. Minimum specific energy required figures from the MAGB survey suggest that kilning might be possible at 2.92 GJ/tonne whereas the maximum recorded figure was 7.60 GJ/tonne. An explanation of the difference between these two figures is largely due to the fact that heat recovery systems were used by those who were most efficient although some small variance must always be allowed because of the different qualities of malt produced. The different malts will vary slightly in moisture content and kilning rate and this clearly affects both the temperature and the length of the kilning cycle. However it is clear that something close to 3.5 GJ/tonne is currently possible as being a best practice and this figure (3.5 GJ/tonne) has been taken as representing best practice which would be available to most maltsters in the industry if they operated with heat recovery and other good practice systems.

299. This data is consistent with Leatherhead Food RA data which suggest 3.5 GJ/tonne is a best practice figure. Average practice, however, may be a little lower than the maltsters data at around 4.0 GJ/tonne. A figure of 4.5 GJ/tonne has been taken as a middle figure between the sets of data.

300. If this data is broadly correct then energy consumption in the malting industry is about 5.99×10^6 GJ, about the same as the Maltsters Survey

suggests. Brewing and malting together are 26.73×10^6 GJ which is about the same as Fuel Purchase Inquiry estimates for 1979.

301. If all firms operated at best practice then energy consumption in the malting industry might be 4.66×10^6 GJ suggesting a potential saving of 1.33×10^6 GJ which, expressed as a percentage of energy consumption in 1980, is 22%.

302. Energy conservation potential in the brewing and malting industry as a whole is shown in Table 35; it is about 22% of total energy consumption in 1980 and represents nearly 6×10^6 GJ.

Energy Conservation Measures in the Malting Sector

303. Considerable discussion is given in the Audit Report on heat recovery systems and heat pumps. This is because energy saving in this industry is dominated by the kilning cycle. There is not much that can be done about electricity consumption most of which goes in motive power for fans and some refrigeration. Nor is it likely that the amount of energy expended on barley drying will be greatly reduced - this will depend in large part on the moisture content of the barley when taken into the store. Energy conservation is essentially about the kilning cycles.

304. A full account is given in the Audit and only a brief note is given below of the kind of measures associated with good practice and investment in conservation associated with the industry. The main points are:

- good housekeeping - this is largely a matter of care and experience in lining the kiln and in ensuring that the kiln is dry before the kilning cycle begins. In some cases a single vessel is used for both steeping, germination and kilning and in these circumstances it is important that pools of water and other moisture are brushed and dried away before the kilning cycle begins. Where separate vessels are used in the kilning cycle the problem is much less serious.

- heat recovery - the major conservation measure is clearly to recover the heat from the kiln exhaust gases and to re-use it for heating the incoming air. It is possible to re-use it for purposes of kilning the next cycle, but this can only take place where two kilns are being used in a back-to-back mode of operation.

Market Developments in Alcoholic Drinks, Soft Drinks and Malting to 2000

305. As the fortunes of the maltings industry are closely tied to developments in brewing, much of this section focuses on likely developments affecting the brewing industry. The consequences for the maltings industry are discussed later. Brewing, of course, is but one form of alcoholic drink and most bigger breweries are also wine buyers and shippers and also producers of soft drinks.

306. The framework within which the brewing industry is planning productive investment and marketing strategies for the next 10 to 15 years has several main strands.

307. <u>Changes in Drinking Habits</u> It is well recognised that beer sales have grown much more slowly than sales of other alcoholic beverages in the last decade. Indeed in the last five years, beer sales have actually fallen by some 10%. The question is whether this trend is simply a cyclical effect caused by the recession from which the industry will recover, or whether some more serious and permanent changes in the social pattern of drinking have occurred which are likely to leave a permanent mark on the industry. Analysis of alcohol sales shows that wines have grown much more rapidly than beer sales, and now account for something like 15% of total alcoholic beverage sales. Lager beers have increased their market penetration at the expense of keg and cask conditional ales and take nearly a quarter of the beer sales market and cider and perrys have also increased sales at the expense of beer and soft drinks. Whisky and other spirits have significant export sales which are currently at a depressed level although there is a long maturation process which affects any assessment of longer term market trends.

308. The pattern of drinking has also changed a little: the take away trade from supermarkets, grocers and off licences, has increased slightly: drinking in clubs and private licence premises (social clubs within large organisations) have all increased while sales at the public house have decreased steadily in the last five to ten years. Thus the nature of the product and the pattern of consumption have changed considerably in the past decade and there is some uncertainty as to where the trade is going.

309. The second observable trend to affect alcoholic drinks and beverages is health consciousness amongst consumers. Sales of health food, natural foods, and low fat, sugar and carbohydrate food and drink, appear to be amongst the more buoyant of the food industry products. Brewers and soft drink manufacturers have recognised the trend and now market an array of products which are calorie reduced, such as 'Diet Pepsi' and 'lite' beers.

310. Most of the brewers assets are in the tied public houses so that brewers may be characterised as property managers, rather than beer manufacturers. In energy terms about half the total energy bill paid directly or indirectly by the main brewers relates to the public houses. So any reduction in trade increases energy costs per unit of turnover because most of the energy used in public houses is used for space heating and refrigeration and air conditioning which is an overhead independent of the number of people in the public house. The need to reverse the trend in public house drinking is therefore critical to brewers. An increase in public house licensing hours may go some way down this path – sales increase of 10% have been estimated [22].

311. This factor has <u>not</u> been included in our estimates but the effect on energy consumption would be significant if drinks sales increased by about 10%.

312. Packaging is also important because the cost of the container is a significant proportion (10-12%) of the final price of a can or bottle of beer; because of the convenience and attractiveness and hence saleability to consumers; and because the kind of packaging may affect the quality of the product [22, 23].

313. A number of developments in packaging may occur in the next decade or so. There is a major development in plastic containers of the one and two litre kind currently used for soft drinks, and it may well be that beer wil

be marketed through off-licenses and other outlets in this form. Much depends upon the quality of manufacturing and hence the cost of the container, but trials are progressing successfully and it may well be that a switch to plastic containers will reduce the cost of the final product without affecting quality. PET (polyethylene terephthalte) has been introduced for soft drinks and could well be used for beer.

314. Another development which is perhaps less probable is the introduction of the bag-in-box foil or plastic container, for cask conditioned ales. The bag-in-box foil container has had a major impact on wine sales, with the three to four litre units, substantially increasing the volume sales of relatively low cost wines. There are doubts, however, about the financial viability of applying this measure to beer - the packaging cost relative to the value of the product and the short shelf life of beer seem to be the two main problems at the moment.

Market Developments in Malting

315. Evidence from the Maltsters Association of Great Britain was not available to the study so the assumptions made here may require to be modified. However, many of the trends affecting the malt whisky and brewing industries will have very similar repercussions for the maltings industry.

Energy Consumption and Conservation in 2000

316. The above points provide a background to the judgements that have been made on the structure of output and SEC over the period to 2000.

317. These are set out in Table 36. SEC is assumed to fall by around 20% in most of the sectors implying that most of the energy conservation potential identified for 1980 will have been realised.

318. This view is based, as with most of the sectors, on the fact that good housekeeping/energy management measures can be achieved quickly and with little investment and much of the conservation potential is of this kind. Where investment is needed the period of 20 years is considered long enough to achieve the replacement of much of the most inefficient plant.

319. No figures were available for how far 'best practice' might improve by 2000 so no estimates of conservation potential in 2000 have been made.

FIGURE 17: Industrial Processes in Grain Whisky Distilling

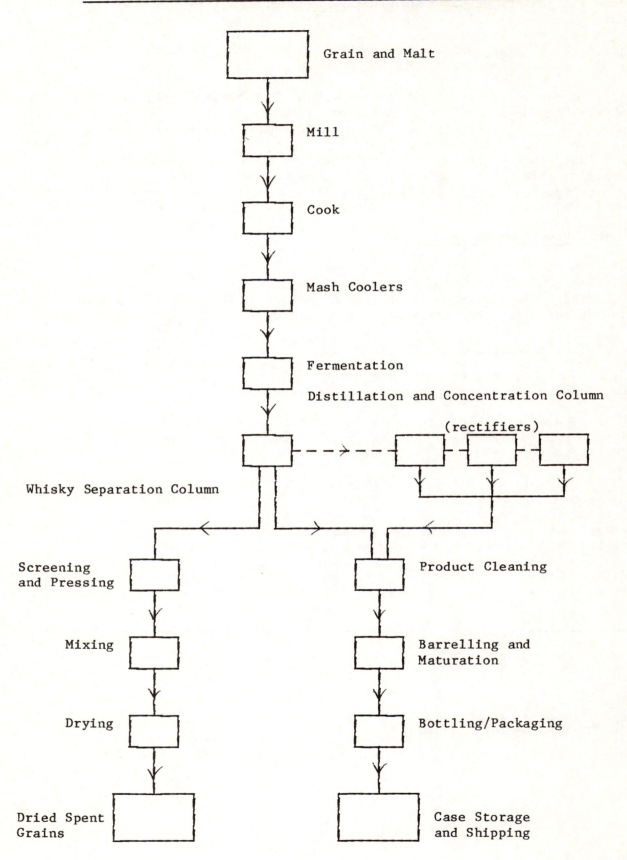

FIGURE 18: Industrial Processes in Brewing

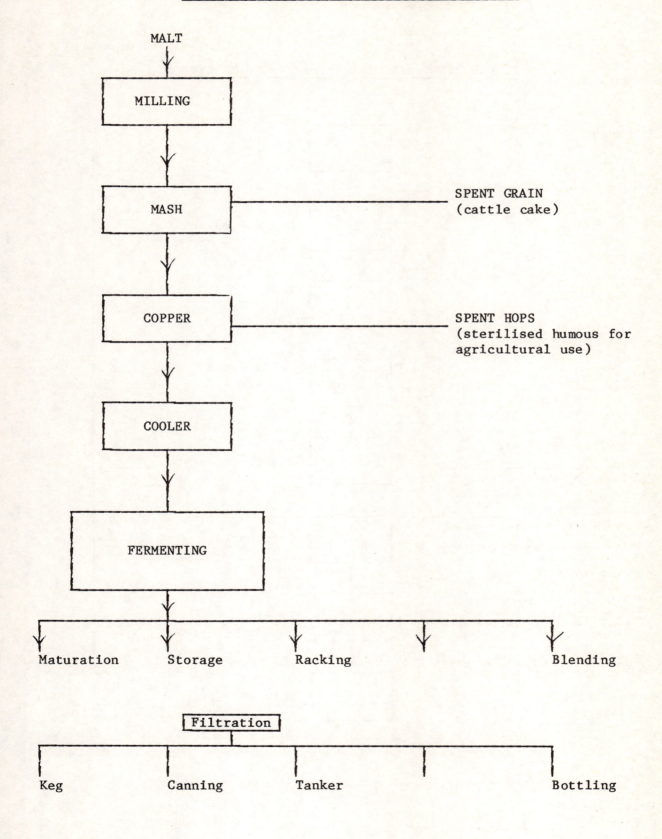

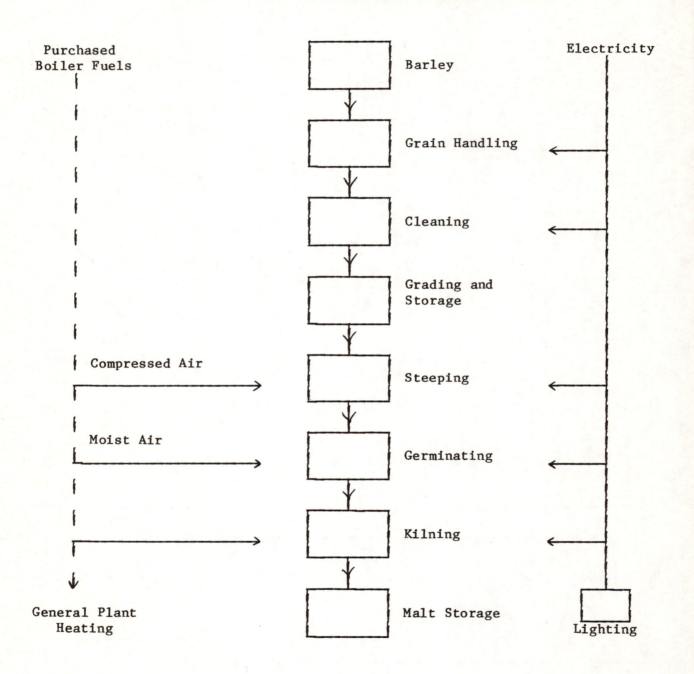

FIGURE 19: Industrial Processes in Malting

TABLE 30: Output and Energy Consumption in Wines, Ciders, Perrys and Spirits, and Soft Drinks

	Output, 10^3 litres	Average Practice,	Best Practice,	Coal,	Oil,	Gas,	Elec,	Total
		MJ/litre		10^6 GJ				
Wines, ciders, perrys	278,250	1.41	1.00	0.01	0.09	0.25	0.04	0.39
Soft drinks	7,035,448	0.60	0.45	0.67	2.70	0.13	0.72	4.22
Malt whisky and	100,205	60.00	55.00	0.12	1.38	3.85	0.66	6.01
Other spirits	569,094	28.00	20.00	0.32	3.66	10.20	1.75	15.93
TOTAL				1.12	7.83	14.43	3.17	26.55

TABLE 31: Energy Utilisation in Soft Drinks, Wines, Ciders, Perrys and Spirits, 1980

	Soft Drinks %	Wines %	Ciders %	Malt Whisky %	Other Spirits %
Process	55-60	55-60	55-60	85	85
of which					
Pasteurisation	30	30	30		
Mashing				15	30
Distillation				80	45
Bottling	30	30			
Space heating		30-50	30-50	10	10
Other	5	5	5	5	5
TOTAL	100	100	100	100	100
% through boiler	90	90	90	80	80
				(except direct fired still)	

TABLE 32: Energy Conservation Potential in Soft Drinks, Wines, Ciders, Perrys and Spirits, 1980

Given by energy consumption at average practice = 26.55 x 10^6 GJ
less energy consumption if all firms operated at best practice = 20.34 x 10^6 GJ
which is 6.21 x 10^6 GJ or 23% of energy consumption in 1980.

TABLE 33: Energy Consumption in Brewing and Malting 1980

	Output, 10^6 tonne	Average Practice, GJ/tonne	Best Practice,	Oil,	Gas,	Coal, 10^6 GJ	Elec,	Total
Brewing*	6.48 +	3.2 +	2.5 +	5.59	6.02	6.85	2.28	20.74
Malting* (excluding brewer grains)	1.33	4.5	3.5	2.39	2.87	0.06	0.67	5.99
				7.98	8.89	6.91	2.95	26.73

* Figures from HM Customs and Excise. Business Monitor data gives 6.48 and 1.18 respectively.
+ Assuming 10^4 hectolitres $\equiv 10^3$ tonne.

TABLE 34: Energy Use in Brewing and Malting 1980

	Brewing	Malting
Process of which kiln	85	90 80
Space Heating	10	5
Other	5	5
	100	100
% through boiler	80	30–40

TABLE 35: Energy Conservation Potential in the Brewing and Malting Sector 1980

Energy consumption if all establishments operated at best practice
= (16.20 + 4.66) = 20.86 x 10^6 GJ
Conservation potential = (4.54 + 1.33) = 5.87 x 10^6 GJ
Expressed as % of energy consumption in 1980 = 22%.

TABLE 36: Output Structure and Specific Energy Consumption in the Drinks Industries, 1980-200

(a) **Structure of Output** (percentage)

	1980	1990 H	1990 L	2000 H	2000 L
Wines, ciders, perrys	2	2	2	2	2
Soft drinks	44	46	44	49	44
Malt whisky	1	1	1	1	1
Other spirits	4	4	4	4	4
Brewing	41	39	41	37	41
Malting	8	8	8	7	8
	100	100	100	100	100

(b) **Specific Energy Consumption** (MJ/litre)

	Average	Best	Average	Average	Best
Wines, cider	1.41	1.00	1.27	1.13	n.a.
Soft drinks	0.60	0.45	0.56	0.54	n.a.
Malt whisky	60.00	55.00	54.00	48.00	n.a.
Other spirits	28.00	20.00	25.20	22.40	n.a.
Brewing	3.20	2.50	2.50	2.30	n.a.
Malting	4.50	3.50	4.05	3.60	n.a.

5.6   THE OILS AND FATS INDUSTRIES

320.   This sub-sector of the industry exhibits some interesting features despite the less than glamorous image of its products. Vegetable oil extraction and refining, which is much the most important in energy terms, is described first.

Vegetable Oil Extraction

321.   Vegetable oils are mainly extracted from imported soya bean, palm kernels and rape seeds although a significant proportion of the raw feedstock is derived from UK produced rape seed. Rape and palm kernel are usually screw extracted first to extract fat. The seed is then crushed and mixed with a petroleum type solvent which dissolves the oil from the seed. The oils are then filtered and the petroleum boiled off in a distillation process, the products of which are oil residues in the form of seed cake; the petroleum vapours are captured and recycled. Figure 20 shows the main processes involved in soya bean processing.

322.   The residues, seed cake or meal must be further prepared before sale to animal feedstock manufacturers. The meal is cleaned, toasted (to destroy enzymes), steam heated to remove any remaining solvent and dried by evaporation. Most crude oil is sold to refineries but some is sold unrefined to animal food manufacturers.

323.   Energy uses in the extraction process are mainly for process use. The distillation (solvent extraction) process operates at temperatures of around 60-80°C (a petroleum solvent) and steam is supplied directly to drying and other processes. Electricity accounts for about 10% of total energy requirements mainly for powered rollers in the cracking mills.

Oil Refining

324.   Most of the crude oil is sent to refineries for refining. Refineries will also handle imported crude oils, palm oil and fish oil, together with soy, sunflower and rape. Refining ensures that the crude oil is cleaned free of fatty acids, is colourless, of appropriate consistency and is deodorised. The energy processes are essentially the provision of steam and hot water at 100-120°C for the neutralising process (caustic refining); and at 180-250°C for hydrogenation and deodorisation. Caustic soda is added to the oil to remove acids and the soaps are then washed out; hydrogenation alters the structure of the oil by passing the oil over a nickel catalyst in the presence of hydrogen to make a 'hard' oil suitable for margarine and similar fats. The deodorising process is also very energy intensive. Vessels containing the oil are heated either by steam to about 180-250°C for a process cycle of around eight hours or Thermex heat transfer liquids such as Dowtherm or Mobiltherm. Winterisation is a process which removes the waxes in oils so that the product does not cloud over in sudden temperature changes from room temperature to refrigerator. It is usual to either hydrogenate or winterise and most UK manufacturers hydrogenate. Energy use in refining is also associated with creation of high vacuum (typically about 5 ml of mercury) for distilling the steam volatiles [24]. Figure 21 shows the industrial processes.

325.   The products of the refinery are fine quality cooking oils like Mazola; less fine oils suitable for biscuit and sweet manufacture; white fats (Cookeen

etc); and oils which pass to margarine manufacture. Margarine is made by adding water, brine, milk products and emulsifiers to the oil, mixing them, and chilling and crystallisation to an appropriate consistency [25].

326. What is interesting but complicated about the industry is that some plants (such as Bibbys at Liverpool) are integrated plants performing all three processes while most of the industries (eg the Unilever plants) operate either extraction plant or refining/manufacture/packing operations. Some margarine plants are manufacturing/packing only.

327. Energy use is shown in Tables 37 and 38. It is easy to describe but difficult to allocate because of the variety of establishments in the industry. Most energy is process energy and most processes involve steam and hot water from oil or gas fired boilers. 60-90% of all energy passes through the boiler. Two variants on this generalisation are:

- <u>Margarine making</u>, where process energy might only be 50-60% of total energy, a high proportion of which would be for refrigeration plant. Space heating, although not important in absolute terms, might account for 50% of the total steam/hot water raised.

- <u>Extraction plant</u>, where almost all the space heating is derived from heat recovery from the solvent extraction process. In some extraction plants sufficient steam is associated with electricity requirements to make some auto generation of electricity economic, eg Van der Bergh at Purfleet.

Meat Rendering and Animal Fats

328. Carcasses are rendered down, compressed and ground by steam heated worm screws in steam jacketed pressure vessels. Fat separates naturally to give fats of given qualities. What determines the quality is the animal and the parts of the animal. Top quality pig fat is lard; top quality cow fat is dripping; and both of these are sold directly to domestic and catering markets. Secondary grades of fat are classed as tallows and quality gradually deteriorates to a dark fat called bone grease. The residue of the carcass is then ground and dried to produce meat and bone meal.

329. Fats may be fractionated (melting out at varying temperatures to provide finer quality fats and tallows) and are then refined (naturalised bleaching and deodorising). Fine lards and drippings go to the food manufacturing industry for pastries, meat products etc; second quality tallows and fats go to soap manufacturers and lower quality tallows go to lubricating oils and greases. Bone grease can be split by hydrolysis and is used in the oleo chemicals industry.

330. Energy use in these processes is essentially the provision of steam at 60-70 psi giving temperatures of around 130-140°C for rendering and much lower temperatures (about 80°C) for fractionating. About 85% or so of all energy is process energy; most of this is passed through boilers. Around 10% of the steam/hot water used is for space heating in this sector of the industry. Table 38 provides the details.

Energy Consumption and Conservation Potential 1980

331. Energy use data is taken from the IETS data. The most important fuel remains oil - taking nearly 45% of all fuels. Electricity is important in margarine and fats refining (12%) but only 8% in total sector energy use. Switching to gas has still some way to go (gas accounts for 22%) and coal is still used in several establishments (accounting for 10% of total fuel use).

332. Data on SEC (GJ/tonne) has been taken from the Leatherhead Food RA visits made under the IETS scheme. Some 12 firms were visited accounting for 73% of employment; 74% of tonnes output in vegetable oil extraction and refining but only 18% of output in meat rendering. The sample may be taken to be full enough to give representative results.

333. This is important because observing the raw data of the visit reports on vegetable oil refining suggest almost a bimodal distribution with a group clustering around 2-5-3.5 GJ/tonne and another group around 9-11 GJ/tonne. Discussion as to why this might be raised several possibilities mostly to do with the technology of production and heat recovery possibilities. But in the absence of more detailed enquiries it was agreed with Leatherhead that the figure selected (8.1 GJ/tonne average or 3 GJ/tonne best) represented a fair view of the evidence. Vegetable oil extraction similarly showed some variance but the figures selected represent the mean and lower quartile positions of the observations. There are lower figures but they do not seem representative of what the industry could achieve.

334. If the energy consumption figures are as indicated in Table 37 the energy consumption in 1980 may be estimated as the product of SEC and output data from Business Monitor. This gives figures set out in Table 37 of 13.7×10^6 GJ. If the 'best practice' figures are right then energy consumption of 8.8×10^6 GJ could be achieved if all firms operated at best practice. This gives energy conservation potential of 4.9×10^6 GJ which, expressed as a percentage of energy consumption in 1980 is around 35%, see Table 39.

Market Developments to 2000

335. The structure of the industry seems likely to undergo further changes in the next 15 to 20 years. Closure of more extraction plants and an increase in imported oil are expected. At present (1983) the oil and fat sector is dominated by Unilever (about 45% of the market); the recently formed Acatos and Hutchinson Group (about 30%) and Bibbys and Croda among the few remaining large independents (about 10%); the remaining 15% of the market is shared among several smaller mills and refineries. Several plants may close in the next decade as output is concentrated in larger mills and refineries. This is the most likely outcome of the rationalisation moves which have occurred in recent years. There is a question of how much extraction may be undertaken in the UK; the growth area (or survival area) is likely to be refining.

336. The market which has remained stable in the past decade is likely to remain stable over the next 15 years or so. Volume output has increased broadly in line with population and with other sectors of the food industry because oils and fats are input to the biscuit, flour confectionery and chocolate and sugar confectionery sectors.

337. Output over the period to 2000 might remain fairly static: the structure
of output may thus be as described in Table 40. There is some uncertainty
about these figures due to the possible effect of imports and, to a lesser
extent, export prospects in the industry. At present imports and exports
represent less than 5% of total sales but in recent years imports,
particularly of animal fats and margarine, have increased at a faster rate
than domestic output. In part this is because of the very large plants which
are located in the centre of the European market (Holland and Germany).
Because of their large size and the margins which they make distributing the
products throughout Europe it is argued by some UK manufacturers that they can
export marginal output at marginal cost to the UK. UK manufacturers find that
exporting is difficult due to the relatively high cost of transport since the
UK is on the very edge of the European market.

338. One important factor affecting domestic sales is the long running battle
between butter and margarine (and other vegetable oil based margarine type
spreads). It is not clear how the health factors (cholesterol, polyun-
saturated fats, etc) will influence sales; nor how the taxes/subsidies
(EEC/CAP intervention prices) will affect relative prices between the
products. Tastes, technology (new additives and mixes in margarine type
spreads) and advertising are other obvious factors which will determine market
shares. The question is not addressed and current market shares are assumed
not to change greatly in the future.

339. Another market development which may well occur on a significant scale
by 1990 is a move to imported part processed oils. In the longer term there
may be very little oil extraction in the UK - extraction and refining of oils
to a neutral oil may be carried out in countries which produce the oil seeds.
Deodorisation is the most complex process and this may be carried out in the
UK. It is too soon to argue strongly that this will occur but changes in the
structure of output may be due to some fall on extraction capacity.

340. To estimate specific energy consumptions in 2000, some view of the
energy implications of technological developments in the industry is
required.

341. Technological Developments to 2000 Incremental rather than radical
improvements are expected in plant and equipment. There seems no major
innovation in the offing in this long established traditional industry.
Innovations in marketing (pure unblended vegetable oils, low cholesterol
margarine, etc) require only minor changes in production technology. So,
improvements in specific energy consumption probably depend on the incremental
improvements brought about by investment in new plant and equipment and better
utilisation of that equipment.

342. Better utilisation is most likely to occur through continuous rather
than batch production. The arguments for and against this move are fairly
close because the benefits of a higher utilisation rate (better quality
control, lower fixed costs) probably produce overall lower unit costs despite
the more sophisticated control gear associated with continuous production and
the increased difficulties in costs associated with maintenance. Energy costs
on balance are probably lower despite the higher electrical loading associated
with control gear. In deodorising, energy costs are much lower in continuous
rather than batch production because of opportunities for heat recovery and
opportunity for continuous use of vacuum plant. Physical refining (in which

the deodorising plant is operated at 250°C to draw off both fatty acids as
well as volatiles) may also be introduced on a wider scale with consequent
energy savings.

343. However, significant improvements up to 20% have been made by some
companies simply by good housekeeping measures (insulation etc) and relatively
low cost bolt on measures such as heat pumps and heat recovery equipment.
This suggests that there is still some prospect of significant savings to be
made in the industry by a move from average practice to best practice over the
period.

344. Significant improvements in energy costs, if not energy utilisation can
be achieved by switching to the use of coal. A number of plants and
refineries have evaluated such a switch; although the paper calculations
suggest significant savings can be achieved, few companies have been prepared
to invest the large capital sums (around one million pounds) required to make
this switch. In the present trading climate informed opinion suggests that
the fuel prices are not likely to move at a rate which will induce significant
switching in the next five years but companies may well do so later in the
decade.

345. The effect of fuel prices to data has been to make companies conscious
of operating costs of new plant and equipment and to insist that replacement
plant is more energy efficient than the old plant. However, running costs are
still of secondary importance compared with the prime requisites of
throughput, reliability, product quality etc, although most firms will say
that running costs are now more important than they were ten years ago. The
relative importance of running costs as against the initial costs is likely to
increase over time as fuel prices rise but it seems in general most unlikely
that plant investment decisions would be brought forward merely in order to
save energy.

346. This is because about 80% of total cost is represented by the raw
materials. Profits therefore in large part depend upon shrewd buying in the
commodities market and shifts in relative prices of rape seed, soya beans
and fish or fish oil (and to some extent palm oil and animal fats) have much
more effect on profits than any other element within total costs. But savings
in energy for a modest outlay must be desirable since they can make a direct
contribution to profits. Process costs are around about 15% of total costs
and energy costs are around about 25% of process costs. Total energy costs
therefore represent no more than 4% of total manufacturing costs.
Conservation activity therefore has a major incentive only if energy costs are
related to profits rather than costs.

Specific Energy Requirements to 2000

347. There is a high degree of uncertainty about any estimate but informed
opinion in the oil refining industry puts a figure of around 15% lower than
present specific energy requirements. No data is available for other sectors,
so similar figures have been assumed. Average practice is assumed to fall
faster than best practice so the difference in 2000 would be significantly
smaller than in 1980 reflecting the more rapid diffusion of operational and
good housekeeping measures compared with improvements due to investment in
plant and equipment.

348. There is little information on rate at which improvements in specific energy consumption will occur. There are no innovations to be introduced and diffused through the industry; there seems to be no future date at which certain techniques or activities might be undertaken if fuel prices were high enough (except fuel switching); so reductions in energy in the vegetable oil extraction and refining is all about the age and efficiency of the plant and, to a lesser extent, operational practices of the good housekeeping kind.

349. It seems not unreasonable, therefore, to assume that the path to 2000 will be a straight line and Table 40(b) is based on this assumption.

FIGURE 20: The Main Processes of Soya Bean Extraction

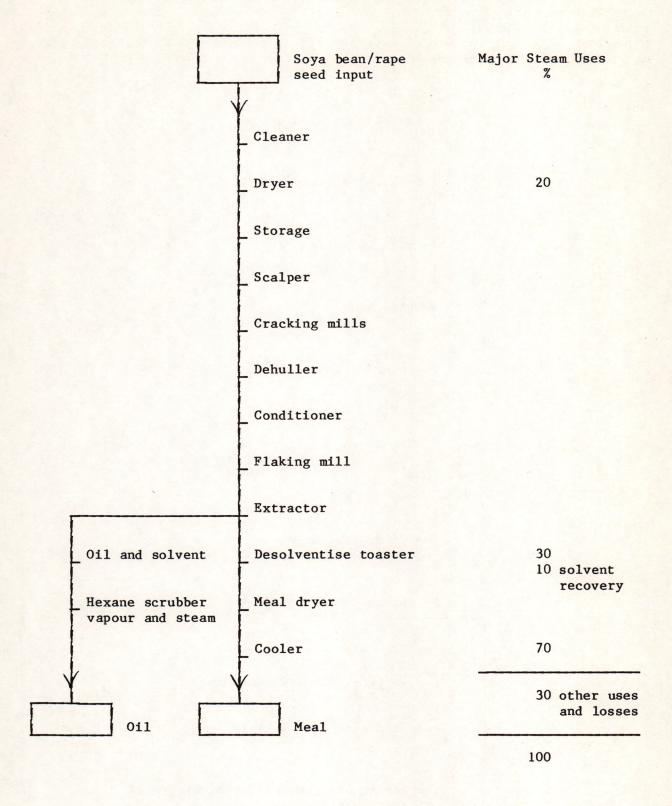

Soya bean/rape
seed input

Major Steam Uses
%

Cleaner

Dryer   20

Storage

Scalper

Cracking mills

Dehuller

Conditioner

Flaking mill

Extractor

Oil and solvent   Desolventise toaster   30
                                                  10 solvent
                                                     recovery
Hexane scrubber
vapour and steam   Meal dryer

                         Cooler   70

Oil   Meal   30 other uses
                                                       and losses

                                                   100

FIGURE 21: Usual Processes in Oil Refining, and Production of Shortening Oils and Margarine

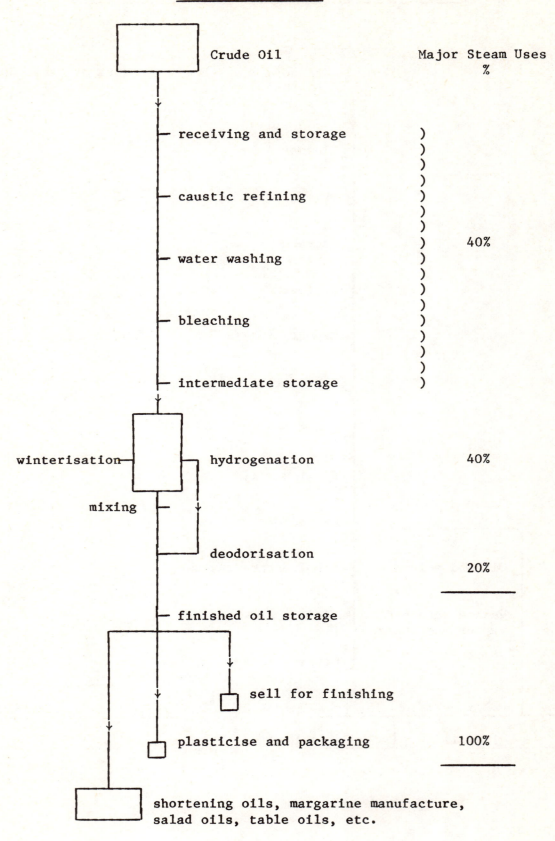

TABLE 37:   Output and Energy Consumption in Margarine, Vegetable Oils and Fats 1980

	Output, 10^3 tonnes	Average Practice, GJ/tonnes	Best Practice, GJ/tonne	Oil, 10^6 GJ	Gas, 10^6 GJ	Coal/ Other, 10^6 GJ	Elec., 10^6 GJ	TOTAL, 10^6 GJ
Margarine & compound cooking fats	522.4	3.94	2.1	0.16	0.02	1.64	0.24	2.06
Oil extraction	737.8	5.0	4.5	1.82	1.41	–	0.46	3.69
Oil refining	642.1	8.1	3.0	3.32	0.22	1.39	0.27	5.20
Meat rendering	242.1	11.4	10.1	1.02	1.39	–	0.35	2.76

TABLE 38:   Energy Use in Margarine, Vegetable Oils and Fats 1980

Energy Use	Margarine %	Veg Oil Extraction %	Veg Oil Refining %	Meat Rendering %
Process	70	95	95	85
of which – refrigeration	30			
Space Heating	25	–	–	10
Other	5	5	5	5
	100	100	100	100
% through boiler	70	80	90	80

TABLE 39:   Energy Conservation Potential in Margarine, Vegetable Oils and Fats Sector 1980

Energy consumption if all establishments operate at best practice = 8.81 x 10^6 GJ
Energy conservation potential = 4.9 x 10^6 GJ
Expressed as % of total energy consumption in 1980 = 35.7%.

TABLE 40: Output Structure and Specific Energy Consumption in the
Oils and Fats Industries, 1980-2000

(a) Structure of Output
(percentage)

	1980	1990 H	1990 L	2000 H	2000 L
Margarine and cooking fats	24	24	25	24	26
Oil extraction	35	34	33	32	31
Oil refining	30	31	31	33	32
Meat rendering	11	11	11	11	11
	100	100	100	100	100

(b) Specific Energy Consumption
(GJ/tonne)

	Average	Best	Average	Average	Best
Margarine and cooking fats	3.94	2.10	3.69	3.43	1.89
Oil extraction	5.00	4.50	4.68	4.35	4.05
Oil refining	8.10	3.00	7.58	7.05	2.70
Meat rendering	11.40	10.10	10.66	9.92	9.09

5.7 MISCELLANEOUS FOODS

350. The range of products manufactured in this sub-sector includes starch and glucose products, salt, coffee and tea packing, yeast products including Marmite and Bovril and a range of other dried powder products. Animal and pet foods are also considered in this section.

351. Most establishments employ steam/hot water which is supplied to the various industrial processes by oil or gas fired boilers. Typically 2/3 of total energy is supplied through the boiler and around 2/3 of total energy is process energy. In most cases the processes include mixing, washing, drying to a stiff concentrate or powder form and some bottling or packing. In some cases, such as tea and coffee packing, circumstances are more like warehouse establishments where the central process is weighing and packing which is not an energy intensive activity but does involve more space heating and some electricity as motive power for machinery. It follows therefore that the steam hot water requirements are substantially less; and that space heating is the major use of boiler fuels. Tables 41 and 42 show the distribution of energy use in the sector.

STARCH AND OTHER FOODS SECTOR

Starch and Glucose Syrup

352. Starch is made from imported maize (although increasing quantities are being derived from wheat) which is soaked and wet milled through stone or steel rollers to produce starch slurry (starch in water). Germ, cake (from starch germ), gluten and maize oil are by-products. The maize germ is sent for refining and the residues are formed into a cake for animal feed. The maize skins are either dumped or partially processed for animal feed. The extracted starch is dried to corn starch or hydrolysed either, in simple terms, by adding acid and boiling the liquor or by adding enzymes to the starch. The hydrolysed liquor is concentrated to a colourless, viscous syrup rather like clear golden syrup which is sold as glucose or various syrups for the baking and sugar confectionery industry. Figure 22 shows the details. In parts of the food industry liquid glucose is known as 'bakers syrup'. This product can also be produced from wheat starch.

353. The main energy using stage is evaporation (typically vacuum evaporation) at around 120°C. But there is also much hot water for cleaning and some warm air drying, particularly to dry the corn starch used in food, paper making and textiles. About 85% of all energy is passed through boilers and most of this is process energy - space heating is small at around 5% or less. In some large companies there is sufficient electrical requirements to install CHP plant.

Farinaceous Foods

354. These products are essentially flour/grain derivatives in powder or dried form. They include self raising flour, noodles, spaghetti and macaroni. Of these, flour is the most important in tonnage and dominates the output statistics in Business Monitor. The process is not particularly energy intensive but there is a high electrical loading for the continuous sieving, mixing and bagging operations.

355. Noodles and other pasta products are made from a flour, water, salt and flavours mix; cooked; extruded to a given shape (including many new assorted shapes for the snack market); and dried in warm air cabinets. Again electricity for motive power (mixes, extrusion) is important as a percentage of process energy (c. 40%) although overall these industries do not consume much energy. There is a requirement for space heating and hot water for cleaning supplied by a small boiler which may also supply steam for the warm air drying cabinets. Perhaps 60% of total fuel is passed through a boiler. Full details are given in Table 42.

Salt

356. This category includes gravy browning and stock cubes such as Bisto, Oxo, Marmite and Bovril, although strictly Bovril is a meat extract. It is a small user of energy; the mixing and packing of salt is an electricity based activity but the other products require some hot water and steam for boiling and cleaning. About 60% of all energy passes through a boiler but a large part of this is used for space heating. Process energy is around 55% of the total most of which is for evaporation and a large part of this is electricity. Although salt is a low energy industry some parts of this sector, particularly Marmite, are much more energy intensive.

Yeast

357. The main output is fresh yeast for bread making. It is produced by fermenting carbohydrates in a highly aerated atmosphere and is sold as a wet filtered cake to the baking trade. Small quantities of brewers yeast and dried yeast for the domestic market are also produced.

358. The main fuel used is electricity for compressing and blowing filtered air through the fermentation vats and for refrigeration of the finished products. Some hot water is also used for cleaning and space heating accounting for around 40% of the total energy used.

Tea and Coffee Packing

359. As the heading implies most process energy is electricity for weighing and packing. Boiler fuels account for around half of the total, much of which goes in space heating.

Other Products

360. A wide range of products is included in this category such as sponge mixes, nuts, flavours and essences, dried soup mixes (canned soups are classified to fruit and vegetable sector) and, most important in energy terms, spray dried coffee.

361. Coffee beans are ground, percolated and the coffee extract then evaporated to about 40% solids before being spray dried or freeze dried. Spray drying is an impressive process. The driers are huge chambers around 40-50 feet in diameter, usually heated by direct firing or indirect firing to around 230°C into which the coffee extract is sprayed. Exhaust gases come out at around 100°C and the dried coffee falls into a hopper for subsequent bagging and transfer to coffee packing sheds/establishments. Freeze drying takes the concentrated liquor at 40% solids; freezes the liquor; and passes it

to a high vacuum chamber. The chamber is heated gently so that the coffee blocks sublime to leave a dry residue which is granulated and packed.

362. For most of the other products a steam based system at less than 120°C from a central boiler is the main source of energy use, much of which is used for space heating. Details of energy use are given in Tables 41 and 42.

Energy Consumption and Conservation Potential, 1980

363. Oil remains the most important fuel (48%) but gas has achieved some penetration (11%) while coal still retains 24% of the market entirely in the starch area. Electricity accounts for about 10% of the total.

364. Data on SEC 'average practice' and 'best practice' have been taken from IETS visit reports held by Leatherhead Food RA. Some 35 sites were visited in the animal foods sector; 34 visits in the starch and miscellaneous food sector accounting for 38% of ouput and employment; the three firms visited in the starch sector accounted for 44% of output. Full details of the sample have yet to be published but the firms visited seem not atypical of the industry. The average SEC and 'best practice' SEC agreed with Leatherhead Food RA are shown in Table 41.

365. If these SEC data are broadly representative of the industry then energy consumption and conservation potential in 1980 are as shown in Tables 42 and 43. Taking the whole sector together energy consumption in 1980 was 11.32×10^6 GJ. If all firms operated at best practice energy consumption would have been about 7.5×10^6 GJ, a conservation potential of 3.8×10^6 GJ or 33% of energy consumption in 1980.

Energy Conservation Measures in Starch and Miscellaneous Foods Sector

366. The Food RA makes several recommendations, including:

- Good housekeeping measures such as better insulation, draught proofing, good management in loading bays and stores.

- Opportunities for heat recovery.

- More efficient use of evaporation processes.

- Improved refrigeration techniques.

COMPOUND ANIMAL FEED AND PET FOODS SECTOR

367. This sub-sector may be categorised partly according to products but mainly according to the degree of drying required in the final product. Canning of pet food is a separate operation akin to fruit and vegetable canning. All establishments are of the type two kind so, as expected, around 70-80% of energy is process energy most of which is supplied by oil fired boilers to give steam and hot water for cooking, drying and cleaning. Energy consumption and fuel use are detailed in Tables 44 and 45.

Compound Animal Foods

368. The mixtures of grains and fats which make up the feedstock need to be
heated to keep the mixture liquid enough for mixing. After grinding and
mixing, the feed stuff is cooked in steam jacketed vessels (or coil heated
vessels) and then strained and dried in warm air cabinets. The dried mixture
is then extruded, pelleted and bagged. Most energy is absorbed in the steam
heated vessels during the cooking phase and the drying phase. These energy
requirements are met by steam or hot water via boilers (nearly 70% of total
fuels are passed through boilers). Electricity is important accounting for
about 1/3 of the total and about half of the total process energy because of
the grinding, extrusion and some air heating in drying cabinets. It is not an
energy intensive process but the output tonnage makes this sector important.

Fish Meal

369. The fish is ground and cooked in steam jacketed vessels where the liquid
fat is separated and the residual mix is then dried. After passing through
a hot air bed the mix is dried in the form of granules or fibrous pieces which
are then fed to a hopper from which they are weighed in bags or sacks.

370. Most of the energy is used in the cooking and drying processes which are
steam based systems supplied by oil fired boilers. Process energy accounts
for about 80% of the total of which electricity is a much smaller proportion
than in compound animal feeds (5%). About 15% total energy is used for space
heating.

Supplements

371. The basic operations are similar to compound animal fats. The only
differences relate to the specialist establishments where finished additives
are mixed with the compound cake. These additives rarely use meat and usually
comprise protein derived from soya bean, reformed protein from oil (Pruteen
the ICI protein) or other protein. This operation is very similar to type
four establishments such as tea and coffee packing which have relatively high
proportions of energy allocated to space heating and correspondingly low
proportions devoted to process energy. Tables 44 and 45 show the pattern of
energy use in this area.

Pet Foods

372. The animal and fish compounds (mostly meat and grains) are mixed,
ground, extruded as lumps and either passed into cans or sold dry. The cans
are then sterilised in a retort at about 130°C. Most pet food canneries have
large scale operations so the retorts operate almost continuously which, as
with vegetable canning, permits a relatively low specific energy per tonne.
Some pet food is not canned but sold in dried form.

373. In the main the processes are steam based raised by oil fired boilers.
Around 80% of energy is process energy with not more than 20% allocated to
space heating. About 80% of energy is passed through the boiler.

374. Some pet food establishments also produce biscuits or biscuit/meat
mixtures. The operations are rather like baking where the mixture is
flattened, cut and baked hard. In these circumstances the ovens are a primary

source of energy and are usually directly or indirectly fired by gas. Again, in such circumstances the proportion of fuel passed through the boiler would be much less than with the steam and water based system.

Energy Conservation Potential in 1980

375. Using the 1980 output data from the Business Monitor and specific energy data supplied by the Food RA, energy consumption in this sector was estimated at 8.61 million GJ. This is about the same as the RA estimated but significantly less than the 1979 fuel price enquiry data.

376. If all establishments operated at best practice then energy consumption might be 6.53×10^6 GJ, offering conservation potential at around 2.08×10^6 GJ, which is 24% of energy consumption in 1980, see Table 46.

Energy Conservation Measures in the Compound Animal Feed and Pet Foods Sector

377. Among recommendations made by the Food RA for the sector are:

- Good housekeeping measures such as better insulation, improved draught proofing of loading bays.

- More effective boiler controls.

- Opportunities for heat recovery exists from the drying processes.

Market Developments to 2000

378. Output in the past decade has been broadly flat and is likely to remain so in all these sub-sectors except for pet food which has not only seen fairly steady growth but is expected to maintain that growth over the next decade. Grain milling is dependent upon the fortunes of bread and flour sales (and hence are likely to remain stable or fall); animal feed depends upon agricultural practices (balance of arable and livestock farming) but is expected to remain broadly stable; while farinaceous foods (pasta products, rice, prepared powders, etc) hopes to maintain its current stable production.

379. The forces acting on these markets are similar:

- high cost raw materials (at least 50% and often nearer 70% of final product price) over which little control may be exercised by the manufacturer;

- the buying powers of retail chains and supermarkets have increased and the proportion of generic foods or own label foods such as flour and rice represent an increasing proportion of grocery sales. Branded goods still account for 75% of all grocery sales (and pet foods for example are dominated by brands) but some areas (eg flour) are under pressure;

- thermo-compression extrusion technology has already been discussed. It is potentially important for pasta products;

- one other factor is the possibility that sweeteners may be based not on sugar products but on glucose derived from maize products. The

technology for producing glucose from maize has been proven for
decades but only recently has it become competitive in price. In the
US, Pepsi Cola Ltd have followed other soft drinks companies in
switching part of their orders for sweeteners to maize based glucose.
The decision is important for both the starch and sugar industries.

In Europe, the CAP protects sugar refiners by imposing taxes on imported maize
which makes the final price of glucose or invert sugar non competitive with
sugar products. If that were to change then European confectionery
manufacture might well follow the lead of the US.

380. Technological Change No new technologies which might significantly
affect output or costs or energy requirement are foreseen in these areas.
Broadly they are all traditional areas of the food industry, apart from pet
food, with long established technologies which are unlikely to change in any
significant fashion. Process plant and equipment will undergo marginal
incremental improvements in efficiency, speed, reliability, but information to
hand suggests no radical innovation which might make any significant
difference to energy consumption. Thermo-compression technology may alter
this view but the effect on energy consumption by 2000 is unlikely to be more
than marginal.

381. The kinds of energy conservation measures appropriate to each of these
sub-sectors available in 1980 were discussed earlier. In general those
industries with large proportions of fuels passing through the boiler will
find major conservation activity in improving boiler efficiencies. It may
also include switching of fuel to coal.

382. It is also the case that even with improvements in boiler efficiencies
significant savings in specific energy requirements depend on the use to which
the steam is put and this is largely in two or three major industrial
processes such as mixing and cooking or evaporation and drying. Most evidence
suggests that process plant equipment suppliers of drying, evaporating,
cooking, steam jacketed vessels, retorts and other equipment for the
production of these food products have long established technologies which are
already moderately efficient and which over the next 10-15 years will merely
make marginal incremental improvements in specific energy requirements. R&D
on process equipment probably will be devoted to faster speeds, larger
throughput, process controls and instrumentation. While this will have some
improvements to running costs, nothing radical might be expected.

383. The net effect on energy consumption therefore would be to see some
marked improvement in average practice SEC between 1980 and 2000 levels as the
1980 average tends towards best practice quite quickly, particularly where
many of the conservation measures are of an inexpensive kind and have to do
with improved boiler efficiencies or with building insulation; which is
particularly important in industries such as tea and coffee, packing salt
manufacture, yeast, etc where space heating is a large proportion (about 40 or
50% of total energy consumption). This is shown in Table 47 where the SECs
for 2000 are mostly about 20% less than the 1980 level. This is most of the
conservation potential identified in 1980. Best practice, however, is more
likely to depend upon re-equipment with new and more efficient plant and
equipment and this is going to be dependent upon replacement investment. The
incentives to re-equip are high but the cash flow and profitability inhibits

such investments. However, no estimates of conservation potential or
improvement in 'best practice' seem robust enough to use in this study.

FIGURE 22: Industrial Processes in Starch, Maize Oil and Syrups Industry

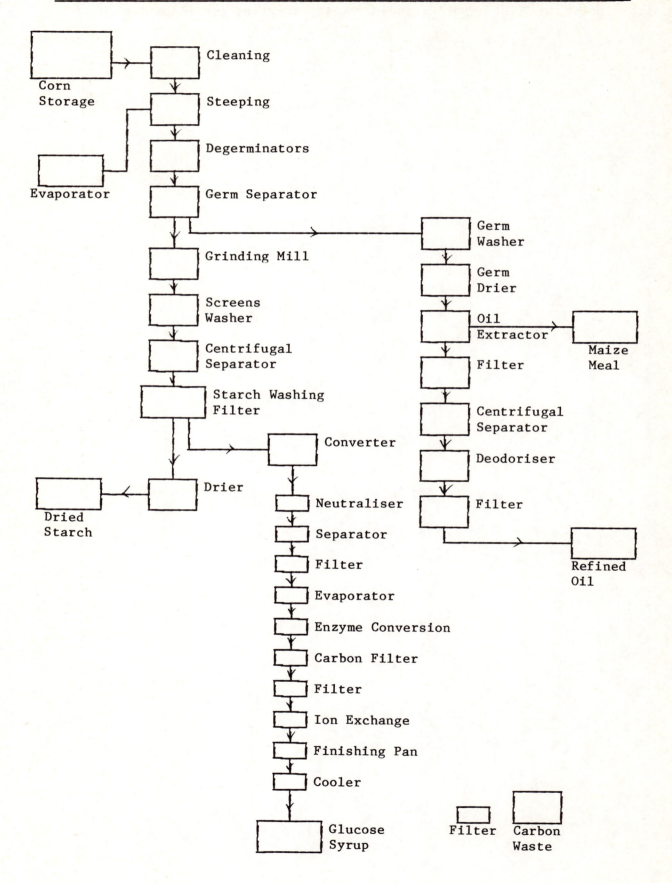

FIGURE 23: Industrial Processes in Pet Food Industry

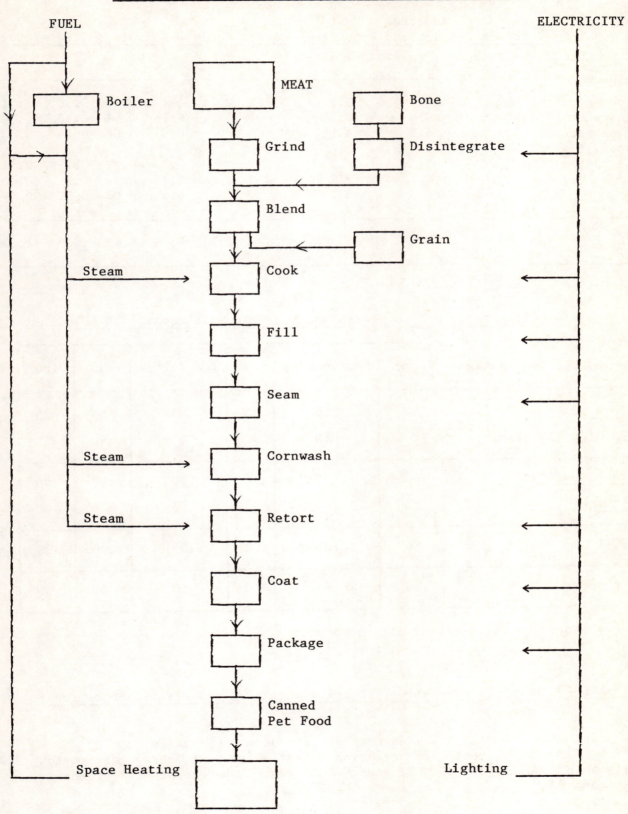

TABLE 41: Output and Energy Consumption in Starch and Miscellaneous Foods 1980

	Output	Average Practice	Best Practice	Other	Coal	Gas	Oil	Elec	Total
	10^3 tonne	GJ/tonne				10^6 GJ			
Starch, glucose	950.5	7.9	5.5	0.82	2.71	0.17	3.43	0.38	7.51
Farinaceous	294.9	4.4	2.1			0.16	0.91	0.23	1.30
Salt & stock cubes	119.8	6.9	4.1			0.01	0.68	0.14	0.83
Tea & coffee	229.8	2.2	1.2			0.23	0.12	0.16	0.51
Yeast	128.0	4.7	3.8			0.18	0.21	0.21	0.60
Other	80.18	7.1	5.7			0.43	0.07	0.07	0.57
TOTAL	1,803.18			0.82	2.71	1.18	5.42	1.19	11.32

TABLE 42: Energy Use in Starch and Miscellaneous Foods 1980 (%)

Energy Use (percentage)	Starch	Farinaceous	Salt	Tea/Coffee	Yeast	Other
Process	90	55	55	45	55	55
of which electricity	10-20	40	40	95	60	25
Space Heating	5	35	40	50	40	40
Other	5	5	5	5	5	5
TOTAL	100	100	100	100	100	100
% through boiler	85	60	60	60	50	60

Some private generated electricity also used.

TABLE 43: Energy Conservation Potential in Starch and Miscellaneous Foods Sector 1980

Energy consumption if all establishments operate at best practice:
Starch: 5.23 x 10^6 GJ, energy conservation potential as % of energy consumption: 30%
Miscellaneous Food: 2.33 x 10^6 GJ, energy conservation potential as % of energy consumption: 38%

TABLE 44: Output and Energy Consumption in Compound Animal Feed and Pet Foods

	Output	Average Practice	Best Practice	Other	Oil	Gas	Elec	Total
	10^3 tonnes	GJ/tonne				10^6 GJ		
Compound foods & concentrate	10,932.8	0.56	0.41	0.06	3.97	0.17	1.92	6.12
Fish Meal	78.02	3.18	2.51		0.24		0.01	0.25
Supplements	92.26	0.43	0.35		0.02	0.01	0.01	0.04
Pet Food	954.71	2.3	1.91		1.37	0.22	0.61	2.20
	12,057.79			0.06	5.60	0.40	2.55	8.61

TABLE 45: Energy Use in Compound Animal Foods

Energy Use (percentage)	Compound Foods %	Fish Meal %	Supplements %	Pet Food %
Process Energy	85	80	85	80
of which Electricity	50			
Space Heating	10	15	10	15
Other	5	5	5	5
% through boiler	65	80	80	80
	100	100	100	100

TABLE 46: Energy Conservation Potential in Animal Compound Foods Sector 1980

Energy consumption in 1980 if all firms operated at best practice
= 6.53 x 10^6 GJ
Energy conservation potential = 2.07 x 10^6 GJ
Expressed as % of total energy consumption in 1980 = 24%.

TABLE 47: Output Structure and Specific Energy Consumption in the Miscellaneous Foods Sector 1980-2000

(a) Structure of Output
     (percentage)

	1980	1990 H	1990 L	2000 H	2000 L
Starch, glucose	)				
Farinaceous	)				
Salt & stock cubes	) 13	11	12	9	11
Tea & coffee	)				
Yeast	)				
Other	)				
Compound foods & concentrates	)				
Fish meal	) 87	89	88	91	89
Supplements	)				
Pet food	)				
	100	100	100	100	100

(b) Specific Energy Consumption
     (GJ/tonne)

	Average	Best	Average	Average	Best
Starch, glucose	)				
Farinaceous	)				
Salt	) 6.27	4.18	5.65	5.02	n.a.
Tea & coffee	)				
Yeast	)				
Other	)				
Compound foods	)				
Fish meal	) 0.71	0.54	0.64	0.57	n.a.
Supplements	)				
Pet foods	)				

5.8   THE TOBACCO INDUSTRY

384.   The tobacco industry cures and prepares imported leaf; cuts the leaf; and prepares it for pipe, cigar or cigarette manufacture; finally the cigarettes and tobacco are packed ready for re-sale.

385.   Energy consumption in the tobacco industry varies widely depending upon the range of functions performed.   Some establishments are integrated works, other specialise in different phases of the industrial processes.

386.   Most imported tobacco is dry and highly compressed.   The bales are loosened and moistened by water sprays and steam to make the leaf pliable for cutting.   The cut tobacco is then dried to a controlled moisture content before manufacture by high speed machinery and final packaging.

Energy Consumption and Conservation, 1980

387.   Electricity for motive power is important (about 25-30% of total energy) because of the high speed machinery.   Other process energy and space heating (which is about 30% of the total) is by steam, raised by oil fired boilers, although most of the drying is performed by indirect heating in separate gas fired dryers.   About 70-80% of all energy is passed through boilers; about 70% of all energy is process energy.   Temperatures are not high - less than 120°.

388.   Some of these proportions vary: where curing is the main activity then 90% of energy passes through the boilers; where manufacture of cigarettes using bought in cut tobacco is the main activity then electricity is over 40% of total energy, and space heating accounts for perhaps 80% of boiler use. Tables 48 and 49 show the details.

389.   Energy consumption in 1980 has been estimated based on average practice data supplied by the Leatherhead Food Research Association based upon visits made by them under IETS activity.   The visits comprise a sample of 13 of the 41 establishments in the industry and accounted for about a quarter of numbers employed and a third of physical output.   Production output for 1980 has been taken from the Business Monitor.   Taken together this provides estimates of 2.076×10^6 GJ.   This is close to the Leatherhead Food RA estimate.

390.   Applying best practice data agreed with the RA gives energy consumption in 1980 of 1.22×10^6 GJ suggesting an energy conservation potential of around 0.85×10^6 GJ which as a percentage of energy consumption in 1980 is around 40%, see Table 50.

Energy Conservation Measures in the Tobacco Sector

391.   The IETS report makes several recommendations which include:

. good housekeping measures - reducing draughts, closing doors, better insulation of pipes and buildings

. more effective use of compressed air

. replacement of boiler plant over 25 years old

. replace burners

.   install exhaust gas analysers

.   heat recovery from gas fired tobacco dryers.

Market Developments to 2000

392.   The output of the tobacco industry fell less than the production of cigarettes during the 1970s because of a move to longer length cigarettes. Output (in tonnes tobacco) increased from 133,000 in 1972 to 149,000 tonnes in 1978 where it has remained broadly unchanged (150,000 in 1980). Output and sales fell back in 1981/2 during the recession (cigarette production has fallen by over 20% since 1980) and a recent survey by Health Education Council suggests that smokers are now a minority of the (adult) population.

393.   The trend in output is probably downward but in a high growth scenario this cannot be certain if income permits people to indulge a long established (or new) habit. Whatever the direction of output the rate of change is likely to be slow.

394.   SEC in Table 51 is shown to fall by about 15% to 2000; this assumes that about half of the potential is taken up and further implies some investment to recover the process energy associated with curing and preparation of the leaf. Recovery of energy for space heating which accounts for most energy in the cigarette manufacturing plants may take some time to achieve because tobacco is not energy intensive when measured in GJ/total purchases (see Table 1).

395.   No estimates of how best practice might change have been prepared so no estimates of conservation potential in 2000 are possible.

TABLE 48: Output and Energy Consumption in the Tobacco Industry 1980

	Output	Average Practice	Best Practice	Oil	Gas	Elec	Total
	10^3 tonne	GJ/tonne			10^6 GJ		
Pipe, tobacco and cigars	14.022	25.1	14.0				0.35
Cigarette tobacco	135.768	12.7	9.0	1.33	0.16	0.58	1.72
	149.790			1.33	0.16	0.58	2.07

TABLE 49: Energy Use in the Tobacco Industry 1980 (%)

	Pipe Tobacco & Cigars	Cigarettes
Process	75	40
of which motive power	30	60
Energy raised as steam of which	80	50
Space heating	20	50
Other	5	10
	100	100

TABLE 50: Energy Conservation Potential in the Tobacco Industry 1980

Energy consumption if all establishments operated at best practice levels
= 1.22 x 10^6 GJ
Energy conservation potential = 0.85 x 10^6 GJ
As % of total energy consumption = 41%.

TABLE 51: Output Structure and Specific Energy Consumption in the
Tobacco Industry 1980-2000

(a) Structure of Output (percentage)	1980	1990		2000	
		H	L	H	L
Pipe tobacco & cigars	9	9	9	9	9
Cigarette tobacco	91	91	91	91	91
	100	100	100	100	100

(b) Specific Energy Consumption (GJ/tonne)	Average	Best	Average	Average	Best
Pipe tobacco & cigars	25.1	14.0	23.50	20.10	n.a.
Cigarettes	12.7	9.0	12.07	11.43	n.a.

REFERENCES

1. Digest of UK Energy Statistics 1981, Department of Energy, HMSO 1982.

2. Energy Audit Series No.3, The Dairy Industry. Issued jointly by the Departments of Energy and Industry, 1978.

3. Energy Audit Series No.8, The Brewing Industry. Issued jointly by the Departments of Energy and Industry, 1979.

4. Energy Audit Series No.15, The Malting Industry. Issued jointly by the Departments of Energy and Industry, 1981.

5. Energy Use in the Food Manufacturing Industry. A G Crawford and C R Elson, Scientific and Technical Surveys No.128. Leatherhead Food Research Association, January 1982.

6. Energy Use in the Meat, Fish, Fruit and Vegetables Processing Industries. IETS Report No.32, Department of Industry, 1983.

7. Energy Use in the Tobacco Industry. IETS Report No.31, Department of Industry, 1982.

8. Energy Conservation Research, Development and Demonstration, Energy Paper 32, Department of Energy, HMSO 1978.

9. 'A Survey of Energy and Water Usage in Liquid Milk Processing.' W F Elsy. Milk Industry 82 (10), 18-23, 1980.

10. 1979 Census of Production and Purchases Inquiry, PA 1002.1. Business Statistics Office, HMSO 1983.

11. Heat Recovery by Heat Pumps in a Dairy. ECDPS project file number 34, Department of Energy. October 1983.

12. Energy Conservation Demonstration Projects Scheme, Demonstration at Magness and Usher Ltd, the use of insulated trucks for wholesale milk delivery, ETSU Report F/18/82/16. 1982.

13. UK Dairy Facts and Figures. Federation of UK Milk Marketing Boards 1982.

14. Innovation in the Food Industry, ACARD, HMSO 1982.

15. Energy Saving in the European Dairy Industry. M E Seligman, Dairy Industries International, June 1982.

16. Energy Saving Techniques for the Food Industry, M E Casper (Ed). NOYES 1977.

17. Business Monitor PQ 218 (fruit and vegetable); PQ 214 (meat and fish) 2nd Quarter, HMSO 1982.

18. Symposium on Biscuit Baking Ovens - Types of Biscuit Ovens, E Humphreys, Simon Vicars Ltd, Leatherhead Food RA 1983.

19.   Symposium on Biscuit Baking Ovens - Power and Efficiency Considerations, W R Mowbray, Baker Perkins Ltd Leatherhead Food RA 1983.

20.   Scottish Grain Distillers. Training Course for Energy Coordination. Derek Sampson and Partners, Glasgow 1982.

21.   Energy Utilisation in Breweries, R W Gordon. Paper to EBC Congress Brewers Society 1981.

22.   Brewing Survey: Financial Times, April 25, 1983.

23.   British Beer: a complex and changing market. C Thurman, Brewers Guardian, June 1983.

24.   Energy Conservation in Oil Refining, B F Brooks. Journal of American Oil Chemists Society, November 1978.

25.   Vegetable Oils and Fats. Unilever Education booklet revised ordinary series no.2.

ENERGY USE AND ENERGY EFFICIENCY IN UK MANUFACTURING INDUSTRY

UP TO THE YEAR 2000

SECTOR 6. ENGINEERING AND ALLIED INDUSTRIES

A V Ward

	Paragraph
Introduction	1
Industrial Processes in Engineering Industry	5
Energy Consumption in the Engineering and Allied Industries	9
Energy Conservation Potential in 1980	29
Energy Conservation Measures	38
Market Changes to 2000	43
Technical Changes to 2000	45
Energy Conservation to 2000	52
Energy Conservation Potential in 2000	56
References	

SECTOR 6. ENGINEERING AND ALLIED INDUSTRIES

Introduction

1.   Engineering and other metal trades accounted for around 23% of energy use in industry in 1980. Because of the particular importance of the non-ferrous metals industries in this study (because of their energy intensity) they have been separated and are treated separately in Sector 1. The engineering and allied industries, less non-ferrous metals, therefore accounted for around 18% of energy use in industry in 1980 on a heat supplied basis.

2.   This figure covers an enormous diversity of products, industrial processes and markets covering some 72 sub-categories within the 1979 Fuel Purchases Inquiry data. These categories range from forgings, pressing and stampings, through cutlery, textile machinery, office machinery to motor vehicles, photographic equipment, clocks and watches. Some aggregation of the 72 subsectors is appropriate to form eight main sectors. These are mechanical, electrical and instrument engineering, ship building, motor vehicles, aerospace, other vehicles and metal goods.

3.   Fortunately, although the products and markets are diverse, the pattern of energy use across the engineering industry is much more uniform. The studies conducted by the Production Engineering Research Association (PERA) and ERA Technology (the Electrical RA) under the IETS scheme [1-7], show clearly and consistently that around 55% (the range is 40-80%) of all energy use across the engineering industry is for space heating. If heat losses in site services are included, this figure rises to around 60%. Process energy is a small proportion (30-35%), mostly electricity for motive power and compressing air but also including some direct heat in those sectors that heat metals during manufacture. The balance of energy use is made up of lighting and other site services which include losses from the steam mains, condensat return, compressed air and probably includes some unidentified auxiliary heating.

4.   The later IETS studies refer directly to the period 1979-1981 whereas most of the earlier studies refer to visits made between 1976-1978. To avoid the problems associated with variable base years, the 1979 Fuel Purchases Inquiry data have been used to apportion total energy consumption between the various fuels. Informed judgement from ETSU, ERA and PERA has been used to establish consistent specific energy consumption figures from the data over the period 1980-2000.

Industrial Processes in Engineering Industry

5.   No account of industrial processes is given in this study because firstly they are not as important in energy terms (process energy being a small percentage of the total) and secondly (and more importantly) because they are complex and diverse (72 subsectors). Thus a flow chart indicating energy use at each point in the industrial process would not only be complex, but would be repetitive given the high proportion of energy use in space heating and site sevices. A brief note, however, is given in the following paragraphs giving the similarities and differences in process energy use in the various sectors of the engineering industry.

6.   The similarities may be seen in the industrial processes associated with metal working. These include:

- forging,

- heat treatment including annealing, carburising, etc.,

- plating,

- paint finishing,

- stove and vitreous enamelling.

All these processes include heating, using direct burners (mostly gas, but some oil or electric) and cooling - typically in immersion baths of hot/warm/ cool water. These processes may be found mainly in mechanical engineering and vehicle sectors but are also present in instruments and electrical engineering products, e.g. stove enamelling of electrical domestic appliances.

7.   The differences in processes and energy use between manufacturing precision instruments like clocks, and assembling vehicles are obvious. The scale of operations and number, kind and order of processes including assembly are different. It may also be noted that the techniques and processes for plating, and paint finishing vary widely in degree, kind and specification between metal windows, tools, vehicles and scientific instruments. Stove enamelling for vehicles, machinery, and domestic electrical appliances will have different specifications.

8.   It follows that no discussion of industrial processes or detailed account of how energy is used in such processes is given here. It is, however, much more important to examine the use of energy for space heating, site services and boiler uses, paying particular attention to heat recovery from process heat in those sectors of the engineering industry that use heat treatment and dry paint or enamel.

Energy Consumption in the Engineering and Allied Industries

9.   Table 1 shows energy consumption in the engineering industries in 1980. Energy consumption has been determined as a product of the number of employees in each sub-sector in 1980 and the specific energy consumption expressed as GJ/employee in that sub-sector. For completeness a measure of energy intensity has also been included in this table, GJ/£k. It is worth dwelling on this question of units of measurement because of the diverse nature of the industries in this sector.

10.   One of the most difficult tasks in this study of the engineering industry has been to decide on the units of measurement. The IETS records of the various sectors of the industry show that there is little doubt that energy consumption is most closely related to the size of the building and the number of shifts operated. So there is no doubt that the best single measure to use is GJ/(m^2 x hours worked). When plotted, such a relationship gives a good fit to the data collected (see [2]).

11. This conclusion is not surprising when it is recognised that over the whole of the engineering industry, space heating accounts for around 2/3 of the total energy used; and space heating, very broadly, is some function of the size of area to be heated. There are exceptions, for example, in the tractor manufacture or jewellery and precious metals subsectors, the energy requirements for process heat in the working and heat treatment of metals is a high proportion (70-80%) of total energy consumption and hence the size of the building is not closely correlated with energy consumption. Similarly, efficient modern buildings or old Victorian two and three storey buildings with uninsulated roof lights etc. distort the picture. But, broadly speaking, across most sectors of the engineering industry, space heating dominates energy consumption and is thus related to the size (and quality) of the building and the number of hours worked (i.e. 1, 2 or 3 shift working).

12. However, it will be recalled that this study is required to project energy consumption to around 2000; and since it is not possible to obtain estimates of the area of the buildings occupied by the engineering industry in 2000, $GJ/(m^2$ hours worked) is not used in this study as a measure of energy efficiency.

13. It will be readily apparent that GJ/tonne of output is not a meaningful measure of energy efficiency either, because of the immense variety of products within the industry. A tonne of bearings has little in common with a tonne of computers or a tonne of ships or aircraft. The IETS data provides no such measure, anyway and hence it is not used in this study.

14. The choice of unit therefore, lies with either GJ/employee or GJ/£k sales. It turns out that the Thrift Scheme data does not include measures of GJ/£k sales because, quite reasonably, this measure is seen as simply another measure of output which has a very weak relationship with energy consumption. So, this study is constrained to use GJ/employee as being the only measure available from the IETS data which can be used both for 1980 and for the year 2000. The measure is also broadly linked with the size of the building because, not unreasonably, the more employees employed the more space is required for them to operate. So, although more variance is observed in the GJ/employee data than in $GJ/(m^2$ hours worked) data both measures do broadly correlate with energy use in buildings. It also seemed a useful measure because forward estimates of employment to 1990 and, by extrapolation, to 2000 are available.

15. However, it turns out that this measure, although adequate for making comparisons across firms in 1980, is much less satisfactory for comparing establishments over time to 2000.

16. This is because specific energy consumption expressed on the basis of GJ/employee is a quotient formed by division of two numbers which can and do vary independently. It will be readily appreciated that if an establishment sheds its labour force quickly then, assuming that the buildings remain heated, the GJ/employee figure may well <u>rise</u> sharply. Equally, if the establishment introduces major energy saving changes such as insulation, ventilation controls, heating controls, heat recovery etc then the energy (in GJ) used will fall and, assuming the labour force remains unchanged, the GJ/employee measure could <u>fall</u> sharply.

17.   Specific energy consumption on the GJ/employee basis could rise or fall over time depending on:

-   . how effectively the firm controls space heating and any process heat recovery that is available;

-   . whether the employment-output relationship recovers. At present labour has been shed at a rapid rate reflecting sharp falls in output. If output recovers there is a question of how much extra output could be produced before more labour is taken on;

-   . whether, in the longer term, robotics, numerically controlled machine tools, micro-processor controls, and the move to continuous production will substitute capital for labour and thus reduce employment per unit output.

18.   If action to improve energy efficiency proceeds at a <u>faster</u> rate than labour is shed (due to both short-term redundancies and longer-term substitution by capital) then lower energy efficiencies in terms of GJ/employee would result. If, for whatever reason, employment was to fall faster than action on energy efficiency then higher GJ/employee figures would result in the future.

19.   There seems no definitive answer to which outcome is the most probable although the best guess available is that energy consumption in GJ will fall more slowly than employment over the period to 2000. This takes the view that major savings in space heating are available to most establishments of around 20-30%, most of which can be achieved without significant use of management time or capital inputs. However, although labour has already been shed at a high rate since 1980 (around 15-18% 1979-81) further expansion in value of output is expected to be accompanied by further falls in employment to 1990. A total fall in employment 1980-90 in engineering as a whole is estimated at around 18-19% by Cambridge Econometrics [8]. Further evidence for this view comes from the Warwick University manpower forecasts [9]. Although the figures in the two studies are different they both point to increases in output accompanied by significant falls in employment. In the period 1990-2000, Cambridge Econometrics suggest [reference 8, Table E11] falls in employment in mechanical engineering (24%), electrical engineering (24%) and metal goods (19%).

20.   Although some off-setting increases in employment might be expected in instrument engineering or electronic engineering it is difficult to see major gains in shipbuilding or motor vehicles. Of the individual sectors only instrument engineering shows an increase in employment in both forecasts to 1990 although vehicles also shows a small increase in the Warwick forecast.

21.   The conclusion to be drawn from this information seems to be that there is a high probability that employment in engineering will fall faster than energy consumption to 2000 even assuming that the whole of the 20% or more o potential saving in energy consumption is achieved by 2000. It follows that the GJ/employee figure will <u>rise</u>. This makes interpretation of specific energy consumption difficult. It seems contradictory that specific energy consumption expressed in any measure should rise by 2000 rather than fall following successful attempts at energy conservation measures particularly

when at least 20% potential saving has been identified. Therefore some other measure of energy efficiency and output must be used.

22. The only other measure of energy efficiency available to this study is GJ/£k sales. This measure is not particularly sensitive because, as already noted, output in the industry is not closely related to energy consumption since some 2/3 of energy consumption is for space heating and therefore is largely determined by degree-days, size and quality of buildings and the number of shifts worked. Output in many cases is only remotely related to energy consumption. Nevertheless, this is the only available measure of energy efficiency which is likely to show a fall between 1980 and 2000. This therefore is the measure chosen.

23. Unfortunately, the IETS data does not record GJ/£k sales for 1980. The measure was therefore constructed from the estimated energy consumption for each sector given in Table 1 and the value of output for the sector from the Annual Abstract of Statistics [10]. Each sub-sector, therefore, shown in Table 1 has a measure of GJ/£k sales for 1980 and this is the basis for the projections of specific energy consumption in the industry to 2000.

24. Table 1 shows energy consumption in the engineering industry in 1980 as 304×10^6 GJ. The Digest of UK Energy Statistics [11] gives 395×10^6 GJ but this includes the non-ferrous metal industries which are addressed separately in Sector 1 and consumed about 91×10^6 GJ. The estimates of total energy consumption in the engineering and allied industries less non-ferrous metals obtained in this study are therefore in good agreement with other estimates.

25. The data shown in Table 2 shows the shares of the main fuels based on the 1979 Fuel Purchases Inquiry data. This source was preferred as some of the IETS data refers to the period 1976/78. The IETS data for some of the sectors, notably instrument engineering, are however more recent (1979/81) and therefore the fuel shares are directly comparable. It is interesting to compare the two sources of data in the other sectors also. The observed pattern was as expected: illustrating a further substitution of gas for oil over the period 1976 to 1979. Electricity remains at much the same percentage in those industries which were observed in both 1976 and 1979 and any major switch to coal is not apparent over the period 1976-1980. Where the data are directly comparable (i.e. both refer to 1979 or 1980), then the shares of fuels recorded are very similar indeed.

26. In mechanical engineering, shipbuilding, airways and aerospace, oil is still the major fuel accounting for around 45% of total energy on a fuel delivered basis; gas accounts for around 35%; and the balance is made up of coal; while electricity takes around 13-18%. Gas has been substituted for oil at a faster rate in metal goods, vehicles, electrical and instrument engineering taking about 40-45% of the market with oil 25-35%. Electricity accounts for 15-30% of energy requirements, in general, with rather more than in the mechanical engineering area.

27. Table 1 includes the percentage of energy used for space heating. It shows that around 55% of fuel is used in space heating over the whole of the engineering and allied industries. The losses of heat from steam mains and condensate return and other site services (which include lighting) account for

a further 5-10%. The balance of around 35-40% is accounted for by process energy use. Of this process energy, much will be direct heat for metal working and paint finishing using gas, but sometimes oil and electricity. There is also a significant use of electricity for powering air compressors which, in many areas of the engineering industry is one of the main consumers of electricity. In some establishments electricity for motive power will be important. There are variations, of course, as the table makes clear: some sectors do little heat treatment and hence have an even smaller proportion of process energy e.g. sound reproduction equipment. Others, like wire and wire products, jewellery and precious metals use significant quantities of heat for metal working.

28.   Motive power does not seem to be a significant overall energy use: of the 43 sub-sectors of the engineering industry in only five did electricity account for more than 30% of total energy use. In these sub-sectors (jewellery and precious metals, cutlery, shipbuilding, electronic computers and radio components) electricity is used for welding and other forms of heat treatment in addition to any motive power requirements. In computers and radio electronic components, electric motors could well be an important use of electricity. Elsewhere in the engineering industry, electricity for motive power seems to be much less important than might be expected.

Energy Conservation Potential in 1980

29.   Using the methodology explained elsewhere in the study, energy conservation potential in 1980 is measured as the difference between estimated energy consumption in 1980 and that energy consumption which would have occurred had all firms operated at best practice. This methodology requires a measure of specific energy consumption for best practice.

30.   Such a measure was obtained from the data recorded on IETS visits. Following inspection of the visit reports, a figure was agreed with officers at PERA and ERA which represented best practice. Such a figure represents an informed view of how much energy a well-managed establishment actually used in 1980 compared with like establishments producing like products. It is a specific energy consumption which all firms or establishments in the sub-sector could, in principle, achieve within 15 years or so. It also represents a level of energy efficiency at which many firms already operate.

31.   The effort required by industry to achieve best practice is not negligible but neither does it imply an overturning of current financial allocations, management time or investment practices. The achievement of best practice will require modest resources and a realisation that space heating, although a relatively small part of total costs, is not trivial in absolute terms. Energy savings can be readily achieved with space heating and these may make a useful contribution to profits. Best practice figures for specific energy consumption reflect conservative judgements about what is reasonable for all the sub-sector to achieve.

32.   In the following paragraphs, the steps taken to establish the best practice figures for each sub-sector are described. The starting point was in each case the data showing mean values of energy consumption per employee in each industry. The IETS visit reports were then examined to ensure that no extreme values have significantly biased the mean values. This figure

represents total energy consumption in the sample of firms within the industry divided by the numbers employed in the sample of firms.

33.   The next step was to observe the range of specific energy consumption values in GJ/employee across the sample of firms visited within each sub-sector. Best practice figures were established by selecting the lowest value which in the judgement of ETSU, PERA and ERA could be achieved, in principle, by all firms. This was not the lowest recorded since in all cases this was associated with some special circumstances or practices such as product mix or location which made it impossible for other firms to emulate.

34.   The procedure involves considerable judgement and no claims to high precision are made. It is the industry view that it is unlikely that the results are seriously in error and that they are probably accurate to about ± 10-15%. Of course, larger samples or more detailed disaggregation combined with more detailed accounting of energy consumption by firms would permit a more accurate measure to be established.

35.   The size of the samples in the IETS visits varied across the industry from 63 visits in the electrical engineering sub-sector to 215 visits in the metal goods sub-sector. These figures represent about 10-20% of all invitations sent out to establishments with more than 25 employees. The samples could be biased towards those firms more eager to improve their energy management. Some of the data is relatively old (1976-1978) and requires updating although later reports cover 1979-1981.

36.   The samples, at the minimum, accounted for around 12% of the total employment in the mechanical engineering sub-sector and, at the maximum, 21% of the vehicles sub-sector. They usually covered more of the larger firms (i.e. those that together accounted for 25-50% of the total employment) and less of the smaller firms. The samples are considered to be adequate for our purpose and the techniques employed rely heavily on judgement of the industries concerned.

37.   The results of this analysis are given in Table 1. If the entire engineering industry adopted best practice the total energy consumption would amount to 241×10^6 GJ which suggests a conservation potential of 63×10^6 GJ or 21% of 1980 energy consumption.

Energy Conservation Measures

38.   A full range of measures are laid out in some detail in the Thrift Scheme reports. Taken together the main points are:

Process energy

.   insulation of metal heating containers including relining or veneering of furnaces with ceramic fibre which will reduce heat losses, and speed start-up times;

.   heat recovery from the exhaust gases of plant such as furnaces, ovens, paint dryers etc. including use of heat pumps to take advantage of exhaust heat recovery;

. use of floating plastic balls to reduce evaporation of immersion
   tanks used in heat treatment and plating.

Space heating

. insulation of buildings;

. reduce excess ventilation;

. strip covers for access bays, loading doors, etc;

. better controls of heating times and temperature.

Boiler

. regular maintenance;

. boiler controls;

. sensing devices.

Site services

. lagging of pipe runs, steam leaks etc;

. air compressor. In many engineering establishments compressed air is
  the largest single use of electricity. Few companies carry out tests
  of the air lines despite the fact that a hole as small as 0.4 mm at
  7 bar gauge wastes energy at the rate of 0.36 MJ per hour;

. more efficient use of motive power. This is not only important for
  compressed air equipment but also for other motor-driven equipment.
  Measures include switching off plant when not in use; sizing motors
  for the individual tasks and levels of utilisation; developing high
  efficiency motors and variable speed drives.

. power factor correction is important. This is because electricity,
  although only around 15-20% of energy on a heat suppled basis, can be
  up to 1/3 of energy costs. The installation of power factor
  correction equipment should be considered wherever the power factor
  is less than 0.9. Maximum demand may be reduced by staggering the
  start-up of plant which draws on a large initial electrical load and
  by scheduling the throughput of work to even out or shed loads where
  possible. Also power factor correction units may save energy as well
  as total costs.

39. The relative importance of these measures are in order, space heating,
boilers and site services. Most of the energy saving activities require
only modest management time and relatively little capital investment
(replacement of boiler plant is a separate question). However, when capital
is in short supply it is likely that energy cost saving ventures have low
priority in management problems.

Developments to 2000

40.   Several factors combine to suggest strongly that not too much should be made of this section of the study. First, it is apparent from paragraphs 9 to 22 that the measurement basis of the data presented in Tables 1 and 3 is rather weak because energy consumption in the engineering industry is not closely related to the value of output.

41.   Second, the wide range of products represented by each of the five main groups within the engineering industry make it very difficult to generalise about prospects to 2000. Yet, in the space available it is not possible to consider each sub-sector in any detail.

42.   Third, and perhaps most importantly, energy consumption in the engineering industry has more to do with the efficient use and conservation of heat in buildings than with technical change in products or production processes. In this sense, this paper can note the main points only and concentrate on the rate at which the measures discussed in paragraph 38 will diffuse through the industry.

Market Changes to 2000

43.   The most recent account of the prospects for the engineering industry over the next decade come from the NEDC [12]. The main points are:

- consumer products (which includes domestic electric appliances) are expected to expand slowly despite the effects of import penetration;

- production engineering equipment, components and specialist machinery have been dependent on demand from UK customer industries which has fallen away in the last decade. Demand has fallen in part due to the UK recession but also to a lack of competitiveness from the supply sector. Two trends in this sector are expected: first a shift to international sourcing of all these categories of equipment and components; and second a trend to install production systems which will permit batch production items to be made available 'with the cost economies normally associated with large volume production'.

44.   It should also be noted that the evidence from the Institution of Mechanical Engineers to the House of Commons Industry and Trade Committee included the point that 'the marketing strength in terms of both quantity and quality of the British machine tool industry is generally extremely poor'. The production index of UK machine tool industry has fallen steadily from 100 in 1975 to 80.9 in 1979 to 45 in 1983.

Technical Changes to 2000

45.   The introduction of robots and microprocessor controls over industrial processes may be expected to bring some small increase in electricity usage; but this could well be offset by the saving of energy in the more closely controlled time, temperature and speeds associated with automatic controlled systems.

46.   Other features of the next 20 years are likely to be mainly extensions and refinements of the points made in paragraph 38. That is:

- improvements in thermal insulation and heat recovery from furnaces and ovens;

- extended use of recuperative burners;

- more efficient drying of the paints and enamels;

- more efficient use of compressed air plant;

- some small extension of electricity as a source of direct heat with associated electronic control systems.

47. Most attention, however, must focus on the extension of more efficient use of space heating systems. The main points have already been made:

- improved heating control by means of thermostats, time controls, etc;

- insulation of buildings;

- improved maintenance of boilers and boiler control systems;

- alternative heating systems where appropriate.

48. These improvements require only modest management time and capital expenditure. Replacement of boiler and process plant is another question. The economic returns on such good housekeeping or bolt-on, or maintenance expenditure can be very high so there is no reason to suppose that companies would prolong such expenditure for many years once they were aware of the potential returns.

49. It is therefore our judgement that perhaps 70% of all potential savings occur by 2000. This is based on the view that engineering firms will need to be low cost producers in order to improve competitiveness and that it will be recognised that savings in energy, although small, actually represent a significant contribution to profits. As this is realised and as the returns to be made on small good housekeeping-type investments are realised action will be taken at a fairly fast rate. This judgement also recognises that not all firms perceive the benefits to be worth the expenditure.

50. One of the most important methods by which this activity might be stimulated is the introduction of a targeting and monitoring scheme. These have been introduced in other industries (see Paper and Board) apparently with some success and a move by the engineering industries would be a major stimulus to efficient use of energy.

51. The major advantages appear to be:

- identification and recognition by senior management of where and how much energy is used in the establishment;

- regular monitoring of the effect of measures so that senior management can see, monthly, the value of savings;

. adjustment of targets furthers the achievement of past targets so clear evidence of gains is available.

The targeting and monitoring scheme seems very effective as a means of establishing energy efficiency measures nearer the top of management priorities.

Energy Conservation to 2000

52. It will be recalled from paragraph 23 and Table 1 that energy consumption in 1980 was recalculated and expressed in GJ/£k sales. The main purposes of this recalculation was to allow the use of estimates of future specific energy consumption to be comparable with 1980 data.

53. The projections of specific energy consumption are also shown in Table 3. These have been estimated by assuming that 70% of the best practice potential in 1980 will be achieved evenly over the period to 2000. This assumption of a constant rate of diffusion is based on the view that most of the savings are in space heating and these measures require both recognition of the savings to be made and the conviction that it is worth acting to achieve them. This point often occurs at the time when buildings or boilers are being refurbished or when production lines are reorganised or when building services are relayed.

54. The changes in specific energy consumption are assumed to be accompanied by general growth in the industry (even if not in terms of employment) and changes in the structure of the industry. Electrical engineering and instrument engineering are both expected to make larger contributions to the output of the industry over the next 20 years, probably at the expense of the mechanical engineering and metal goods manufacturing sectors.

55. Table 4 shows the fuel use in 2000 for the whole industry. This table shows the effects of some small switching to coal and gas and increased use of electricity. These are very tentative estimates, however.

Energy Conservation Potential in 2000

56. There are no estimates of best practice in 2000 because there is no way that the area of buildings occupied by the engineering industry can be estimated; and there is no measure of by how much best practice in 1980 might improve by 2000.

Table 1. Employment, Output, Energy Consumption and Energy Use in the Engineering and Allied Industries, 1980

	Numbers Employed, 000s	SEC GJ/employee		Energy Consumption, 10⁶ GJ		Gross Output, £M 1980 prices	SEC GJ/£k gross output (average)	Space Heating %
		(average practice)	(best practice)	(average practice)	(best practice)			
Agricultural machinery	27	86	69	2.32	1.86		8.93	58
Metal working	61	56	45	3.42	2.74		4.00	70
Pumps and valves	83	103	75	8.55	6.22		10.84	51
Construction machinery	39	134	105	5.23	4.09		5.42	75
Industrial engines	26	67	54	1.74	1.40		2.95	66
Textile machinery	20	70	56	1.40	1.12		5.10	58
Handling equipment	58	83	66	4.81	3.83		3.97	68
Office machinery	20	89	72	1.78	1.44		12.36	62
Misc. machinery	199	80	60	15.92	11.94		14.14	63
Fabricated steel	135	92	74	12.42	9.99		8.38	65
Other mechanical engineering	189	86	70	16.25	13.23		2.93	57
Total mechanical engineering	857			73.84	57.86	19,170	3.85	
Photographic	11	48	38	0.53	0.42		2.05	55
Watches and clocks	9	78	64	0.70	0.58		9.59	50
Scientific instruments	94	43	34	4.04	3.20		2.75	70
Other instruments	26	45	36	1.17	0.94		3.73	65
Total instrument engineering	140			6.44	5.14	1,782	3.61	
Electrical machinery	126	75	60	9.45	7.56	1,809	5.22	66
Insulated wire	40	155	124	6.20	4.96	863	7.18	66
Telegraph equipment	68	55	25	3.74	1.70	940	3.978	63
Radio & elec. comp.	121	25	20	3.03	2.42	1,456	2.08	68
Gramophone records	17	230	184	3.91	3.13	135	28.96	10
Broadcast receiving & sound repro equipment	43	40	22	1.72	0.95	523	3.29	80
Electronic computer	44	73	58	3.21	2.55	1,022	3.14	
Elec. capital goods	101	30	20	3.03	2.02	1,652	1.83	70
Elec. accessories	20	100	80	2.00	1.60	709	2.82	68
Domestic appliances	59	90	72	5.31	4.25	944	5.63	43
Other electrical	80	87	70	6.96	5.60	793	8.78	66
Total electrical engineering	719			48.56	36.74	14,526	3.34	

Cont/d

Table 1. Employment, Output, Energy Consumption and Energy Use in the Engineering and Allied Industries, 1980
(continued)

	Numbers Employed, 000s	SEC GJ/employee (average practice)	SEC GJ/employee (best practice)	Energy Consumption, 10⁶ GJ (average practice)	Energy Consumption, 10⁶ GJ (best practice)	Gross output, £M 1980 prices	SEC GJ/£k gross output (average)	Space Heating %
Shipbuilding	⎫	69	⎫			⎫	⎫	62
Ship repairing	⎬ 149	122	⎬ 82			⎬ 2,070	⎬ 7.27	84
Boat building	⎮	53	⎮			⎮	⎮	86
Marine engineering	⎭	161	⎭			⎭	⎭	72
Total shipbuilding and marine engineering	149	101	82	15.05	12.22	2,070	7.27	
Motor vehicles	424	164	131	69.54	55.54	10,443	6.66	56
Tractors	34	134	115	4.56	3.91	948	4.81	30
M/cycles, cycles	12	45	38	0.54	0.46	143	3.78	57
Aerospace	198	91	73	18.02	14.45	3,890	4.63	50
Railways	44	130	104	5.72	4.58	344	16.63	61
Total vehicles	711			98.38	78.94	15,768	6.24	
Engineers' small tools & gauges	62	71	55	4.40	3.41	718	6.13	56
Hand tools	17	173	138	2.94	2.35	190	15.47	31
Cutlery	10	53	42	0.53	0.42	148	3.58	47
Bolts, nuts	28	127	102	3.56	2.86	336	10.59	46
Wire	34	262	209	8.91	7.11	917	9.72	34
Can & metal box	28	135	108	3.78	3.02	680	5.56	38
* Jewellery & precious metal	21	79	45	1.66	0.94			27
Metal goods	305	118	97	35.99	29.58	3,819	9.42	43
Total metal goods	505			61.77	49.69	8,452	7.31	
TOTAL ENGINEERING AND ALLIED INDUSTRIES				304.04	240.59			

- 13 -

Table 2. Fuel Use in the Engineering and Allied Industries, 1980

Sub-Sector	Percentage of Energy Consumption Supplied by Each Fuel, %				
	Gas	Coal	Oil	Elec.	Other
Agricultural machinery	42	1	36	12	9
Metal working	55	1	26	14	4
Pumps and valves			50		
Construction machinery					
Industrial engines	22	4	50	19	5
Textile machinery	17	22	45	15	1
Handling equipment	34	4	45	18	0
Office machinery			42		
Misc. machinery					
Fabricated steelwork	34	6	42	18	0
Other mechanical engineering	36	8	32	20	4
Photographic equipment	23	12	38	27	–
Watches and clocks	10	–	62	28	
Scientific instruments	38	3	32	27	
Other instrument engineering	31	3	39	27	
Electrical machinery	34	6	39	21	–
Insulated wire & cables	44	9	24	23	
Telegraph & telephone equip.					
Radio & elec. comp.	44	1	20	34	–
Gramophones					
Broadcast receiving & sound repro equip.					
Electronic computer	39	0	23	38	–
Elec. capital goods	32	1	46	21	–
Elec. accessories	48	0	32	20	–
Domestic appliances					
Other electrical engineering	37	32	37	26	–
Shipbuilding	24	1	43	31	1
Ship repairing	12	1	67	19	1
Boat building	8	–	75	15	2
Marine engineering	18	–	61	19	1
Motor vehicles	40	12	31	17	0
Tractors					
M/cycles, cycles					
Aerospace	34	7	42	17	0
Railway	17	25	48	9	1
Engineers' small tools	38	–	32	29	1
Hand tools	51	2	28	20	–
Cutlery	41	2	24	33	–
Bolts, nuts	52	2	20	26	0
Wire	53	2	27	17	–
Cans & metal boxes	53	–	28	18	–
Jewellery & precious metals	37	–	18	45	–
Other metal goods	35	5	41	18	1

Table 3. Output Structure and Specific Energy Consumption Assumptions
in the Engineering and Allied Industries, 1980-2000

(a) Structure of Output (percentage)

	1980	1990 H	1990 L	2000 H	2000 L
Mechanical engineering	31	30	29	29	28
Instrument engineering	3	3	3	3	3
Electrical engineering	24	25	26	27	28
Ship building	3	3	3	3	3
Motor vehicles	17	17	17	17	17
Aerospace	6	6	6	6	6
Other vehicles	2	2	2	2	2
Metal Goods	14	14	14	13	13
TOTAL	100	100	100	100	100

(b) Specific Energy Consumption, GJ/£k

	Average	Best	Average	Average	Best
Mechanical engineering	3.85	3.02	3.56	3.27	na
Instrument engineering	3.61	2.88	3.35	3.10	na
Electrical engineering	3.34	2.53	3.05	2.77	na
Ship building	7.27	)	6.79	6.31	)
Motor vehicles	6.66	)5.09	6.19	5.72	)na
Aerospace	4.63	)	4.31	3.99	)
Other vehicles	7.55	)	7.09	6.64	)
Metal goods	7.31	5.88	6.81	6.31	na

Table 4. Fuel Use in the Engineering and Allied Industries in 2000

		Percentage of Energy Consumption Supplied by Each Fuel, %			
Sector	Fuel:	Gas	Coal	Oil	Elec.
Mechanical engineering		35	20	15	30
Instrument engineering		40	6	23	31
Electrical engineering		38	15	15	32
Ship building)					
Motor vehicles)					
Aerospace) vehicles		30	25	15	30
Other vehicles)					
Metal goods		11	35	20	34
TOTAL		28	24	16	32

REFERENCES

1.   Energy Use in the Shipbuilding and Marine Engineering Industries. IETS Report No.4, Department of Industry 1979.

2.   Energy Use in Some Sectors of Mechanical Engineering Industry. IETS Report No.21, Department of Industry 1980.

3.   Energy Use in Cans, Cutlery, Engineers Small Tools, Hand Tools, Nuts, Bolts, Wire and Wire Products. IETS Report No.25, Department of Industry 1981.

4.   Energy Use in Motor Vehicles, Aerospace and Railway Industries. IETS Report No.26, Department of Industry 1981.

5.   Energy Use in Three Sectors of the Electrical Engineering Industry. IETS Report No.29, Department of Industry 1982.

6.   Energy Use in the Instrument Engineering Industry. Unpublished IETS Report No 50, Department of Trade and Industry. 1984.

7.   Energy Use in the Metal Goods Industry. IETS Report No 51. Department of Trade and Industry. 1984.

8.   Cambridge Econometrics Forecast Number 82/3, June 1982.

9.   Warwick University: Review of the Economy and Employment, Spring 1982.

10.   Annual Abstract of Statistics 1983, Edition No.119 CSO HMSO 1983.

11.   Digest of UK Energy Statistics 1981, Department of Energy, HMSO 1982.

12.   NEDC Paper 83(14). Summary of the Sector Working Party Assessments.

ENERGY USE AND ENERGY EFFICIENCY IN UK MANUFACTURING INDUSTRY

UP TO THE YEAR 2000

SECTOR 7. THE TEXTILE, LEATHER AND CLOTHING INDUSTRIES

A V Ward

Subsector	Paragraph No
Introduction	1
7.1 Wool Industry	21
7.2 Cotton Industry	33
7.3 Knitted Goods Industry	55
7.4 Textile Finishing	64
7.5 Jute and Carpet Industries	69
7.6 Miscellaneous Textiles	79
7.7 Synthetic Fibres	93
7.8 The Footwear, Leather and Clothing Industries	103
7.9 Summary	111
References	

7. THE TEXTILE, LEATHER AND CLOTHING INDUSTRIES

THE TEXTILE INDUSTRIES

1. The textile industries in 1980 together accounted for about 5% of the total energy consumed by industry. Although not a major energy user, the textile industries were some of the first to be examined under the Industrial Energy Thrift Scheme (IETS) operated by the Department of Industry. The first IETS report was on the woollen and worsted industry [1] followed by several others of the textile industries [2-8].

2. The textile industries have been divided up for the purpose of the present study into seven sectors:

    woollen and worsted,
    cotton spinning,
    cotton weaving,
    textile finishing,
    carpet and jute,
    miscellaneous textiles,
    man-made fibres.

These major sectors require further subdivision into subsectors so that energy use can be adequately described and the energy conservation potential can be assessed.

3. The individual textile industries are markedly different from each other and together produce an immense range of finished products. The industries include the spinning and weaving of woollen, worsteds, cotton and blended fibres (including man-made fibres) of many kinds; knitted goods; flax; rope and twine; carpets and jute; and textiles. The products range from gauzes to heavy duty industrial fabrics. Final products may have various finishes including:

    . waterproofing

    . oil proofing

    . stain resistance

    . shrink resistance

    . abrasion resistance

    . laminates with paper, board or metal stiffening.

4. The industries are made up of establishments which fall into two general types. In the first type of establishment, energy use is dominated by a process use of steam or hot water or both. Some of the steam or hot water is used for space heating but this is not usually particularly important compared with process energy use. In the second type of establishment the reverse situation applies and energy use is dominated by space heating and process energy use is usually unimportant.

5. Preparation and spinning of wool in the woollen and worsted industries provide a good example within the textiles industry of establishments which are dominated by steam/hot water distribution. The wool is washed and dried,

three, four, or even five times, in the preparation of spun and dyed yarn from raw wool. The combination of fairly high temperatures, several wet processes and the associated drying makes it clear why the industry uses over two thirds of its energy as process energy, most of it raised through boilers for steam and hot water. The steam and hot water is allocated between three or four rather different processes together with associated space heating and there are few studies which permit identification of how much steam and or hot water are used in each process.

6.   In contrast many establishments in the cotton spinning and knitting sectors perform only the single function of spinning or knitting. This is not particularly energy intensive and process energy accounts for only about 40% of total energy requirements with the remainder being taken up for space heating. In spinning cotton, air temperature and humidity is maintained at somewhat higher levels than would be required purely for comfort reasons but, nevertheless, the energy is used for space heating albeit at somewhat generous standards. In other sectors of the industry there are broad percentage splits of between 60-40 to 40-60 for process energy and space heating. The differences depend largely on how much finishing of the fabric or yarn or carpet is required since this in turn largely controls the number of wet processes involved.

7.   Process heat is therefore largely a function of the quantity and frequency of hot water usage which in turn depends on how much finishing is carried out in the establishment. It is clear that there is quite a diverse range in the pattern of energy use between the various establishments. These aspects are treated in some detail in later paragraphs of this Appendix.

8.   It seems useful to make three further general remarks before turning to a detailed treatment of each of the textile industries. First, most of the data on energy use and energy consumption has been taken from the IETS studies conducted by the Shirley Institute and the Woollen Industries Research Association (WIRA) [1-9]. These studies were conducted over a period 1977-1981 and some results, particularly those relating to fuel use, may be out of date. They have been compared with the Digest of UK Energy Statistics 1982 [10] and Fuel Purchases Inquiry data for 1979 [11] and a brief note on the changes in fuel use if any are made in each section. In general, the most noticeable change has been the continued switch from oil to gas, though oil is still the most important fuel in many of the textile industries.

9.   The second point is that there is a close relationship between the textile industries and laundries, in that similar processes, such as cleaning and drying are carried out in the two industries. This is recognised in the ECDPS which covers both industries. Only textiles are covered in this study, however.

10.   The third general remark refers to the problem of measurement. The IETS studies offer several measures: GJ/site; GJ/tonne; GJ/employee; and GJ/(m^2 x hours worked)*. The 'best' measure depends heavily on the kind of establishment and the nature of the product.

---

* 'm^2 x hours' is often used as a measure of the rate of occupied floorspace of a building, to primarily determine energy used for space heating purposes.

11. Where establishments are dominated by process use of steam or hot water or both, then GJ/tonne output is a satisfactory measure of energy efficiency or of the specific energy consumption. This covers most establishments in woollen and worsteds industry and textile finishing; few establishments in cotton spinning; but many establishments in cotton weaving, knitted goods, carpets, and household textiles where a significant number of finishing activities are carried out.

12. However, GJ/tonne as a measure of energy efficiency raises the problem of converting between the various official statistics of output which are expressed in tonnes, linear metres (of various widths), square metres (of various densities), dozens of pairs, or numbers of pieces (knitted garments). Although it is possible to convert these measures to weight,

- it is complex (many different weights and widths),

- liable to error because the proportions of output lying within a band of < 68 grams/m^2 or > 150 gram/m^2 is not always recorded, and

- given the variety of finished goods, it is not always helpful to compare GJ/tonne for all goods between gauzes and carpets.

Nevertheless, there may be a case for carrying out a detailed account of those areas of the textiles industry for which this measure is useful. Illustrative examples of GJ/tonne are shown in Table 1.

13. As it is too difficult to obtain energy efficiency measures based on GJ/tonne or GJ/(m^2 x hours) worked because such a measure is not available for the year 2000, an alternative remaining measure is GJ/employee. This measure is useful where energy use in the establishment is dominated by space heating because there is some broad correlation between the size of the building and the numbers employed; the measure acts as a proxy for GJ/(m^2 x hours) worked which is the 'best' measure for such establishments. But output bears only a loose relationship to numbers employed as labour is shed and products move to a higher technology/value added kind so that <u>tonnages</u> may fall but <u>value</u> rises.

14. Numbers employed can also give odd results when energy consumption in similar sized firms is very different because of the nature of the building (very old or very new) or because the industrial processes happen to be unusually labour intensive (making up garments) or capital intensive (dyeing/finishing). Nevertheless GJ/employee has been the measure used to assess energy consumption in 1980.

15. It was also thought that GJ/employee had the advantage that employment forecasts were available for 1990 and, although less refined, for 2000. It turns out that these forecasts are not only too aggregated (textile fibres and textiles n.e.s.) but show steady <u>falls</u> in employment to 1990 and beyond. This has peculiar implications for the estimation of Specific Energy Consumptions for 2000.

16. If employment falls more rapidly than energy consumption then the measure GJ/employee could <u>rise</u> to 2000; if energy consumption falls faster than employment, then the measure could <u>fall</u> to 2000. The details are

similar to those set out fully in Appendix 6 dealing with the Engineering and Allied industries. The main point to make is that the interpretation of the GJ/employee measure is difficult. An alternative approach based on the £ value of sales was employed as described in the following paragraphs.

17. The 1980 estimates of energy consumption based on GJ/employee were converted to GJ/£k sales by dividing by the value of output for each subsector in 1980 prices abstracted from the Business Monitor. Output value data for 1980 could then be used as a basis for forecasting specific energy consumptions to 2000.

18. For consistency this approach has been adopted for all sectors of the textiles industries. It is not ideal for some sectors such as woollens, but it seems the best way forward in what is a difficult area.

19. A general point relates to energy conservation measures. In four of the six industries considered here space heating accounts for over 2/3 of energy requirements and hence the opportunities for energy conservation are all to do with boiler efficiencies, insulation and control systems within the building. It seems sensible therefore to bring together all the conservation measures at the end of the discussion of the individual industries. Where process energy is a particularly significant proportion (woollen and worsteds and textile finishing) discussion of energy conservation measures are taken after the description of the industrial processes. This is because many of the conservation measures relate to heat recovery or improved technology and investment associated with those processes. We proceed then to an account of each of the textile industries.

20. Finally, Volume I (The Main Report) contains actual estimates of energy consumption in 2000 for the Textiles sector. Also contained is a discussion of the scenarios adopted in this study.

7.1 WOOLLEN AND WORSTEDS (SPINNING AND WEAVING) INDUSTRIES

21.   A basic definition of a worsted cloth is one made from fine quality
combed wool. This means that short fibres are removed by the combing process
before spinning so that the fibre distribution of the remaining wool will
enable a fine yarn to be spun. Fibre length is critical in the worsted
industry because short fibres are not suitable for the process machinery.
Fibre length is much less critical in the rest of the woollen sector.

22.   Within the industry the number and sequence of operations varies
according to the end product: material is recombed for high quality suitings
while for hand knitting yarns this operation is not necessary. Dyeing may be
carried out on tops, yarns, or fabric according to requirements. The energy
requirements will vary according to the nature and number of these processes:
so, to obtain an accurate assessment of energy consumption per tonne of
fabric, it is necessary to classify the final product and aggregate the
energy consumption for each process involved in its production. However,
given the wide variety of final products, from gauzes to heavy industrial
fabrics, and available data, this is not a practicable way of examining
energy consumption in the industry. In the woollen and worsted industries
four main sets of activities may be distinguished [1] and these are detailed
in the following paragraphs.

Wool Combing

23.   The raw wool is processed to produce what is called a top. This is a
ball made from a rope of long wool fibres parallel to each other and of
uniform length and diameter. Combing sorts out the short fibres (which are
sold off to the woollen industry) to leave a long rope of uniform raw wool.
The processes involved are:

- scouring (washing);

- carding (aligning the fibre in a smooth rope);

- combing (to remove short fibres);

- finish and gilling (which mixes and aligns the fibres to produce a
  top);

- synthetic fibre tops do not need scouring or back-washing for their
  production.

Worsted Spinning

24.   The tops from the wool combing processes may be dyed to give a coloured
yarn. The dyed tops are usually re-combed and back-washed and remade into
tops of coloured yarn. The spinning operations then attenuate and mix the
tops to produce a fine yarn. Twisting and winding operations produce yarns
in which single ends are twisted together (two fold yarn) and are then wound
on to a suitable 'package'. Some yarns are then dyed ('packaged dyeing') at
this stage. The yarn may also be set in an autoclave as part of the spinning
sequence to reduce snarling and twisting, when unwound. An autoclave is a
steel container in which the contents are subjected to steam under pressure

(often used for sterilisation in the food industry) rather like a pressure cooker.

Woollen Yarn

25.   The woollen industry has a different set of operations from the worsted industry because of the great variety of material with which it works. The main objective of the mill is to produce a yarn with a given performance at a competitive price; so much effort is put into selecting and blending a variety of new wool, re-claimed fibre (often non-wool fibre) and man-made fibres to produce the required yarn. The woollen components of the blend are scoured before mixing, and the fibre may be dyed prior to blending. The prepared blend is then carded to produce slubbing from which the yarn can be spun. Twisting and winding operations produce the required yarn. For carpet yarn, hank dyeing is normally preceded by scouring.

Fabric Manufacture

26.   The initial warp and weft preparation processes set up the loom for weaving specific fabrics. The fabric is again scoured and uncoloured fabric will be dyed at this stage. The finishing processes vary considerably depending on the fabric. All impurities such as grease, oil, dirt are removed and the structure of the fabric is then prepared by coating or impregnating the fabric as required (water-proofing, flame-proofing, backing, etc).

27.   Worsted fabrics are usually scoured again, dyed - if still uncoloured - and are passed to a tenter or stenter dryer. This machine stabilises the dimension of the fabric by holding the edges while the fabric is heated and dried. The fabric is then cropped to remove surface fibres and pressed. An additional high temperature decatising process (setting the fabric by steaming in an autoclave) is often given to worsted fabrics to reduce the possibility of shrinkage during garment pressing operations.

28.   It is apparent from this description of the main processes involved in the production of woollen and worsted fabric that process energy is a large part of the total energy requirements of the industry as is to be expected of a type two industry. Estimates from the Shirley Institute put the figure at around 70%. An example from the Audit [9] indicates the detail necessary for estimating the allocation of energy within the industry. Most woollen fabrics are blends: a 55/45 polyester/wool cloth for mens' suiting was examined. It was woven from a 2/30 worsted count (a measure of the thickness of the yarn) to a weight of 350 g/m^2. The total energy requirement was 281.5 GJ/tonne which can be broken down as follows:

Polyester fibre production	36%
Spinning including top dyeing	42%
Weaving	5%
Finishing	17%.

Pattern of Energy Use in Woollen and Worsted Industry 1980

29.   Table 2 shows the pattern of fuel use using the UK Digest and the 1979 Fuel Purchases Inquiry. These are preferred to the IETS data because they are more recent than the IETS surveys.

30.   Energy use in most areas of the wool industry is dominated by process energy (typically over 70% shown in Table 3) for which steam and hot water are supplied via boilers. Oil accounts for 41% of fuel consumption, coal 28%. These figures are similar to the IETS data for 1977 [1] although it is estimated that oil now takes a lower proportion. Electricity accounts for 17% of the total but is much higher in the wool spinning mills.

31.   The allocations of energy use between process energy and space heating shown in Table 3 were taken from the IETS visit reports where possible. They also were derived in part from the judgements of the Shirley Institute and WIRA. They are not precise but they reflect the practice which has been observed in the establishments visited. The same is true for the figures dealing with the quantities of boiler fuels.

Conservation Measures

32.   The implications of the observed pattern of energy consumption for potential energy conservation measures are clear and the options are:

- boilers and boiler controls,

- water usage,

- heat recovery from the various stages of woollen and worsted manufacture.

A full account of possible action is given in the Thrift Schemes [1] and Textiles Audit [9] but the main points to record are:

Boilers and boiler controls

- fit sensors to monitor oxygen,

- improve maintenance,

- check burners,

- lag pipe runs, check and correct steam traps.

Water usage

- controls over temperature,

- reuse where practicable.

Heat recovery

- heat recovery from water discharged from dyeing and wool scouring processes despite the high dirt or short fibre content of the effluent,

- heat recovery from dryer exhaust air,

- scope for direct gas firing for drying,

- examine scope for heat pumps.

7.2 THE COTTON SPINNING AND DOUBLING INDUSTRIES

33.   Cotton yarn varies in thickness and weight and the most suitable method for defining a yarn is in mass per unit of length. The industry uses two such measures. These are:

(i)   <u>TEX</u> which is measured in grammes per kilometres, and

(ii)   <u>Cotton count</u> which is a measure of the number of hanks of 840 yards length which weigh 1 lb.

The IETS study suggest that there are broadly four types of yarn used in the UK:

. carpet yarns and waste yarns,

. condenser type yarns,

. medium counts - suitings,

. fine counts - shirts, blouses, gauzes.

The Yarn Manufacturing Process

34.   The process of spinning is a joining together of short fibres (staple) by joining them from a fibrous mass and twisting them together. Spun yarns are produced from natural fibres, such as cotton, flax and wool, from man-made rayon fibres, from synthetic fibres such as polyester and from fibre blends. Spinning of man-made and synthetic fibres is clearly a different process from spinning of a natural fibre. Man-made fibres are spun by the extrusion of material through a spinneret into a coagulating bath or into a cooling atmosphere. The fibres are continuous filaments and yarns are referred to as continuous filament yarns. Filaments may be cut into short lengths, known as staple, for conventional spinning, either alone or in blend with other fibres.

35.   The preparation of the fibre for spinning depends upon the type of fibre. The raw material arrives at the mill usually in large bales; layers from several bales are fed into a bale breaking machine, thus blending the fibres, and the opening process reduces the material to a soft fluffy mass from which any impurities such as sand and leaf are removed (this is called trash). The material then moves to other machines which continue to loosen the fibres and remove further trash before the fibres are pulled by air suction against revolving cages where they form a sheet. The sheets are thick webs of fibres which are then rolled into laps which are similar to large rolls of cotton-wool, but with the fibre still entangled in the form of small tufts.

36.   The next process is carding, which disentangles the material and loosens the last traces of trash. The carding cylinder is covered uniformly with short wires with a density of several hundred points per square inch, which comb and tease out fibres from the lap so that the cylinder surface is covered with a uniform layer of fibres more or less parallel to each other. At the output end of the carding machine the fibres are stripped from the main cylinder to give a thick web, which is then fed through a tapered hole

and consolidated into a sliver or rope of fibres which is gently coiled into cylinder containers ready for further processing.

37.   The slivers are then drawn into a uniform linear yarn with parallel fibres by a draw frame. When fine cotton yarns are being prepared the fibre of the sliver may be further arranged by combing in which the cotton fibres in the sliver are better spaced and from which weak short fibres are removed. When a satisfactory sliver has been prepared, it can be further drawn down on a flyer frame to produce a much finer strand, or roving.

38.   Strictly speaking, spinning is the twisting together of the parallel-ised fibres to produce a yarn. However, the complete operation of the spinning frame involves drawing out the roving, twisting it and winding the twisted yarn on a bobbin. On modern machinery all these processes are carried out continuously by ring, flyer and bobbin and cut spinning methods.

39.   The maximum production speed of these spinning methods is limited by the fact that each twist requires a turn of the package. Open end or break spinning has been developed in recent years, in order to maintain higher processing speeds and further automation. The open-end spun yarns are often weaker than those spun conventionally and in practice it has been found that yarns up to about 30 cotton count (20 TEX and coarser) can be spun more economically on the open-end system than on the ring system though the open-end spun yarns require more twist than ring frame yarns for satisfactory use. UK production of yarn is currently estimated at about 85% ring-spun and 15% open-end yarn.

40.   A further stage of processing known as doubling is used where yarn strength is an important factor in the final product. Typical yarns which are doubled are sewing threads and quality fine yarns used in light shirt, or blouse fabrics. The process is simply that of twisting or plying yarns together, and is carried out on a simplified spinning frame [see 2 and 3].

The Pattern of Energy Use in the Spinning Industry 1980

41.   Table 2 shows the pattern of fuel use used in the above industrial processes based upon the UK Digest and the 1979 Fuel Purchases Inquiry data. There is no clear distinction between mills which spin only and mills which both spin and conduct finishing processes, such as dyeing. IETS data shows that when spinning only is carried out, much of the energy is in the form of electricity for motive power to drive the spinning frames, accounting for over 40% of energy usage. The table shows some 36% of the fuels used is oil, and the balance is made up of coal and gas. These fuels are used primarily to raise steam and hot water in boilers mainly for space heating purposes, although there is some washing of the final product.

42.   The IETS data shows that mills which conduct both spinning and finishing activities use much more water as the yarn is washed and dried several times in the course of finishing. It follows that boiler fuels are much more important in these mills and oil, which was the main boiler fuel used at the time of the survey, accounted for nearly 70% of total fuel used. Electricity accounted for only 16% of fuel use although in absolute terms it was about the same as in spinning only mills. The data from the UK Digest and the Fuel Purchases Inquiry gives 'average' results although it is possible to detect some move towards substituting gas for oil as gas now accounts for some 5% of fuel use as opposed to 1-2% in the IETS Report.

However, no distinction can be made within the 1979 data between spinning mills and mills which both spin and finish.

43.   Over the cotton spinning sector as a whole about 70% of energy was used for space heating; about 30% for process energy most of which is electricity for driving the spinning frames and air conditioning and a small balance was used for lighting and site services. The heat generated by the spinning frames is significant and reduces the space heating load in winter and increases the load on air conditioning in the summer. Site service losses have been included in the space heating figures.

44.   Where finishing trades are involved the requirement for steam and hot water is increased and the balance of space heating and process heating (and particularly the balance of electricity and oil within process energy) will alter. However, overall, the cotton spinning sector probably retains this balance of 70/30 between space heating and process energy.

45.   It is difficult to establish a useful measure of energy efficiency as the earlier discussion made clear. There are particular difficulties in using GJ/tonne measures in cotton spinning because of the very wide differences in the quality of the yarns produced. Quality or fineness of the yarn is measured in tex or in cotton counts which reflect the mass of yarn produced per unit length of yarn. Given very different qualities of yarn varying between sewing cottons and gauzes and heavy industrial yarns it is difficult to interpret a GJ/tonne figure. Unless very clear relationships can be established between the tonnage and the 'count' it remains difficult to reconcile the wide variance in GJ/tonne of yarn output which are 'averaged' in Table 1.

COTTON WEAVING AND FINISHING

46.   It is clear from the IETS visits that the sector is divided into those firms which weave only; those which produce industrial fabrics (which would include both weaving, spinning and finishing); and those which engage in finishing and garment making as well as weaving. Industrial fabrics include canvas, filter cloths, tile cords, and heavy fabrics for paper making and base cloths and for fabric coating sectors.

47.   Table 3 shows that where finishing is a significant part of the activities, then process energy is proportionally higher as a percentage of total energy requirements. Where weaving only occurs then the power looms use only electricity and hence most energy is used in space heating. It follows therefore that some account of industrial processes is important to understand those processes which utilise energy.

Industrial Processes

48.   A fabric is woven by the interlacing of two sets of threads, the warp and the weft. The warp lies parallel to the edges of the fabric and the weft is inserted at right angles to the warp. The threads to be used for weaving may be spun yarns of continuous filaments. Spun yarns must be cleared of any weak or uneven lengths and this may be done either at the spinning mill or by the weaver.

49.   The yarn which is to become the warp has to be prepared for the loom by first winding hundreds of threads simultaneously onto a cylindrical core in

such a way that they can be drawn off as a sheet of parallel threads. Each of these sheets is treated with size for protection against abrasion in the loom and is then rewound in parallel with a number of others to make a single broad sheet.

50.   In the loom the threads of the warp are drawn apart to allow the weft to pass between them. The weft may be inserted either as thread wound on a shuttle or as a thread transported by mechanical or fluid dynamic means in a shuttleless loom. The latter has the advantage that long continuous lengths of weft may be used because the whole supply of weft does not have to pass between the threads. In some looms with shuttles the supply of weft on the shuttle is replenished automatically.

51.   Weaving is a relatively slow way of producing fabric. Typical rates for fabric 36 inches wide are 3-4 inches/minute (8-10 cm/minute). In contrast, production by knitting may well be 4-5 times as fast as that of weaving. However the different methods of forming fabrics produce different characteristics and woven fabrics are usually much more stable dimensionally than many knitted fabrics.

52.   Finishing and the production of industrial fabrics is considered elsewhere, in the textile finishing sector. It is enough to note that most of the processes involve some coating impregnation or additional wet processing. The material may well have to be prepared (washed) so that the steam heating, washing and drying processes absorb significant quantities of energy.

Energy Use in Cotton Weaving and Finishing 1980

53.   Where weaving only is carried out in the mills, process energy accounts for around 30% of total energy requirements. This is mostly electricity which powers the looms together with air conditioning. Space heating is clearly important (around 70%) of the total and oil is still the major fuel used to fire boilers. Gas and coal are much less important - although gas seems to have made some impact in this sector accounting for about 22% compared with 8-9% at the time of the IETS visits in 1977.

54.   The relatively high temperature and humidity required for spinning operations means that it is difficult to distinguish between process energy and space heating in the use of boiler fuels. Temperatures between 20-28°C and humidities between 38-55% are typical conditions necessary for travel free spinning of most fibres. Normal space heating temperatures might be between 18°-21°C so certainly some proportion of heat and most of the humidity requirement might be allocated to process energy use. Air conditioning systems are designed to avoid pollution of the working atmosphere. There are legal limits for pollution control. Air conditioning systems either exhaust to the external atmosphere or have been recycled through filters. Where filters are used high powered fans are needed to overcome the pressure drop across the filters. Electricity is used predominantly to drive electric motors fitted to process machinery (opening units, card drawn frames and spinning frames) - the process heat generated makes a significant contribution to the energy required for space heating particularly if losses through openings are minimised. Energy conservation measures in both cotton spinning and weaving are mainly associated with buildings and space heating and are discussed in paragraphs 126-129.

7.3 THE KNITTED GOODS INDUSTRIES

55. Energy consumption in the knitting industry varies directly with the degree of water used in the industrial processes. Thus mills which operate knitting machines only are not very energy intensive and what energy is used is primarily space heating as the electrical drives for the machines absorb only modest quantities of electrical energy. Finishing involves wet processes typically washing dyeing and drying which consume much more energy (around 2-4 times as much). Thus in the IETS visits two groups of establishments may be identified - those that knit only and those which both knit and finish the goods.

56. Industrial activities may be divided into two broad categories:

   (1)   fabric knitting and

   (2)   garment knitting

which are shown separately in the output and trade statistics. Fabric is produced by both warp and weft knitting and is sold to be cut and made up. The knitting of garments is always by the weft knitting route and garment statistics appear either as pieces produced or by monetary value. It follows that within this sector it is difficult to achieve a GJ/tonne measure of energy efficiency which is meaningful. This is important because there are over 500 weft knitting establishments but only around 50 warp knitting units. These latter units are typically much larger than the weft knitting units because warp knitting machines need yarns on beams before they can be knitted and so require an intermediate warping activity. The machines are built in 84,126 and 168 cm width units and knit 100-3,000 metre lengths of fabric. Specialised handling equipment is needed to load machines with beams and unload the fabric so the plants are much larger. Because of the problems of handling long and wide fabrics, a finishing plant is commonly found on the same site and the end product is produced in the form of wrapped rolls of finished fabric which can be man-handled and are ready for garment making up.

57. Weft knitting machines need from 1-96 cones of yarn which come from the manufacturer ready for use to produce a fabric or garment. Knitted garments or fabrics can be easily man handled and removed from the kitting machine so the industry lends itself to activities in small units.

58. Garments and fabrics may then be further processed on site or forwarded to the textile finishing or garment manufacturer for making up. The main categories of the weft knitting industry include:

   .   hose and half-hose

   .   fully fashioned knitwear

   .   cut and sew knitwear.

59. Two types of knitting machines are in use; rotary continuous action machines for the manufacture of seamless hose, halfhose, and panty hose and reciprocating action straight bar machines which are employed exclusively in the small fully fashioned hose sector and for the fully fashioned knitwear

market. There appears to be little difference in the energy requirement of such machines. Fabric invariably requires wet processing and or finishing and this may be provided either on the same site or conveyed to specialist finishers. In the knitwear sector it is rare to have wet finishing facilities so extensive use is made of commission dyers and finishers.

60. A variety of dyeing machines are used depending upon the particular type of goods being produced. Dyeing processes vary in temperature requirement. At temperatures above 100°C as in the case of dyeing polyester, machinery must be pressurised. For fabric dyeing, atmospheric winch machines or pressurised partially or fully flooded jet machines are used. For articles like hose, half hose and garments most dyeing machines are of the atmospheric side paddle or rotary drum type. Yarn is usually dyed in pressurised machinery. Drying operations are energy intensive and vary considerably with the type of goods processed. They are generally dryed and heat set on stenters where the fabric is passed through a series of hot air chambers with careful control of temperature.

61. Scouring to remove knitting lubricants and other impurities from fabrics usually precedes dyeing and in some cases where natural fibres are used or where discolouration has occurred in heating processes, fabric may need bleaching. In addition to dyeing, colours may be applied by conventional printing or by transfer from pre-printed papers.

62. It follows that the energy requirements largely depend upon the degree of dyeing and finishing and this is reflected in the distinction made between knitted goods only and knitting and finishing indicated in Table 3.

Pattern of Energy Use in Knitting Industries 1980

63. Energy use in this sector is determined by the relative proportions of knitting and knitting combined with finishing establishments. Overall a significant proportion of finishing activity was found so that electricity use accounted for only 23% of total energy requirements. Most of this electricity of course will be used for the knitting machines. Finishing trades employ wet processes for which steam and hot water will be raised from boilers and these are fired by oil (21%) and gas (50%) with coal using 6% of total energy. A comparison with the IETS data suggests that substantial substitution between gas and oil has occurred. As in other sectors, the IETS Report notes that measures to control space heating have the greatest conservation potential (see paragraphs 126-129) but heat recovery from processes in finishing plant (eg from dye liquors) and better process control can offer nearly as much in energy-saving terms.

7.4 TEXTILE FINISHING

64.   Output of this sector includes yarns and fabrics, both woven and knitted, which require to be bleached, dyed, printed and to which a finish eg foam backing, water proofing etc, is applied. The range of finishing processes is immense and as energy consumption largely depends upon the kind of finishing used it is one of the most difficult areas to establish average and best practice data.

Industrial Processes

65.   The basic processes involved in producing finished yarns are broadly common to fabric yarn and fibres. The processes comprise:

- Preparation

  The fabric or yarn is cleaned usually by soaking in alkaline liquors containing detergents. The fabric may then be brushed, cropped to remove loose fibres and bleached again, typically in hot alkaline solutions of hydrogen peroxide to improve the colour. A further treatment in strong caustic soda solution may be applied to influence dye absorption and to enhance translucence of cotton fibres. When synthetic fibres are present in the fabric they may be heat set at this stage.

- Dyeing and printing

  The second stage is complicated by the presence of two different fibres such as in fabrics blended from cotton and polyester where two dyes and two stages of dyeing may have to be used, thus incurring a higher requirement for energy. Recent development in dyeing techniques however have made simultaneous dyeing of some types of fabric possible. An additional reason for large energy consumption is that fabrics containing polyester are dyed under pressure and at temperatures of around 130°C. The traditional method of printing with engraved rollers is still used but newer methods of printing using screens, either flat or in roller form, is increasing.

- Application of finish

  The third and final stage applies a finish. These vary immensely and will include one or more of the following:-

  - Resins for mimimum aftercare or for shrink proofing
  - Softeners to modify handling characteristics
  - Soil release agents
  - Flame retardants
  - Water repellents
  - Moth proofing or rot proofing finishes.

  These finishes are applied by controlled addition of the requisite chemicals.

The Pattern of Energy Use in Textile Finishings 1980

66.   As Table 3 shows, wet processes dominate energy use in textile finishings. Most energy passes through the boiler for steam and hot water raising with oil accounting for about 1/2 of total fuel use. Gas accounts for around 30% which is higher than in most other areas of textiles and electricity accounts for about 10% of total energy use which again is rather lower than elsewhere in textiles. This is a little surprising as there is a good deal of machinery in the sector which of course is powered by electric motors. However the proportions of wet processing to power machinery are such that electricity, despite its importance, accounts for only a small proportion of total fuel requirements. Coal retains about 10% of the market. Comparison with the IETS fuel use data suggest some shift towards substituting gas for oil.

67.   Table 3 shows that process energy dominates energy use in this sector although it should be noted that stack losses (20%) and the exhaust air and water vapour (a further 15%) have been included in process energy. Energy requirements for space heating as a percentage of total energy consumption is small at around 15%. A detailed picture of the industrial processes and the allocation of steam and hot water across the many different processes (see list of 21 different processes in Appendix 2 of [9]) is available in some extended energy surveys and from Shirley Institute data.

68.   A special use of heat is for 'setting' the material before dyeing by passing it at 210°C across a stenter which stretches the material (at correct tension) to the required width before drying and (after drying) the material is heat fixed on a stenter at 200°C to fix the dye. So, energy use associated with stenters is important. Detailed accounts of energy use are not available. The IETS report notes only that:

Net production	50%
Exhaust air	8%
Water vapour	7%
Stack losses	20%
Space heating and miscellaneous uses	15%

The Report also lists the following areas of opportunity for energy conservation: conversion to direct firing of bakers and dryers, heat recovery from wet processes and insulation of process plant, in addition to those mentioned in paragraphs 126-129.

7.5 CARPET AND JUTE INDUSTRIES

69.   The jute industry is related to the carpet industry because much of its output is used as backing for carpets. Polypropylene is increasingly used for carpet backing as well as for bags and sacks which represent the traditional output of the jute industry.

Carpets

70.   The industry is traditionally separated into the woven and non-woven sectors, a classification based on the method of fabrication. Non-woven floor coverings have a much higher rate of fabrication than woven and can be produced at a much lower cost per square metre.

Woven Carpets

71.   Competition based on the relatively high prices of woven carpets has caused this sector to decline steadily in the last decade as a proportion of total output of carpets. The principal fibres used in the facing pile i woven carpets are wool (50%) acrylic (29%) and polyamid (15%). Blends of these fibres, particularly wool with polyamid, are frequently used. The main process in manufacturing carpets is weaving although some mills also carry out a certain amount of spinning, dyeing and finishing of the fibres and yarns. The woven carpet is then backed by a secondary fabric and held by a latex binder. The non-weaving methods are relatively new and do not require traditional skills in carpet making.

Jute

72.   Major changes may be observed in the jute industry over the last two decades as polypropylene has been substituted for jute due to the unstable price of the latter material. Polypropylene has proved to be an excellent substitute for jute for most products (bags, sacks and similar containers).

73.   The manufacturing processes associated with jute are similar to those of other textile products - spinning the fibre into yarn; weaving the yarn; fabricating, dyeing and finishing although for many industrial purposes the material is not bleached or dyed.

74.   The manufacture of polypropylene yarn and the weaving and preparation of the fabric are more diverse. Fibre may be made by extrusion (molten polymer forced through spinnerets) or by fibrillating slit tapes or film. Yarn may be spun by conventional textile spinning methods or by relatively simple twisting. Fabric is usually woven using traditional textile materia machinery.

The Pattern of Energy Use in the Carpet and Jute Industries 1980

75.   As indicated in Table 2 the main fuel used in the jute industry is oil (nearly 45%) and the balance of boiler fuels is split fairly evenly between gas and coal - coal is important in the carpet and jute industry accounting for over 20% in the carpet industry. The 1979 fuel purchase data shows similar figures to the IETS data although, surprisingly, in the jute industry in 1979 there was less gas and more oil than that recorded for 1978 the year to which the Carpet and Jute Thrift report relates.

76.   Electricity usage is a high proportion – 30% in jute because of the
extrusion and spinning machinery for polypropylene yarn. Electricity is a
smaller proportion of total energy in the carpet industry (21%) because
proportionately more finishing (dyeing) occurs thus increasing the proportion
of boiler fuels for the total energy used.

77.   The main fuel used in the carpets industry is oil (34%) while gas and
coal account for 24% and 21% respectively. Nearly all this fuel is passed
through boilers for steam and hot water, most of which is used for process
(drying, washing, backing and fusion drying).

Energy Conservation Measures

78.   Full details are given in the Thrift Scheme report but the main points
relating to process energy are:

- heat recovery from hot effluent via plate or tube heat exchangers;

- maintenance of instrumentation on machinery particularly jute
  machinery, where dust reduced the efficiency of electric motors;

- insulation of hot water based tanks such as dye tanks and carpet
  backing plant.

7.6 MISCELLANEOUS TEXTILES

79. Miscellaneous textiles comprises three sectors: narrow fabrics; made up textiles; and other textiles.

80. A narrow fabric is classified as being a woven, knitted or braided fabric not exceeding 30 cm in width. Speciality yarns used in the manufacture of these fabrics are also included in this sector. The principal products are:

- covered rubber thread

- elastic thread

- woven and knitted elastic fabric

- tapes and webbings

- cut edge ribbons

- woven and printed labels

- non elastic braids

- boot, shoe, corset and similar laces

- trimmings including fringes and twist cords, tassels, etc.

- bias binding.

81. The major processes involved in the narrow fabrics industry are the production of covered rubber thread which involves the winding of thread, usually cotton, onto a rubber core, weaving, knitting, braiding, dyeing, printing and cutting. A wide range of synthetic and natural fibres are used in these processes to produce the products noted above.

82. Household textiles cover a range of products such as towels, mats, table cloths, sheets, quilts, sleeping bags, handkerchiefs etc.

83. The major processes employed by the industry are, weaving, dyeing, printing, finishing, and making-up. The major fibre used in the manufacture of these products has been cotton, and where products require absorbent properties as in towels, cotton is still used. The introduction of synthetics, however, has resulted in blends or 100% synthetics and this is used for many products such as bed spreads and sleeping bags.

84. Canvas goods include a range of products including sacks and bags, sailcloth tarpaulins, tents and marquees, which are heavy fabrics using blends of synthetic fibres, as well as natural fibres such as cotton and jute. Sacks and bags are now usually made from polypropylene, while tents, marquees and tarpaulins, include polyester, nylon, as well as natural cotton.

85. Asbestos is used in a wide array of products, including process fibre and rope lagging, asbestos yarn, asbestos cloth, sheeting, belting and

friction materials such as brake and clutch linings, joints and gaskets. For the purpose of presenting a complete view of the processes in textile industries, the following paragraphs cover the asbestos industries. However, their energy use and energy conservation potential are dealt with in greater detail under other building materials elsewhere in the report.

86. The major processes employed by the asbestos industry are spinning, weaving, blending, carding, roving, lapping and braiding. Asbestos is a mineral material which is a short staple, requiring blending with small quantities of other textile fibres with a longer staple length, to enable successful spinning to take place. The amount of blended fibre is carefully controlled so that the required properties of asbestos are not impaired. Asbestos does not dye readily but colour is not usually important in the end uses to which it is applied.

87. Typically, asbestos yarns and fabrics are used as safety clothing (flame resistant), laundry and dry cleaning press covers, conveyor belting, dust filters, brake linings, and seals.

88. Miscellaneous other textiles cover an array of products for filling materials, PVC coated fabrics for upholstery and a wide range of non woven products, from carpet underlays to medical and household applications. Typically the fibres are chemically bonded fibres, wet laid, dry laid or spun bonded, coated fabrics, needleloomed carpeting, and upholstery hair fibre filling.

The Pattern of Energy Use in Miscellaneous Textiles 1980

89. In most of these industries process energy accounts for around 60% of total energy requirements because of the need for significant dyeing, washing and finishing operations to the woven textiles. Nevertheless space heating requirements are not insignificant (around 25-40%) and most heat is raised through boilers.

90. The fuels used based on the UK Digest and 1979 Fuel Purchases Inquiry are mainly oil and gas with some small percentage of firms (around 10%) still using coal. Electricity is significant at over 20% used primarily for motive power for spinning and weaving machinery.

Energy Conservation Measures

91. There are significant opportunities for heat recovery from process use, including heat from dye liquors and washing water before discharging to waste and heat recovery from the drying ovens.

92. However, it is somewhat surprising that in the Thrift Scheme Report for miscellaneous textiles, the most important sources of conservation in all these industries, accounting for half of the total savings identified, are improved space heating controls and good housekeeping. It was noted repeatedly that thermostats were either incorrectly set or not available at all; that time controls were rarely fitted; that simple energy management practices associated with process controls were not observed; and that the building and boiler maintenance could easily be improved in order to save quite significant quantities of energy. All of these practices could be achieved with little management time and less expenditure.

7.7 SYNTHETIC FIBRES

93. Production of synthetic fibres falls into two categories:

  (i) Regenerated fibres for which the raw material is cellulose derived from wood or short cotton fibre waste, called linters. Wood pulp is dissolved in a solution containing caustic soda and carbon disulphide and the cellulose is generated by extrusion into an acid bath. Cotton linters are processed to give cellulose acetate which can be dissolved in organic solvents, such as acetone for spinning into fibres. Both processes require considerable energy and chemical inputs and requires substantial plant with chemical recycling and heat recovery for efficient operation.

  (ii) Synthetic fibres including nylon, polyester, acrylic and polypropylene are derived from oil based products requiring considerable additional expenditure of direct energy during chemical processing, polymerisation and spinning into filament form. Synthetic fibres account for a major share in the UK man made fibre industry, as they have many desirable properties both in their continuous filament form and when cut into short staple for blending with natural and regenerated fibres. The production of synthetic fibres varies between manufacturers but there are three basic routes, which use polymer chips or salts. These are:

  (a) dry spinning in which a polymer solution in a solvent pours through a series of tiny holes in a spinneret. On contact with warm air the solvent evaporates and the polymer solidifies into a continuous filament yarn;

  (b) wet spinning in which the polymer solution is forced through the holes into another solution where the polymer is coagulated into a continuous filament yarn;

  (c) melt spinning in which a solid polymer is melted and poured through the holes of the spinneret into cool air and the reaction solidifies the polymer into a continuous filament yarn.

94. After extrusion through the spinneret, fibres intended for staple are subsequently drawn to improve their strength and orientation in tows of up to several million filaments and treated with an appropriate finish prior to cutting and baling. Where continuous yarn is required, only a small number of filaments (5-50) are extruded from each spinneret and each yarn must be drawn, processed and wound individually. This increases machinery and energy requirements compared with staple yarns.

95. The fibre or yarn is supplied to the textile industry either in filament or staple form and it can be textured or left untextured. Filament yarns are taken from the spinning process and wound onto a textile package. Where staple fibre is supplied the production from one hundred or more spinnerets is collected in rope form, crimped and cut into the desired length of spinning on conventional short staple spinning machinery.

96.   Several processes may be applied to fibres either before production or after the yarn has been spun to produce appropriate characteristics required by the weaver or knitter. These modifications include adding substances to the polymer such as dye, changing the size and shape of the holes in the spinneret and varying the condition of extrusion. Texturing changes the physical characteristics of the yarn to give increased or (decreased) bulk and some degree of stretch. The normal form of texturing is a false twist, which applies a temporary twist to the filaments which is then set permanently into the yarn by a heater. POY (partially oriented yarn) is an increasing part of the output of synthetic fibre because high speed spinning can often provide sufficient orientation to avoid the drawing stage.

The Pattern of Energy Use in Synthetic Fibres 1980

97.   As might be expected energy use is dominated by process heat. About 80% of total energy requirements is used for process heat most of which is supplied through a boiler to give steam and hot water. In some large integrated plants some CHP provides a proportion of electricity. The major fuel used is oil (69%) but coal accounts for 10% of the total. Little gas is currently used (7%). Electricity requirements, some of which as noted above are met by CHP, account for 14% of total energy input. These proportions are taken from the 1979 Fuel Purchases Inquiry and are somewhat different from those found on the IETS scheme visits. There, no coal was found in any of the seven plants visited, and (including the oil equivalent of purchased steam from adjacent sites) nearly 80% of the fuel use was oil. Gas however accounted for some 13% which is twice the quantity noted in the 1979 Fuel Purchases Inquiry. The site visits were carried out in 1981.

Energy Conservation Measures

98.   The principal source of energy loss in processes was radiated heat from the spinning heads. This could be significantly reduced simply by insulation of the machine heads and mouldable insulation is available which could achieve a saving of around 15%.

99.   There are opportunities for changing from batch processing to continuous processing but the capital investment involved is high and it requires some confidence that the increased throughput from a given number of hours operation can be solved. In current market conditions there is little incentive to move to continuous production.

100. Heat recovery is possible from washing and drying operations. Washing removes processed impurities from the filament yarn or staple fibre and waste water from the washing ranges is discharged to the drain in large quantities at temperatures of around 60°C. Heat exchangers could raise the temperature of incoming warm water to the level required for washing. Plate type heat exchangers suitable for this application could recover up to 70% of the waste heat. Further large savings are possible by recovering waste heat from drying ovens. The dryers are usually continuous steam heated hot air ovens, and it is feasible to obtain heat recovery from exhaust air by using heat exchangers, although this operation is limited to filament tow because staple fibre will foul the heat exchangers' surfaces and increase fire risk. Suitable systems are available which have nominal heat recovery factors of around 60%.

101. Space heating is a small part of total energy requirements but
nonetheless is one of the easiest to control. Improved insulation of
buildings, control on boilers and improved maintenance and control of
thermostats around the buildings were all noted which could lead to
significant reduction in space heating energy requirements.

102. The cost of energy in the synthetic fibre sector is around 7% of total
production costs. Good housekeeping could save 1-2% without much
expenditure. Heat recovery from the process heat involves some bolt-on
investment but nonetheless should give significant returns.

7.8 THE FOOTWEAR, LEATHER AND CLOTHING INDUSTRIES

Introduction

103. This section is based solely on the IETS report on footwear, leather and fur industries [12]. No visits or other information were available to supplement this evidence so the conclusions are much less robust than other areas in the textiles sector. Nor is there any account of the industry or industrial processes.

Energy Consumption and Conservation in Footwear, Leather and Clothing Industries 1980

104. The fuel use data (percentage use of fuel) was taken from the 1979 Fuel Purchases Inquiry and the proportions have been applied to the 1980 estimates of total energy consumption. Oil accounted for some 54% of energy consumption with gas 22% and electricity between 9% and 14% depending on the sector. These figures are similar to those presented in the IETS report although in the intervening period up to 1980 gas has increased slightly at the expense of oil. Table 7 gives the detailed estimates.

105. Like many other areas of the textile industry the pattern of energy varies widely depending upon the degree of process heat used. The leather sector is dominated by leather tanning which has a very high proportion (65%) of process energy while the leather goods subsector (which is small in absolute terms) uses energy mainly for space heating. In both sectors the heat is supplied by steam or hot water raised through a central boiler.

106. The footwear and clothing subsectors are mainly concerned with making up garments and footwear. This largely comprises cutting out, making up with sewing machines together with some cleaning, polishing and pressing of the completed article. Process energy, therefore, comprises mainly electricity for motive power with some small quantity of heat for steam pressing and cleaning if garments are given a final clean and press. In general, space heating dominates energy use in this sector.

107. The methodology used to prepare these estimates is as follows: an SEC figure was obtained from the 1979 Fuel Purchases Inquiry (energy consumption divided by net output in GJ/£k sales) which was then applied to the net output figure for 1980 to give an estimate of energy consumption in 1980. Table 7 sets out the details and shows that total energy consumption was about 15.3×10^6 GJ in 1980.

108. No evidence is available on which to estimate a 'best practice' SEC so the estimate of energy conservation potential published in the IETS report has been used. They suggest a potential saving of 33% of energy consumption observed in footwear and 30-53% in leather industries in 1977/78. These figures are assumed to apply for 1980 and form the basis of the improvement in SEC which is further assumed to progress smoothly to 2000 in the high scenario when all of the 30% improvement should have been achieved. Most of the improvements are expected to occur in space heating so these gains could well be achieved ahead of 2000 (perhaps by 1990) but it seems best to be conservative about rates of penetration. Table 9 shows the assumptions made concerning specific energy consumption through to 2000. Thus, the energy conservation potential in 1980 in leather, leather goods, clothing and

footwear is estimated to be about 4.3×10^6 GJ. No estimates for energy conservation potential in 2000 have been made.

109. Energy conservation measures include: heat recovery and/or heat pump drying in leather tanning - low temperature heat pump drying could save as much as 23% of a tannery's total energy; improved space heating controls - saving 3-9% of total energy use, depending on the establishment; boiler replacement and building insulation - these two measures together could save nearly 15% of energy used in the footwear and clothing sector, where space heating is important.

110. The estimates of fuel use in 2000 are based on a number of judgements. The assumption is made that a further switch from oil to gas will occur. There will also be some small penetration of coal particularly in the leather tanning area which uses significant quantities of boiler fuel for process energy. Elsewhere, coal is assumed to penetrate only slowly because of the proportion of energy used for space heating. There seems no reason why electricity should increase in absolute terms because it is not apparent that any direct use of electricity for space heating or for process energy is likely to develop. However, in a relative sense because much of the conservation occurs in boiler fuels it may be expected that electricity would improve its share of the market. Table 10 provides the details.

7.9 SUMMARY

Energy Conservation Potential in the Textile Industry

111. Conservation potential is measured as the difference between energy consumption as estimated for 1980 and the energy consumption which would occur had all firms been operating at best practice levels. This methodology requires some assessment of best practice in each of the sectors within the textile industry.

112. The data was obtained from the summary reports, and, where necessary, inspection of the original visit reports made by consultants employed for the IETS scheme visits. These are held at the Shirley Institute and discussions with staff at the Shirley Institute provided an agreed set of average practice and best practice data.

113. It should however be noted that assessing energy consumption, particularly best practice or average practice energy consumption in some areas of the textile industry is extremely difficult. This is because for those establishments which have a high process energy requirement, that requirement is set by the characteristics of the yarn or cloth, which in turn is determined by the expected market. Sales are often 'ordered' in the sense that clients (typically those in clothing and retail establishments) order yarn or cloth or fabric of certain characteristics and it is these characteristics which determine the nature, kind and frequency of the processes used and hence the energy requirements.

114. Production related energy requirements are further complicated by the high energy costs associated with finishing. One of the points not made thus far is that the industry is highly fragmented horizontally: that is, specialist processes are carried out by quite separate establishments rather than being integrated in a single works. This means that most finishing establishments and some weaving establishments do not own the cloth they process but do it on commission to specifications laid down by the owner (often the spinner or weaver who in turn is producing the yarn or cloth for a known client). In these circumstances it is difficult to establish close control over the industrial processes because they are deliberately varied in order to achieve characteristics in yarn, cloth, colouring, finish etc, to specifications which are not within the manufacturers' control.

115. This makes it very difficult to compare the energy efficiency in terms of GJ/tonne or GJ/employee for different production runs because they will vary quite significantly because of the differences associated with the finish or other characteristics of the yarn or cloth being processed. Monitoring and targeting and other means of establishing better energy management is thus a formidable task in the textiles industry.

116. The Shirley Institute have carried out a number of studies of energy use in finishings and other areas of the textiles industry and have achieved some success in analysing the determinants of energy use. The number of variables which determine gigajoule per tonne or some other measure of energy use have been reduced to output, degree-days (which affects both incoming air and water temperature as well as determining space heat requirements), the quantity of water used, and the length of production run. Time series data for the plant shows quite significant seasonal variations in output as well

as clearly affecting energy requirements due to the degree-days factor. It is also possible to isolate the production and non-production (space heating) requirement, albeit rather crudely. The time series data available can be fitted to these variables and regression analysis allows some assessment of the relative importance of the variables used. The fits are quite good and some modest prediction of energy use for given levels of output, temperature and other variables have been carried out at a number of establishments.

117. The first step taken was to identify 'average' practice as the mean energy efficiency on a GJ/employee basis calculated by Shirley Institute or WIRA for each industry group. The data for individual mills or establishments were checked to ensure that the industry averages were not significantly distorted by any atypical mills with extremely high or low energy efficiency values. In almost all cases, the published mean values were acceptable.

118. The results of this inspection and discussion are shown in Table 1 where, for each subsector of the textile industry, average practice expressed in gigajoule per employee is shown. Gigajoule per tonne is important in a number of the textile sectors and average practice is shown on this basis also. The number of employees in each sector in 1980 is given for information.

119. Establishing 'best' practice was more difficult for two reasons: variety in the mix of output and the sample sizes available. The full sample of firms in each subsector was observed using the GJ/employee or GJ/tonne measures and the records of those firms with the lowest values were examined. It usually turned out that they were not representative of the subsector for some reason. They perhaps performed only one function or carried out little in the way of finishing operations or were unusually small or large establishments. It turned out from the judgement of ETSU, Shirley Institute and WIRA that in most subsectors, firms or establishments with values around the 25-30 percentile were not significantly different in product mix, range of functions, technologies, size and location from most of the other companies in the sample. In principle all the firms in the sample could achieve these levels of energy efficiency. These values could then be taken as a measure of 'best' practice. These judgements have been examined within the industry and found to be sound and are presented in Table 1.

120. The degree of precision in the estimates of 'best' practice is difficult to assess. Within the sample, the range of values is such that the error is likely to be $\pm$ 10-12% or better. Overall it is most unlikely that any figures are in error by more than $\pm$ 25% in the judgement of the Shirley Institute, WIRA and ETSU.

121. The sample sizes vary between seven visits in synthetic fibres to 111 visits in carpets and jute. Mostly, the samples include about 50-60 sites. The sample is self-selecting: firms are invited to participate in the IETS scheme but had positively to elect to do so. Most did not, so the samples of firms or establishments are typically around 15-20% of those with more than 25 employees. The samples may be biased towards the more energy conscious. In all other respects, ie size, location and production, the samples appear to be representative. They typically account for about 15-40% of total output or employment in their respective subsectors.

122.   The results of this analysis are therefore judged to be soundly based and to have adequate coverage. They accord well with the industrial experience of the Shirley Institute and WIRA. So, despite some weaknesses, the results are taken to be a reasonable account of energy consumption and use in the industries in 1980, with an uncertainty of about $\pm$ 10-12%.

123.   The effort that would be required by the industry to achieve 'best' practice is not negligible. Neither, however, does it imply a revolution in either technology, investment, management practices or closures. What would be required is what would be expected to happen over the next 15-20 years:

   . a replacement of older plant,

   . some increases in management attention to energy matters, particularly in buildings.

A number (20-25%) of all establishments already operate at or below this level of energy consumption per unit of output. The achievement of 'best' practice by most firms will take time and some may not achieve it, but it is possible.

124.   Energy consumption in 1980 is estimated as 105×10^6 GJ while energy consumption if based on best practice would be 80×10^6 GJ. The energy conservation potential in 1980 therefore, is estimated as 25×10^6 GJ which, expressed as a percentage of energy consumption in 1980, is 24%.

Energy Conservation Measures in the Textile Industry

125.   Much of the energy conservation measures associated with process heat have already been discussed under those sectors which use substantial quantities of energy for process use. For completeness it may be noted that such measures largely are of the following kind:

   . heat recovery from waste water (particularly from washing, scouring and dyeing effluents);

   . heat recovery from drying ovens;

   . improved insulation of process plant particularly vats, drying ovens, hot water tanks, etc;

   . improved process management - minimising start-up times, turning off process equipment when not in use, particularly ovens;

   . improved process controls, eg moisture meters to control the degree of drying.

The rest of this section refers to conservation measures associated with space heating.

126.   The proportion of energy devoted to space heating varies considerably from less than 20% in high process using industries such as synthetic fibres to 80% in type four establishments such as cotton spinning establishments. All establishments had boilers and in some of the sector IETS reports in

around 40% of sites visited, boiler replacement was recommended. Other measures were:

. lagging pipe runs;

. condensate return;

. improved installation of the boiler and boiler house;

. sensing equipment to maintain efficiency;

. improved and more regular maintenance;

. insulation of the building.

127. One other general recommendation which is not made forcibly in the earlier Thrift Scheme Reports but features in some of the later Thrift Reports is the importance of targeting and monitoring systems. Much was made of these systems in the Paper and Board Report and to a lesser extent in Food, Drink and Tobacco. They are seen to be of major importance in the textiles industry and there are good reasons why this should be so.

128. The two major advantages of targeting and monitoring schemes are first, that although senior management time is required in the establishment of the scheme, once running, management time is not required. Secondly, it is relatively inexpensive to establish and operate a targeting and monitoring scheme. Meters and other control and measuring devices are usually already available and operating. Labour costs are modest depending upon the size and complexity of the operations in recording and writing monthly accounts of energy costs and energy usage.

129. The major benefit associated with targeting and monitoring, apart from the actual observed savings which occur, lies in the regular reporting of energy use and costs to senior management. This keeps energy matters in a high profile position as far as senior management is concerned and thus improves the chances of more major items of plant and equipment which may be sought on the capital budget for energy conservation purposes. Overall, the important point about targeting and monitoring is that it raises the level of energy consciousness of senior management.

Recent Developments in the Textiles Industries 1975-83

130. It is well known that the textiles industry has suffered substantial reductions in output and employment in the last decade. 'Rationalisation' of the industry in the 1950s failed to prevent further rounds of cutbacks during the 1970s and the current recession has only reinforced the trends towards a smaller industry with more efficient, and usually more capital intensive units of production.

131. The severity of the recent changes are shown in Table 4. The industry has lost over half its spindles (spinning capacity) since 1971; and weaving capacity has fallen by nearly one third. Comparing output figures in 1975 and 1981 (both years of trough in the short-term business cycle) spinning yarns fell by 44% while weaving output fell by 40%; while in the woollen and worsted sectors output fell by around a third.

132. The NEDC report [13] which reviews the most recent state of UK manufacturing industry suggests several reasons for the downwards movement in the textiles industry. These include:

. very depressed UK consumer demand;

. competition from imports particularly from those countries which are suppliers of raw cotton, woollens and synthetic fibres;

. the trend for supplying countries to process or semi-process materials up to spun yarn or even woven cloth and for UK to engage in finishing and/or weaving. The finished cloth is often re-exported for garment making and re-imported as clothing;

. lack of capital investment (or perhaps unwise investments in Wales, Northern Ireland, etc) in areas of textiles which, even with subsidised plant and machinery, cannot compete effectively on design, quality or price with imported goods;

. the better performance of high value added products as opposed to volume production such as basic cotton yarn. So NEDC recommend concentration on finishings, knitted goods, high quality fabrics and high quality finishings. It may be that very little spinning (including synthetic fibres) will be carried out in the UK by 2000.

133. These points provide a necessary link between the picture of the industry as given by the 1980 base used in this study and the view ahead to around 2000 which is presented in the next section. Output has fallen significantly since 1980 so gigajoule per tonne measures may well show little change (or look even worse) compared with the 1980 figures despite some energy savings. This phenomenon can be found in other industries such as brewing which have sought to establish targets and have failed to meet them because output in tonnes or hectolitres has fallen as fast or faster than energy consumption. In the same period employment has fallen rapidly as labour has been shed, accelerating a trend already apparent over the past decade, so gigajoule per employee figures may also show a higher level in 1983 than in 1980. The problem of the interpretation of specific energy consumption data is a serious issue already mentioned.

Developments in the Textile Industries to 2000

134. Perhaps the first problem is to establish the likely path of output and output structure and employment in the textiles industry to 2000. Most current estimates of future activity conclude that the industry will see further falls in capacity and employment, certainly to 1990, and probably to 2000. There is also general agreement about the direction of change towards an industry composed primarily of high value added, high quality, finished textiles with emphasis on the specialist, high technology finishes. The future of spinning of textile fibres of all kinds is particularly at risk.

135. The area which has suffered most seriously in the period 1975-81 has been cotton textiles. The industry has seen its sales fall by around a third; wool textiles has fallen by around 25%; while knitting and clothing have fared best, albeit falling over 10% in the same period. The fall in

employment in the same period has been somewhat higher at 35%; 37%; and 21% and 23% respectively.

136.   The NEDC Sector Summary paper reports that 'competition from the newly industrialising countries based on a low wage cost and working practices and conditions which would not be acceptable in the UK; tariff and non-tariff barriers which virtually close the southern hemisphere to UK clothing and textile exports making it difficult to penetrate the US markets; and subsidies to some European textile industries; poor use of design and design management and inadequate fashion leadership'. This effectively sums up the problems in textiles and the report then recommends some things which might be done to maintain the industry over the next decade or so. The NEDC paper recommends:

- extending and consolidating the move up-market to specialist and/or high value added product ranges;

- specific products manufactured in short runs and targeted to specific markets;

- a move away from mass marketing strategies to closer links with retailers for the supply of limited runs to order;

- further improvement in production efficiency using microprocessor controls and more advanced production systems in order to achieve the benefits of continuous production while maintaining the flexibility of batch output.

137.   In terms of the structure of output this is taken to mean further falls in cotton textiles and synthetic fibres particularly on the spinning side; less severe effects on woollen textiles; and further concentration on weaving and finishing particularly the enormous range of finishes produced by the specialist finishing industry. The position of other textiles such as jute, carpets, household textiles, etc is uncertain.

Technical Change in the Industry

138.   With one, perhaps two, exceptions most technical change foreseen over the next two decades represents improvements, refinements and gradual diffusion of practices already in operation in the industry. In no particular order of importance these are as follows.

(i)   Foam based finishes

139.   In most finishing processes the chemical additives for fire proofing, water repellent properties, shrink proofing, etc are added wet and the material is impregnated, sprayed or otherwise processed in a fully wet form. Clearly this then has to be dried out which is an energy intensive process. Current R&D work at the Shirley Institute is attempting to achieve the same quality and durability of finish by applying the chemical additives as a foam. This substantially reduces the drying time for the completion of the fabric. Indeed about half the energy consumption in the finishings industry is associated with drying, so substantial savings might be expected were the process to be technically satisfactory. The first extension of this process would be to weaving, where many of the threads are sized before the weaving

process commences, in order to strengthen the yarn to resist abrasion in the loom.

Temperature and time controls in washing

140. Various processes are currently in use in a demonstration phase for closer control over the large number of washing processes which occur, particularly in the wool industry. Using an array of microprocessor controls for temperature, time and moisture content, it has proved possible to control much more accurately the volume of water required for washing and the degree of moisture at the end of the drying process. Sensing of moisture content is important in the drying stage: a too dry fabric will need to be moistened before the next process commences. This attempt to optimise the washing process has saved up to 50% of energy associated with the washing processes in some mills. It is too early for the technology to be proven but further trials and demonstrations are likely to confirm the promising early results.

Low temperature and combined processes

141. Many dyeing processes and some finishing processes are currently carried out in hot water conditions with the associated washing, fixing and drying processes which together make dyeing and other finishing processes very energy intensive. Trials have shown that some dyeing and finishing processes can be carried out in low (or lower) temperature conditions. It is also the case that certain two-stage dyeing processes (where the materials are a mixture of natural and synthetic fibres) can be combined in a single dyeing and finishing process. Naturally this has the effect of saving energy costs, in some cases up to two thirds of current process energy costs. If the trials are successful and can be extended to adjacent areas potential savings could be at least 25% and perhaps as much as 70% in a large number of finishing mills.

Electric motors

142. It was argued that there was some scope for improving the efficiency of electric motors but the major saving would come from matching the tasks carried out with the size of the motor. At present a large number of looms and spinning machines have individual motors which are not appropriately sized for the task carried out.

Friction losses

143. Another area which may prove useful is the possibility of reducing friction losses in belt drives. Energy consumption rises by the square of the speed with which the frame operates but there is some question whether the speed achieved by many frames is really required by the firm. Rates of production may be unnecessarily fast in relation to the total flow of work. By reducing the speed of one part of the system, little effect would occur in overall productivity rate but significant energy savings might occur at the weaving stage. This is still under investigation.

144. For certain processes, particularly drying, there has recently been a move by the textile industries to seek to emulate the chemicals industry (with which of course it has very close links). That is, certain parts of the process tend now to be fully utilised and then stopped. Thus all the

energy costs and overheads are absorbed in one full output session and the
plant is then mothballed for a time lag period, until further work sufficient
for another flat out production run has been generated. Once ready, the
plant is then fully utilised until such time as the next particular batch of
output has been completed. High energy users such as stenters and certain
dyeing equipment fall into this category. Such a process may not be
practicable at present because output levels and demand prospects do not
justify such practices. But it may be possible, if work can be scheduled in
order to fully utilise these highly energy-intensive processes.

Fuel Prices

145.   Fuel prices were regarded by the Shirley Institute as fairly important
as a stimulant to improved energy efficiency but by no means the key to
effective action. Clearly fuel prices have a greater incentive in the high
energy using finishing trades but otherwise are much less important. It is
also the case that in the high energy finishing trades much of the energy use
is process energy and improvements here are largely determined by the
investment cycle. The factors affecting investment are clearly complex but
appear to be dominated by:

  . quality of the yarn or fabric;

  . productivity factors, notably speed of output;

  . then, and only then, energy costs which are now more important than
    they used to be in the factors affecting investment.

146.   Even when pressed it proved very difficult for officials at the Shirley
Institute to assess by how much an increase in fuel prices would act as an
inducement or incentive to become more energy efficient either by changing
energy management practice or by stimulating investment which would
substitute capital for operating costs. As far as they were aware fuel price
increase in real terms had only a marginal effect on investment decisions and
energy management practices. What might be more significant would be
relative changes in fuel prices. But past variations betwen gas, oil and
coal had been such as to cause uncertainty and until such time as it was
fairly clear which fuel would be consistently relatively cheaper no
significant switching could be expected. In the long term there was expected
to be some shift towards coal for steam raising. It was unlikely that
significant savings could or would be made in electricity consumption and, if
the fuel proved relatively less expensive, there may be some increase in
absolute terms in electricity consumption particularly in certain drying
processes.

Energy Conservation to 2000

147.   Specific energy requirements as measured in GJ/£k is shown for each of
the categories in the textiles industry in Table 5. These are based on the
IETS visit Reports which estimated energy savings in the industry ranging
from 14% in the knitting industry to 22% in carpets. The high and low
scenario SECs are upper and lower estimates based on these IETS Report
figures. For those sectors of the textile industry where space heating is
the major form of energy much of the potential is assumed to be achieved by
1990 leaving a small balance to be achieved over the remaining decade. Where

the energy consumption in the sub-sector is dominated by process energy then the potential savings accrue in a linear fashion. These assumptions are rather general and the measures somewhat crude but are good enough for the purposes of this exercise.

148. Table 6 shows the estimates of the pattern of fuel use reflecting the judgement that most of the energy conservation potential will fall on boiler fuels and hence most of the significant changes occur in the relative importance of the electricity and boiler fuels. There has also been assumed to be some small switching to coal of around 15%; but this is no more than a suggestion.

149. Discussions at the Shirley Institute suggest that even those mills operating at or near best practice in 1980 could still show some significant improvement by around 2000 — an improvement put at something around 15%. This, therefore, has been taken to be the improvement in best practice across all the categories by 2000 and it is considered to be real and likely to materialise.

TABLE 1: Output, Employment, Specific Energy Consumption and Energy Consumption in the Textiles Industry 1980

Subsector	Output, £M 1980 prices	Employment, thousands	SEC, (average practice) GJ/tonne	SEC, (average practice) GJ/employee	Energy Consumption, (average practice) 106 GJ	SEC, (best practice) GJ/employee	Energy Consumption, (best practice) 106 GJ	SEC, GJ/£k
Woollen & Worsted								
Wool spinning	) 877.7	) 65	50	150	) 13.39	120	) 10.75	) 11.59
Wool combing	)	)	40	410	)	370	)	)
Woollen mills	)	)	120	150	)	113	)	)
Carpet yarn	) 277.1	)	160	210	)	168	)	)
Weaving & finishing	)	)	10	220	)	176	)	)
Cotton								
Spinning only	) 530.7	) 42	20	130	) 6.09	108	) 4.62	) 11.48
Spinning & finishing	)	)	57	160	)	120	)	)
Weaving only	) 550.5	) 36	30	132	) 5.98	122	) 4.96	) 10.86
Industrial fabrics	)	)	26	178	)	142	)	)
Weaving & finish	)	)	38	189	)	150	)	)
Knitted Goods								
Knitting only	) 1257.5	) 100	44	44	) 8.50	33	) 8.0	) 6.76
Knitting & finish	)	)	90	188	)	140	)	)
Textile Finishing*								
Fibre/yarn only	) 380.3	) 41	45	490	) 21.42	294	) 12.85	) 56.37
Fabric processing	)	)	60	560	)	336	)	)
Carpets & Jute								
Tufted carpet	) 588.8	) 30	38	408) 311	) 9.45	326) 238	) 7.24	) 16.05
Woven carpet	)	)	72	214)	)	150)	)	)
Jute	) 99.6	) 6	7	176	) 0.97	140	) 0.77	) 9.74
Synthetic Fibres	) 693.3	) 30		1127	) 33.81	902	) 27.06	) 48.79
Miscellaneous								
Narrow fabrics	) 577.7	) 12	NA	88	) 1.06	67	) 0.81	) 9.22
Household textiles	)	) 23	NA	53	) 1.22	41	) 0.94	)
Canvas goods	)	) 6	NA	70	) 0.42	58	) 0.35	)
Other textiles	)	) 9	22	292	) 2.63	220	) 1.98	)

* This represents not total value but value added at this processing stage only.

TABLE 2: Percentage of Total Energy Consumption Accounted For By
Each Fuel in the Textiles Industries, 1980 (percentage)

Subsector	Fuel: Oil	Gas	Coal	Elec.
Woollen & Worsted	41	14	28	17
Spinning of Cotton	36	5	16	43
Weaving of Cotton	41	22	7	30
Knitted goods	21	50	6	23
Textile finishing	49	29	12	10
Carpets	34	24	21	21
Jute	43	14	14	29
Synthetic fibres	69	7	10	14
Narrow fabrics	36	29	14	21
Household textiles	44	34	11	11
Canvas goods	42	42	0	15
Other textiles	22	50	3	25

TABLE 3: Energy Use in the Textiles Industries 1980

Subsector	Percentages of Fuel Consumption, %		Percentage of Fuel Through Boilers, %
	Process	Space Heating	
Woollen & Worsted			
Wool spinning	50	50	80
Wool combing	90	10	80
Woollen mills	70	30	80
Carpet yarn	60	40	80
Weaving & finishing	70	30	
Cotton			
Spinning only	20	80	65 (significant elec. motive
Spinning & finishing	50	50	power)
Weaving only	30	70	70 (significant elec. motive
Weaving & finishing	60	40	power)
Knitted Goods			
Knitting only	45	55	80
Knitting & finishing	80	20	80
Textile Finishing	85	15	90
Jute & Carpets			
Tufted carpet *	25	75	70
Woven carpet +	25	75	70
Jute	30	70	55 (high elec. motive power)
Synthetic Fibres	80	20	80
Misc. Textiles			
Narrow fabrics	70	30	70
Household textiles	60	40	80
Canvas goods	60	40	80
Other textiles	50	50	70

* Nearer 5-95 if no backing to carpet.
\+ Reverse proportions if latex backs.

TABLE 4: UK Output and Employment in Textiles Industries 1970-82

	1971	1973	1975	1977	1979	1980	1981
Synthetic Fibres							
Production, 10^3 tonnes	613.1	730.8	562.5	551.8	596.3	449.7	394.7
Employment, thousands				40	37	32	27
Cotton							
Production of single yarn, 10^3 tonnes	215.9	207.8	170.7	175.4	164.4	124.0	93.9
Production of doubled yarn, 10^3 tonnes	85.0	82.1	68.4	69.8	58.0	49.1	39.8
Employment, thousands				56	47	42	33
Total spindles, millions	2.9	2.5	2.3	2.2	1.9	1.7	1.2
Cotton, million linear metres	1038.8	956.3	908.0	865.4	832.0	664.1	550.8
Loom activity, thousands of looms	57.8	49.9	42.8	37.5	32.9	26.2	19.8
Wool							
Wool & tops, 10^3 tonnes	134	150	124	128	111	100	87
Woollen yarn, production, 10^6 kg	131	135	112	106	105	81.2	71.6
Worsted deliveries, 10^6 kg	83.8	88.6	67.0	72.8	60.5	52.8	51.6
Woollen & worsted fabrics, 10^6 square metres	185.8	192.4	151.4	150.1	137.8	118.2	97.1
Blankets, 10^3 tonnes	28.5	30.6	24.7	22.2	18.2	15.0	9.9
Carpets, 10^3 tonnes					75	65	56

TABLE 5: Assumption of Output Structure and Specific Energy Consumption
in the Textile Industries 1980-2000

Sector	1980	1990		2000	
		H	L	H	L
(a) Output Structure (percentage)					
Woollen & Worsted	20	20	20	19	19
Cotton	18	17	17	16	16
Knitted Goods	22	23	23	23	23
Textiles Finishing	6	7	7	7	7
Carpets & Jute	12	12	12	13	13
Miscellaneous Textiles	10	10	10	11	11
Synthetic Fibres	12	11	11	11	11
	100	100	100	100	100
(b) Specific Energy Consumption (GJ/£k)					
Woollen & Worsted	11.59	10.43	10.72	9.27	9.85
Cotton	11.17	9.88	9.88	8.93	8.93
Knitted Goods	6.76	5.96	6.04	5.47	5.63
Textiles Finishing	56.37	50.74	52.14	45.10	47.91
Carpets & Jute	15.15	13.41	13.41	12.12	12.12
Miscellaneous Textiles	9.22	8.26	8.53	7.30	7.84
Synthetic Fibres	48.79	43.91	45.13	39.03	41.47

TABLE 6: Percentage of Total Energy Consumption Accounted For By
Each Fuel in the Textiles Industries, 2000

Subsector Fuel:	Coal	Gas	Oil	Elec.
Woollen & Worsted	45	17	17	21
Cotton Spinning	25	15	15	45
Cotton Weaving	15	20	35	30
Knitted Goods	12	45	18	25
Textile Finishing	15	40	37	8
Carpets	25	30	27	18
Jute	20	25	33	32
Narrow Fabrics	20	30	27	23
Household Textiles	15	25	43	17
Canvas Goods	10	45	27	18
Other Textiles	15	40	25	20
Synthetic Fibres	15	15	54	16

TABLE 7: Output, Specific Energy Consumption and Energy Consumption in the
Leather, Leather Goods, Footwear and Clothing Industries in 1979 and 1980

Sector	Gross Output in 1979, £M	Energy Consumption 1979, 10^6 GJ	SEC, GJ/000£	Gross Output 1980, million	Energy Consumption 1980, 10^6 GJ
Leather, Leather Goods	420	4.29	10.21	420	4.29
Footwear and Clothing	4963	11.30	2.28	4827	11.00
TOTAL	5383	15.59		5247	15.29

TABLE 8: Fuel Use in the Leather, Leather Goods, Clothing
and Footwear Industries 1980 in 10^6 GJ

Sector	Oil	Gas	Coal	Elec.	TOTAL
Leather, Leather Goods	2.38	0.73	0.72	0.46	4.29
Clothing and Footwear	5.77	2.29	0.99	1.95	11.00
TOTAL	8.15	3.20	1.71	2.41	15.29

TABLE 9: Output Structure and Specific Energy Consumption Assumptions for the Footwear, Leather and Clothing Industries 1980-2000

	1980	1990		2000	
		H	L	H	L
(a) Output Structure (percentage)					
Leather, Leather Goods	8	8	8	8	8
Footwear and Clothing	92	92	92	92	92
	100	100	100	100	100
(b) Specific Energy Consumption (GJ/£k)					
Leather, Leather Goods	10.21	8.45	9.04	7.15	8.17
Footwear and Clothing	2.28	1.89	1.89	1.60	1.60

TABLE 10: Percentage Fuel Use in the Leather, Leather Goods, Clothing and Footwear Industries, in the year 2000

(High and Low Scenarios)

Sector	Oil	Gas	Coal	Elec.	TOTAL
Leather, Leather Goods	20	45	20	15	100
Clothing and Footwear	25	45	10	20	100
TOTAL	24	45	13	18	100

REFERENCES

1.   Energy Use in the Woollen and Worsted Industry. IETS Report No 1. Department of Industry, 1978.

2.   Energy Use in the Spinning and Doubling (Cotton System) Industry. IETS Report No 6. Department of Industry, 1979.

3.   Energy Use in the Weaving (Cotton System) Industry. IETS Report No 13. Department of Industry, 1979.

4.   Energy Use in the Knitting Industry. IETS Report No 16. Department of Industry, 1979.

5.   Energy Use in Textile Finishing. IETS Report No 20. Department of Industry, 1980.

6.   Energy Use in Carpets and Jute. IETS Report No 30. Department of Industry, 1983.

7.   Energy Use in Manmade Fibres. IETS Report No 41. Department of Industry. 1983.

8.   Energy Use in Miscellaneous Textiles. IETS Report No 44. Department of Trade and Industry. 1984.

9.   Energy Audit Series No 17. The Textile Industry. Issued jointly by the Department of Energy and the Department of Trade and Industry. July 1983.

10. Digest of UK Energy Statistics 1982. Department of Energy. HMSO 1983.

11. 1979 Census of Production and Purchases Inquiry. Business Statistics Office. HMSO 1983.

12. Energy Use in the Footwear, Leather and Fur Industries. IETS Report No 22. Department of Industry, 1980.

13. NEDC Paper 83(14). Summary of the Sector Working Party Assessments, April 1983.

Printed in the UK for HMSO, Dd.736732, C15, 11/84, 5673, 4607.